NEUROENDOCRINE CONTROL

NEUROENDOCRINE CONTROL

Edited by
KINJI YAGI and SHO YOSHIDA

A HALSTED PRESS BOOK

JOHN WILEY & SONS
New York – Toronto – London

Published in North America and Europe by Halsted Press, a Division of John Wiley & Sons, Inc., New York.

Printed in Japan.

Library of Congress Cataloging in Publication Data

Yagi, Kinji.
Neuroendocrine control.

"A Halsted Press book."
Includes bibliographical references.
1. Neuroendocrinology. 2. Neurophysiology.
I. Yoshida, Sho, joint author. II. Title.
[DNLM: 1. Endocrine glands—Physiology. 2. Neurophysiology. WK102 Y12n 1973]
QP187.Y3 612'.4 73-12315
ISBN 0-470-96951-2

PREFACE

Various neuroendocrine control systems as well as neural and endocrine control systems are responsible for adaptive behavior and reproductive functions in higher vertebrates, including human beings. Though neuroendocrine control systems vary widely in anatomical features and functions, fundamental common problems in need of solutions do exist.

About three years ago one of us (K. Y.) organized a study group to carry out integrated investigations of such fundamental problems. Financial support for this group study was obtained from the Ministry of Education for the period from 1970 to 1972. The problems we studied and the investigators concerned are as follows: (1) Development of rhythmicity in the brain (Drs. N. Tsukahara, Y. Arai, and I. Yamashita); (2) Properties of neurosecretory cells (Drs. S. Iwasaki, T. Akiyama, and R. Yamada); (3) Neural control of endocrine glands (Drs. N. Watari, S. Yoshida, T. Hiroshige, K. Yamamoto, and M. Kawakami); (4) Feedback effects of peripheral hormones on the central nervous system (Drs. T. Kobayashi, T. Tamada, and K. Yagi); and (5) A synthetic study of analytical investigations including simulation of neuroendocrine control systems by mathematical models (Dr. S. Inoué).

The members of this group are the most active researchers of neuroendocrinology and neurophysiology in this country. They met as a group once or twice a year to enthusiastically criticize and encourage each other. As a result, they produced many findings which provide some evidence for and against hypotheses not yet proved. It was, therefore, decided to publish the results of this integrated study in the form of articles by members of the group and their colleagues in the hope that this volume would be of particular value to the advancement of neuroendocrinology.

This group study program was supported by the Research Grants nos. 0803 and 60803 and the Grant-in-Aid for Publication No. 001 from the Ministry of Education of Japan. The editors wish to thank Miss Yukiko Sawaki, Department of Physiology, Jichi Medical School for her invaluable assistance throughout the editorial work.

January 1973

Kinji Yagi
Sho Yoshida

LIST OF CONTRIBUTORS

KAZUO ABE

Department of Physiology, Hokkaido University School of Medicine, Sapporo, 060, Japan

TOYOHIRO AKIYAMA

Department of Physiology, Jichi Medical School, Minamikochimachi, Tochigi-ken, 329–04, Japan

YASUMASA ARAI

Department of Anatomy, Juntendo University School of Medicine, Bunkyo-ku, Tokyo, 113, Japan

SHIGEO ARAKI

Department of Obstetrics and Gynecology, Gunma University School of Medicine, Maebashi, 371, Japan

YUTAKA ASANO

Department of Psychiatry and Neurology, Hokkaido University School of Medicine, Sapporo, 060, Japan

TAKEHIKO BANDO

Department of Biophysical Engineering, Faculty of Engineering Science, Osaka University, Toyonaka, Osaka, 560, Japan

HIDEKI EBIHARA

Department of Obstetrics and Gynecology, Gunma University School of Medicine, Maebashi, 371, Japan

TSUTOMU HIROSHIGE

Department of Physiology, Hokkaido University School of Medicine, Sapporo, 060, Japan

SHOJIRO INOUÉ

Division of Biocybernetics, Institute for Medical and Dental Engineering, Tokyo Medical and Dental University, Chiyodaku, Tokyo, 101, Japan

SHIZUKO IWASAKI
Department of Physiology, Tokyo Medical College, Shinjuku-ku, Tokyo, 160, Japan

MASAZUMI KAWAKAMI
Department of Physiology, Yokohama City University School of Medicine, Minami-ku, Yokohama, 232, Japan

FUKUKO KIMURA
Department of Physiology, Yokohama City University School of Medicine, Minami-ku, Yokohama, 232, Japan

TOSHIKAZU KIYOHARA
Department of Biophysical Engineering, Faculty of Engineering Science, Osaka University, Toyonaka, Osaka, 560, Japan

TAKURO KOBAYASHI
Department of Obstetrics and Gynecology, Branch Hospital, University of Tokyo School of Medicine, Bunkyo-ku, Tokyo, 112, Japan

SEIICHI MATSUMOTO
Department of Obstetrics and Gynecology, Jichi Medical School, Minamikochi-machi, Tochigi-ken, 329–04, Japan

RYUSUKE MAYUZUMI
Department of Obstetrics and Gynecology, Gunma University School of Medicine, Maebashi, 371, Japan

HIROSHI MINAGUCHI
Department of Obstetrics and Gynecology, University of Tokyo School of Medicine, Bunkyo-ku, Tokyo, 113, Japan

TAKASHI MOROJI
Department of Psychiatry and Neurology, Hokkaido University School of Medicine, Sapporo, 060, Japan

TOSHIO MURASE
The 3rd Department of Internal Medicine, Faculty of Medicine, University of Tokyo, Bunkyo-ku, Tokyo, 113, Japan

FUMIHIKO OKADA
Department of Psychiatry and Neurology, Hokkaido University School of Medicine, Sapporo, 060, Japan

TOSHIKAZU SAITO
The 3rd Department of Internal Medicine, Faculty of Medicine, University of Tokyo, Bunkyo-ku, Tokyo, 113, Japan

Yoshiro Saito
Department of Psychiatry and Neurology, Hokkaido University School of Medicine, Sapporo, 060, Japan

Shoichi Sakamoto
Department of Obstetrics and Gynecology, University of Tokyo School of Medicine, Bunkyo-ku, Tokyo, 113, Japan

Youko Satow
Department of Physiology, Tokyo Medical College, Shinjuku-ku, Tokyo, 160, Japan

Yukiko Sawaki
Department of Physiology, Jichi Medical School, Minamikochi-machi, Tochigi-ken, 329–04, Japan

Takashi Sekiguchi
Faculty of Engineering, Yokohama National University, Minami-ku, Yokohama, 232, Japan

Katsuo Seto
Department of Physiology, Yokohama City University School of Medicine, Minami-ku, Yokohama, 232, Japan

Taro Tamada
Department of Obstetrics and Gynecology, Gunma University School of Medicine, Maebashi, 371, Japan

Takashi Tamura
Department of Obstetrics and Gynecology, University of Tokyo School of Medicine, Bunkyo-ku, Tokyo, 113, Japan

Ei Terasawa
Department of Physiology, Yokohama City University School of Medicine, Minami-ku, Yokohama, 232, Japan

Nakaakira Tsukahara
Department of Biophysical Engineering, Faculty of Engineering Science, Osaka University, Toyonaka, Osaka, 560, Japan

Yoshiji Tsukui
Department of Obstetrics and Gynecology, Gunma University School of Medicine, Maebashi, 371, Japan

Kinji Yagi
Department of Physiology, Jichi Medical School, Minamikochi-machi, Tochigi-ken, 329–04, Japan

Ritsuji Yamada
The 1st Department of Internal Medicine, Faculty of Medicine, University of Tokyo, Bunkyo-ku, Tokyo, 113, Japan

Kiyoshi Yamamoto
Department of Physiology, Institute of Endocrinology, Gunma University, Maebashi, 371, Japan

Itaru Yamashita
Department of Psychiatry and Neurology, Hokkaido University School of Medicine, Sapporo, 060, Japan

Kosuke Yamazaki
Department of Psychiatry and Neurology, Hokkaido University School of Medicine, Sapporo, 060, Japan

Sho Yoshida
Department of Internal Medicine, Jichi Medical School, Minamikochi-machi, Tochigi-ken, 329–04, Japan

Nakazo Watari
Department of Anatomy, Nagoya City University Medical School, Nagoya, 467, Japan

CONTENTS

Preface.. i
List of contributors iii

BIORHYTHM IN THE CENTRAL NERVOUS SYSTEM

The Properties of the Reverberating Circuit in the Brain
N. Tsukahara, T. Bando, and T. Kiyohara 3
Introduction 3
Mutual monosynaptic excitation between the nucleus interpositus (IP) and nucleus reticularis tegmenti pontis (PN)...... 5
Dynamic properties of the reverberating circuit 11
Model 16
Conclusion 22
References 25

Sexual Differentiation and Development of the Hypothalamus and Steroid-induced Sterility
Y. Arai 27
Introduction 27
Development of the acyclicity of GTH regulation in the male.. 29
Masculinization of the female hypothalamus by androgen...... 33
Localization and sexual differentiation of the neural substrates for regulation of ovulation 36
Possible mechanisms of hypothalamic masculinization......... 41
Persistent diestrus syndrome—a pharmacological effect of neonatal treatment with sex steroids........................ 45
References 48

Circadian Rhythm of Glucocorticoid Secretion
T. Moroji, K. Yamazaki, F. Okada, Y. Saito, Y. Asano, and I. Yamashita 57
Introduction ... 57
Circadian rhythm of pituitary—adrenocortical function in patients with central nervous system diseases ... 61
Correlation between circadian rhythm of plasma corticosteroid concentrations and daily variation in contents of transmitters in the brain ... 65
Conclusion ... 74
References ... 74

NEUROSECRETION

Electrical Characteristics of the Membrane in Neurosecretory Neurons
S. Iwasaki and Y. Satow 85
Introduction ... 85
Ultrastructure of the X-organ—Sinus gland system of crayfish ... 86
Electrical characteristics of the membrane in the secretory neuron ... 89
Conclusions ... 104
References ... 105

Divalent Cations and Release of Humoral Substance from Axon Terminals
T. Akiyama 111
Introduction ... 111
Method of estimation of the amount of transmitter released ... 112
Experiments with GABA ... 113
Effect of changing [Ca] and [Mg] on the release of transmitter ... 118
References ... 133

Secretory Mechanism in Adrenal Medullary Cells
R. Yamada 137
Introduction ... 137
Subcellular pools in rat adrenal gland ... 139
Influence of secretion upon subcellular pools of catecholamine in rat adrenal medulla ... 142
Influence of biosynthesis upon subcellular pools of catecholamine in rat adrenal medulla ... 147
Comparison of subcellular pools of catecholamines in pheochromocytomas and neuroblastomas to pools in the adrenal medulla ... 152

The role of subcellular catecholamine pools in adrenal medulla 157
The role of the subcellular catecholamine pools in pheochromocytoma and neuroblastoma 159
References 159

NEURAL CONTROL OF ENDOCRINE GLANDS

Innervation of Pancreatic Islet Cells
N. Watari 165
Introduction 165
Nerve supply in the pancreatic islets 166
Morphological alterations of the nervous elements in some experimental conditions 172
References 172

Control of Vasopressin Secretion
S. Yoshida, T. Saito, and T. Murase 189
Introduction 189
Osmoregulation and ADH 190
Volume regulation and ADH 194
References 200

Role of Brain Biogenic Amines in the Regulation of ACTH Secretion
T. Hiroshige and K. Abe 205
Introduction 205
ACTH-Releasing potency of brain biogenic amines 207
Effect of brain amine depletion on ACTH release 214
Effects of intraventricular injection of monoamines 216
Possible role of biogenic amines in ACTH release mechanisms 220
References 223

Nature of the Hippocampal Function in Relation to Gonadotropin Secretion
M. Kawakami, K. Seto, F. Kimura, and E. Terasawa 229
Introduction 229
Hippocampal connections with the hypothalamus 232
Characteristics of hippocampal electrical activity 233

The role of the hippocampus in controlling gonadotropin secretion in the female ... 234
The role of the hippocampus in controlling gonadotropin secretion in the male ... 247
Electrophysiological aspects of the hippocampal function ... 248
Conclusion ... 251
References ... 252

FEED-BACK EFFECTS OF PERIPHERAL HORMONES ON THE ACTIVITY OF HYPOTHALAMO-HYPOPHYSIAL SYSTEMS

Neural Control of Human Fetal Adenohypophysis
T. Tamura, H. Minaguchi, T. Kobayashi, and S. Sakamoto 259
Introduction ... 259
Materials and methods ... 260
Results ... 261
Discussion ... 266
References ... 269

Interrelationships between the Levels of Ovarian Steroids and Luteinizing Hormone
T. Tamada, Y. Tsukui, S. Araki, and S. Matsumoto 271
Introduction ... 271
Interrelationships between estrogens and LH ... 273
Urinary excretion of estrogens and gonadotropins in subjects treated with clomiphene ... 282
Interrelationships between progestogens and LH ... 287
The significance of LH on estrogen secretion during follicular maturation ... 291
References ... 294

Feedback of Estrogen in the Hypothalamic Control of Gonadotrophin Secretion
K. Yagi and Y. Sawaki 297
Introduction ... 297
Estrogen receptor cells in the hypothalamus ... 299
The effects of estrogen on preoptic and hypothalamic unit activity ... 300

Neural connections between hypothalamic nuclei and the median eminence 308
Unit activity in the preoptic and hypothalamic areas of cyclic and castrated female rats 313
Conclusion 317
References 318

CONTROL MECHANISM OF NEURO-ENDOCRINE SYSTEM

Biological Factors Controlling the Synthesis and Release of Adenohypophysial Hormones
K. Yamamoto 329
Introduction 329
Principles, procedures, and limitations of the method 331
Variations in pituitary functions under several biological conditions 335
Summary and comments 356
References 358

Neuroendocrine Control Systems and Systems Analyses of the Brain-Hypophyseal-Gonadal Mechanism
S. Inoué and T. Sekiguchi 361
Systems approaches to neuroendocrine systems 361
Control mechanism of the brain-hypophyseal-gonadal system .. 370
Analysis of dynamic estrogen—uterine weight interrelationships 385
References 391

BIORHYTHM IN THE CENTRAL NERVOUS SYSTEM

THE PROPERTIES OF THE REVERBERATING CIRCUIT IN THE BRAIN

NAKAAKIRA TSUKAHARA, TAKEHIKO BANDO, and TOSHIKAZU KIYOHARA

This article deals with the physiological and theoretical properties of the reverberating circuit between the cerebellar nucleus and a part of the pontine nuclei in cats. Experimental evidence suggesting the existence of a reverberating circuit between the nucleus interpositus and nucleus reticularis tegmenti pontis is given. The dynamics of the reverberating circuit was investigated in the surgically and pharmacologically isolated neuronal circuit. The dynamic properties of the reverberating circuit were characterized with the regenerative behavior. Above and below threshold, there were two stable states, the excited state where many constituent neurons were active and the resting state where all neurons were inactive. Furthermore, the time course of the neuronal recruitment depended upon how many cells were excited initially. These two properties were successfully reproduced by a mathematical model proposed in this article.

INTRODUCTION

In the last decade, the synaptic connection of the neuronal networks in the brain has been the subject of extensive study in the field of neurophysiology. Thanks to the progress of cell physiology in general, the detailed cell properties of the constituent neurons in the networks have been defined using intracellular microelectrode techniques. However, there are large gaps in our understanding between the mostly known physiological properties of a single neuron on one hand and the largely unknown overall dynamic properties of an entire network on the other. The normal function of the nervous system can only be achieved by all networks as internally interrelated dynamic wholes, and yet we are not at present at the stage to relate the overall activity of such a network to the properties of its constituent neurons. Although this aspect stands as a crucial point in integrative neurophysiology, progress seems to be hampered by the lack of adequate conceptual and analytical tools for handling it.

Fortunately, however, considerable information is at hand concerning the interconnection in and around the cerebellum, which has been a subject of growing interest in neurophysiology as well as in neuroanatomy for these ten years (Eccles, Ito & Szentagothai, 1967). The authors tried to take advantage of this information and chose the network made up of the cerebellar nucleus and pontine nucleus for an experimental subject. As a first step in the study of such a problem, an attempt was made to define the physiological properties of the constituent neurons using microelectrode techniques (Tsukahara & Bando, 1970; Tsukahara, Korn & Stone, 1968).

In order to study the network properties, the network in question was isolated from the rest of the brain. It was found that such a network exhibited a regenerating reverberation (Tsukahara, 1972; Tsukahara, Bando, Kitai & Kiyohara, 1971). To relate the properties of the network to those of the constituent nerve cells, a mathematical model was introduced. This model reproduced fairly satisfactorily the main features of the network. Since the problem is of importance in our understanding of neuronal networks in general, it may also provide a strong background for interpreting neural functions, such as neuroendocrine control.

In order to provide sufficient background for the present analysis, a brief account of the current concept of cerebellar physiology will be given below. The cerebellar Purkinje cells, which are the only output system from the cerebellar cortex, are in their entirety inhibitory in nature and exert a direct inhibitory control upon all their target cells. As for the cerebellar nuclei, which are excitatory, Purkinje cells represent a complex feedforward inhibitory system, which regulates their activity and modulates the excitatory action of the nuclear cells on the brain stem nuclei. The nuclear cells themselves receive a steady background depolarization via the collaterals of the cerebellar afferents, which generates a tonic firing of these neurons and thus represents a good substratum for inhibitory action of the Purkinje cells (Eccles *et al.* 1967).

In order to account for these tonic discharges of cerebellar nuclear neurons, the presence of reverberating circuits through the cerebello-nuclear neurons has been suggested but not yet proven. In view of the available histological evidence, the cerebellonuclear-pontonuclear-cerebellonuclear connection is a possible candidate for these rever-berating circuits (Eccles, 1967; Forbes, 1922; Lorente de Nó, 1938).

In the following account, a demonstration and an analysis of a reverberating circuit in the brain are presented using the example of the cerebello-pontine reverberating circuit, which has been found recently in the author's laboratory, and further the problems relevant to the neuronal mechanisms of the neuroendocrine systems are highlighted.

MUTUAL MONOSYNAPTIC EXCITATION BETWEEN THE NUCLEUS INTERPOSITUS (IP) AND NUCLEUS RETICULARIS TEGMENTI PONTIS (PN)

For this experiments cat were anesthetized with pentobarbitone sodium. Decerebrated cats were also used; they were immobilized with flaxedil (galamine triethiodide) and artificially respirated. A conventional recording system was used for intracellular recording. Recording from the nucleus reticularis tegmenti pontis (Tsukahara *et al.* 1970), nucleus interpositus of the cerebellum (Tsukahara, Korn & Stone, 1968), and magnocellular red nucleus (Tsukahara, Toyama & Kosaka, 1967) was made by the same procedures as reported previously. Picrotoxin was injected intravenously (10–5 mg/kg) after immobilization by flaxedil. Further details of the experimental procedures are given in previous papers (Tsukahara, Fuller & Brooks, 1968; Tsukahara *et al.* 1970; Tsukahara *et al.* 1971).

Cells were sought in the dorsal pontine region, in the area of the nucleus reticularis tegmenti pontis (PN). Histological investigation has shown that relatively large cells are concentrated in this region. When a cell was impaled, action potentials appeared that responded to the brachium pontis (BP) (Fig. 1B), the cerebellar hemisphere (CH) (Fig. 1C) and sometimes IP stimulation (Fig. 1D and E) with short and fixed latencies. These potentials were produced without any prepotentials with brief and fixed latencies, 0.9 msec, by BP stimulation. Therefore, it is apparent that they are produced antidromically through the pontocerebellar fibers.

In the samples identified in this way, stimulation of the IP induced an excitatory postsynaptic potential (EPSP) (Fig. 1F) with brief latencies, an average of 1.19 msec from 32 PN cells. The observation was made after the antidromic spikes were completely deteriorated. This result is in accordance with the histological investigation showing that the descending branch of the brachium con-

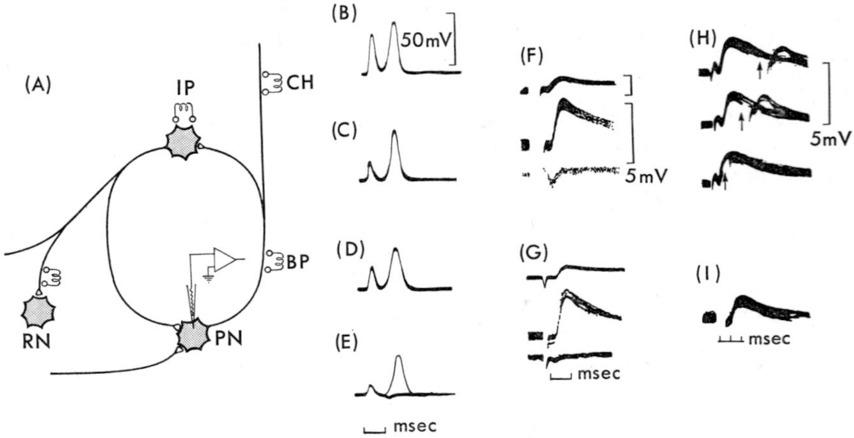

Fig. 1. A, Diagram of the experimental arrangement for study of the pontine nucleus (PN). Stimulation was applied to the cerebellar hemisphere (CH), brachium pontis (BP), nucleus interpositus (IP), and red nucleus (RN). B–E, Antidromic spike evoked in a PN cell by BP (B), CH(C), and IP (D, E) stimulation; dc recording. Time mark of 1 msec and voltage calibration of 50 mV apply from B to E. F, EPSP induced in a PN cell by stimulating the IP. Uppermost traces are dc recordings of intracellular potential. Middle traces are ac recordings with high gain of the intracellular potential. Lowermost traces are extracellular fields corresponding to the middle traces. G, An EPSP produced in the same cell as F by stimulating the RN. Upper, middle and lower traces are the same as in F. H, Conditioning of the IP-induced EPSP, control in I, by a preceding RN stimulation. The arrows indicate the onset of IP stimulation. Voltage calibration 5 mV in H and I applies to all ac traces.

junctivum ends in the dorsal pontine nuclei. Since the IP innervates the magnocellular red nucleus (RN) and the ventrolateral nucleus of the thalamus (VL) monosynaptically by axon collaterals, it may be asked whether IP-PN fibers are collaterals of IP-RN-VL fibers.

When stimulation was applied to the RN where the majority of the fibers of the brachium conjunctivum end or pass through, EPSPs were produced in PN cells with short latencies. Figure 1G illustrates an example of an RN-induced EPSP in the same PN cell as in Fig. 1F. The average latency of the RN-induced EPSPs in 40 PN cells was 0.87 msec. The collision experiment between RN and IP stimulation showed that these IP- and RN-induced EPSPs indeed originated from the same line of fibers (Fig. 1H). It was thus concluded that IP- or RN-induced EPSPs in PN cells were both mediated by the axons of the brachium conjunctivum as diagrammatically illustrated in Fig. 1A and that RN-induced EPSPs in PN cells were mediated by

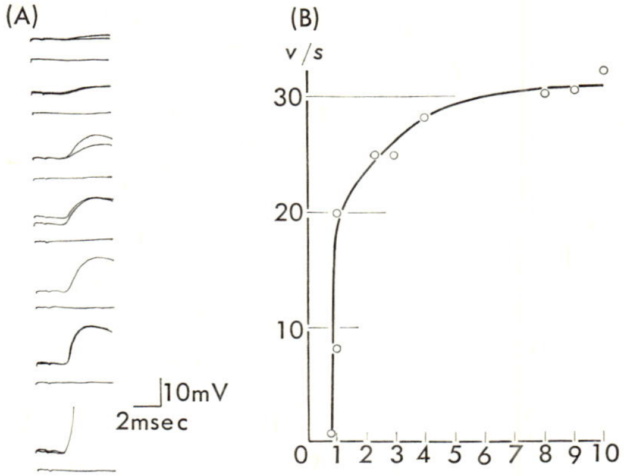

Fig. 2. A, Upper traces, EPSPs induced in a PN cell by IP stimulation with increasing stimulus intensities. Lower traces, extracellular controls. B, Relationship between maximum slope of the EPSP and stimulus intensity.

the axon-reflex pathway by stimulating one branch of the fibers of the brachium conjunctivum at the RN. The short latencies of the RN-induced EPSPs in PN cells, 0.87 msec, indicates that these EPSPs are produced monosynaptically.

To obtain further details of the synaptic connection from the IP to PN, the convergence of IP axons onto a single PN neuron was estimated in the following way. Figure 2A illustrates an IP-induced EPSP in a PN cell with increasing stimulus intensities from the top to bottom traces. The maximal rising slope of the EPSP is measured on its curve corrected for the extracellular field potential and is plotted in Fig. 2B as a function of the stimulus intensity. Saturation occurs above approximately 8 V of IP stimulation. This would indicate that all of the IP axons converging onto the PN cell are excited. The steepest rising slope of the maximal EPSP was almost 50 times as large as that of the unitary EPSPs. The rising slope of just threshold EPSPs for initiating spikes was 35 times as large as that of the unitary EPSPs. The size of the unitary EPSP was 0.8 mV in this PN cell. This would indicate approximately 50 IP axons converge onto a single PN cell.

As can be noted in Figs. 1F and G or 2A, the RN- or IP-evoked EPSPs had a relatively short summit time. In several PN cells, the decay time course of the IP-induced EPSP was analyzed by plotting the amplitude of the EPSPs on the semilogarithmic coordinates against time. The slope of this straight line gives the time constant

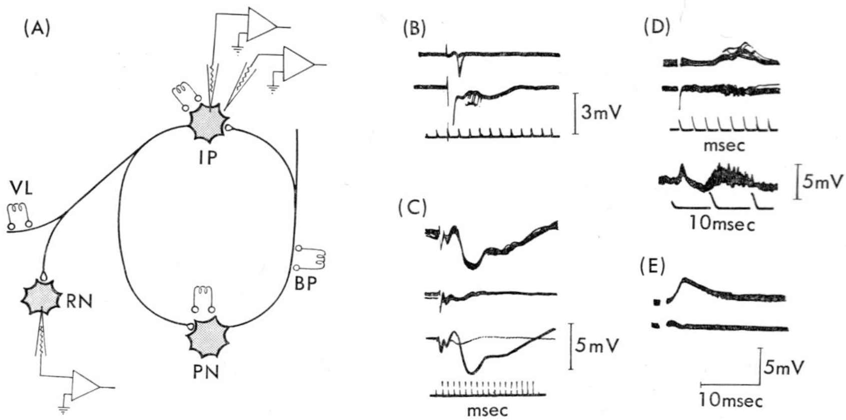

Fig. 3. A, Diagram of the experimental set-up for study of the pontine influence on the nucleus interpositus. The brachium pontis (BP), nucleus interpositus (IP), ventrolateral nucleus (VL) of the thalamus, and nucleus reticularis tegmenti pontis (PN) could be stimulated. The intracellular and extracellular activity of neurons of the IP nucleus and the red nucleus (RN) were recorded. B, Upper trace, antidromic activation of an IP neuron from VL. Lower trace, orthodromic activation of the same IP neuron by BP stimulation. C, Upper trace, intracellular potentials generated in an IP neuron by BP stimulation. Middle trace, extracellular potential generated just outside the same cell by the same stimulus. Lowermost trace, superimposed tracings of the intra- and extracellular potentials. D, Intracellular potentials induced in an RN cell by BP stimulation. The corresponding extracellular field generated outside the same cell. E, Upper trace, EPSP induced in an RN cell by PN stimulation. Lower trace, extracellular field potential recorded just outside of the cell.

of 7 msec. These experiments establish that there is a monosynaptic excitatory connection from the IP to PN.

In order to test the possible synaptic connection from the PN to IP, microelectrodes were inserted into the region of the nucleus interpositus of the cerebellum. Single units in the IP were identified by their antidromic activation after stimulation of the ventrolateral nucleus (VL) of the thalamus as shown in the upper trace of Fig. 3B. Extracellular recordings have shown that stimulation of the ipsilateral BP evoked in the IP a burst of spikes with longer and more variable latencies as shown in the lower trace of Fig. 3B.

The minimum latency of the BP-evoked spikes was quite short, between 0.9 and 3.9 msec (mean 1.8 msec), in the 45 units identified as IP neurons. It was thus presumed that the BP-evoked response was produced orthodromically. In order to confirm the excitatory action from BP to IP, intracellular recordings were performed in a few

IP neurons. As would be expected, there was an EPSP by BP stimulation. As illustrated in Fig. 3C, the EPSP is followed by a hyperpolarization, which is presumed to be due to the IPSP produced by the cerebellar Purkinje cells activated by the mossy fiber pathway via the cerebello-pontine afferents, since the cerebello-pontine fibers terminate as mossy fibers (Sasaki, Kawaguchi, Shimono & Yoneda, 1969).

Since IP neurons excite RN cells monosynaptically, we may expect, according to the results mentioned above, that the RN cells are activated disynaptically by BP stimulation. We may also expect that stimulation of the PN induces axon reflex activation of the fibers of the brachium conjunctivum; consequently, monosynaptic EPSPs are produced in RN cells. These were further tested by intracellular recording from red nucleus neurons. After BP stimulation, EPSPs were recorded in RN cells as indicated in Fig. 3D. In a sample of 41 cells, the latency of these EPSPs varied from 1.6 to 4 msec. In view of the monosynaptic latency of EPSPs in RN cells induced by IP stimulation, 0.9 msec, BP-induced EPSPs are presumed to be produced di- or polysynaptically.

Further evidence that the BP-induced activity reaches the red nucleus after a synaptic relay in the IP is given by the following experiments (not illustrated). Conditioning stimuli were applied to the cerebellar cortex and the test stimuli to the BP, while their effects were observed in the RN cell intracellularly. As would be expected, the EPSP generated in the RN cell by BP stimulation was completely suppressed when preceded by cerebellar cortex stimulation at short stimulus intervals. This result was taken to indicate that the BP-induced EPSPs in RN cells were mediated via the target nucleus where the cerebellar inhibitory action was exerted. In the light of the previous investigations on the cerebellar efferent system, this nucleus was presumed to be the IP. As shown in the intracellular records of Fig. 3D, stimulation of the BP produced a series of potential changes in the RN neurons in the following sequence (1) a depolarization (EPSP), already described, (2) a hyperpolarization lasting 100 msec or more, (3) a strong depolarization. This sequence of membrane potential changes corresponds to the EPSP-IPSP sequence of IP neurons as illustrated in Fig. 3C.

As would be expected from the diagram in Fig. 3A, an EPSP was produced in an RN cell with brief latency by PN stimulation (Figs. 3E and 4B). When the EPSP induced by PN stimulation was combined with that of IP, a collision phenomenon occurred. Figure 4C

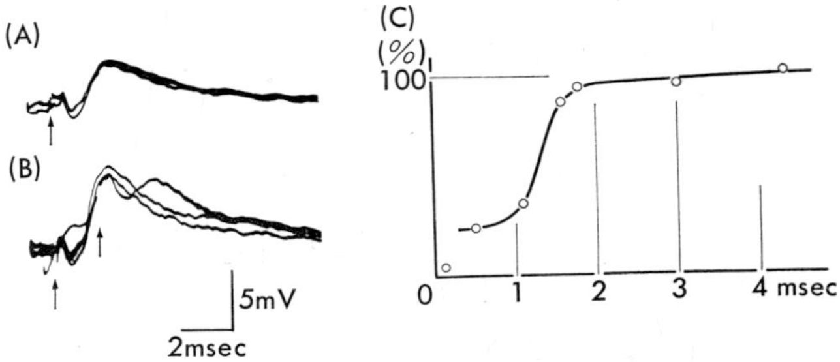

Fig. 4. Intracellular potentials induced in an RN cell by IP and PN stimulation. A, EPSP induced in an RN cell by stimulation of the IP. B, IP-induced EPSP preceded by PN stimulus. Arrows indicate the onset of the stimuli. C, Graph of percent reduction of the control EPSPs plotted against time after PN stimulation.

plots the peak amplitudes of the test IP-evoked EPSP as a function of time interval between PN and IP stimulation. At intervals of less than 2 msec the testing EPSP was depressed. The reduction of the testing EPSP was similarly observed when examined with the IP-PN sequence. According to the previous postulate (Tsukahara *et al.* 1967; Toyama, Tsukahara, Kosaka & Matsunami, 1970), the depression of the testing EPSP is explained by blockage of presynaptic impulses due to impulse collision and refractoriness when they travel along the same line of fibers in opposite directions. Furthermore, the testing EPSP reduces to 20% of the control, which indicates that approximately 80% of the IP axons actually end in the PN. These results substantiate that RN neurons are good monitors of IP-PN activity. This is discussed below in Section III.

In summary, the results demonstrate that there is a mutual monosynaptic excitatory connection between the IP and PN. From the latency measurements, the loop time of this presumed excitatory loop may be estimated in the following way. (1) Tl=0.9 msec (average of 33 cells) PN to BP, as measured from the antidromic latencies of PN cells by stimulating BP. (2) T2=1.8 msec (average of 45 cells), spike initiation of IP cells stimulating BP. (3) T3=1.2 msec (average of 32 cells), onset of EPSP in PN cells stimulating IP. (4) T4=0.5 msec, from onset of EPSP to the spike initiation of the PN cell and this was estimated as 0.5 msec based on the assumption that the spike potentials are initiated at the peak of the EPSP. The sum of these four values, 4.4 msec, is the average time for an impulse starting from one cell to return to that cell.

DYNAMIC PROPERTIES OF THE REVERBERATING CIRCUIT

If the IP-PN excitatory connection actually constitutes a reverberating circuit, one may expect a long-lasting depolarization to be produced in the neurons of these nuclei when one of these nuclei or their pathways are stimulated. Since the IP innervates the PN and also the magnocellular red nucleus (RN) monosynaptically by its axon collaterals, a longlasting depolarization may also be produced in the neurons of the RN by stimulating the IP or PN or their pathways. On the other hand, the Purkinje cells of the cerebellum are known to have strong inhibitory action upon the neurons of the IP (Ito, Yoshida, Obata, Kawai & Udo, 1970), and they are excited by the pontocerebellar mossy fiber input (Sasaki *et al.* 1969). Therefore

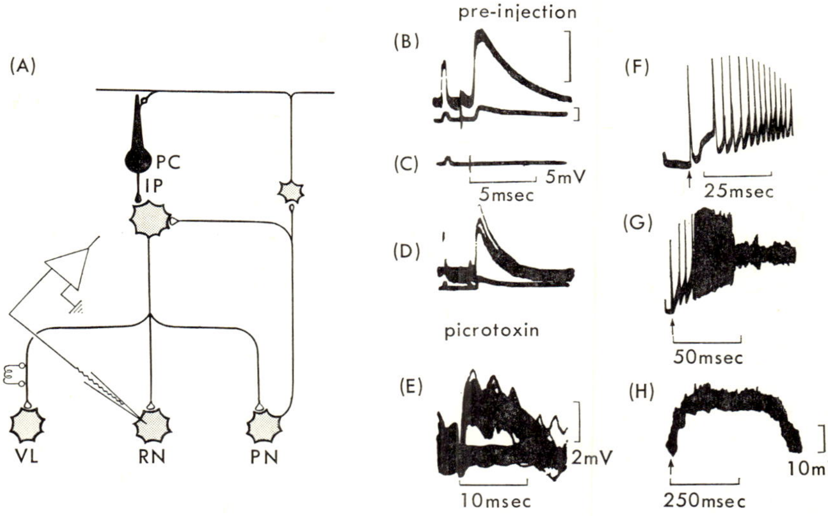

Fig. 5. A, Diagram of the experimental arrangement for study of the red nucleus (RN) neurons. Stimulation was applied to the ventrolateral thalamic nucleus (VL). Related anatomical structures are the Purkinje cell (PC), nucleus reticularis tegmenti pontis (PN), and nucleus interpositus (IP). B–D, Potential changes of red nucleus cells induced by VL stimulation before picrotoxin injection. B and D, Intracellular EPSP. Upper traces are ac recordings of intracellular potentials. Lower traces are dc recordings corresponding to the upper traces. C, Extracellular record corresponding to the dc recording in B. Time mark of 5 msec in C applies to B and C. Voltage calibration of 5 mV in B also applies to C and D. E–H, EPSPs and spike potentials induced in an RN cell by VL stimulation after picrotoxin injection. Arrows in F–H indicate the moment of VL stimulation.

this Purkinje inhibitory action would prevent the reverberation of impulses by acting at IP neurons. Fortunately, it has recently been found that picrotoxin blocks the inhibitory action of Purkinje cells upon the Deiter's neurons (Ito, Highstein & Fukuda, 1970; Obata, Takeda & Shinozaki, 1970). We may also expect that this is the case for the cerebellar nuclei. Therefore, an attempt was made to demonstrate the existence of the impulse reverberation by use of this pharmacological action of picrotoxin.

Figure 5 illustrates the transmembrane potential changes of RN cells induced by VL stimulation before (Fig. 5B–D) and after (Fig. 5E–H) picrotoxin injection. After VL or BP (Fig. 6C and G) stimulation, a prominent depolarization was produced with a burst of spike potentials. As shown in Fig. 5H, the depolarization attained a plateau after about 150 msec, lasting for several hundreds of milliseconds, and then returned to the resting state. In this case, the entire time course was about 500 msec. Multiple bursts of spikes were also observed extracellularly after picrotoxin injection. It is unlikely, therefore, that this depolarization was caused by injury to the cell impalement. A short depolarizing pulse applied through the microelectrode initiated only a single spike and never a burst of spikes (Fig. 6B). This would exclude the possibility that this depolarization is due to an after-depolarization associated with spike potentials. Since the synaptic noises increased during the depolarization phase (Fig. 6G), it is suggested that this prolonged depolarization is caused by repetitive bombardments of the EPSPs.

As for the origin of the inputs involved in triggering the EPSPs under picrotoxin, the known inputs to RN cells, the IP and the cerebral cortex via the internal capsule were taken into account (Tsukahara, Fuller & Brooks, 1968; Tsukahara & Kosaka, 1968). An attempt was made to eliminate the cerebral input to RN cells by sectioning the internal capsule through which the cortico-rubral fibers descend. This procedure also eliminated the excitatory input to RN cells via the cerebrocerebellar excitatory loop which was theorized to exist (Tsukahara, Korn & Stone, 1968). In this preparation, picrotoxin-induced depolarization still persisted.

There is also a possibility that these excitatory bombardments are produced by some loops via the spinal cord. Therefore, the spinal cord was sectioned at the second cervical segment. This procedure did not abolish the picrotoxin-induced depolarization of RN cells. Finally, there is a possibility that the picrotoxin-induced depolarization is produced by some excitatory loop via the brain stem. The

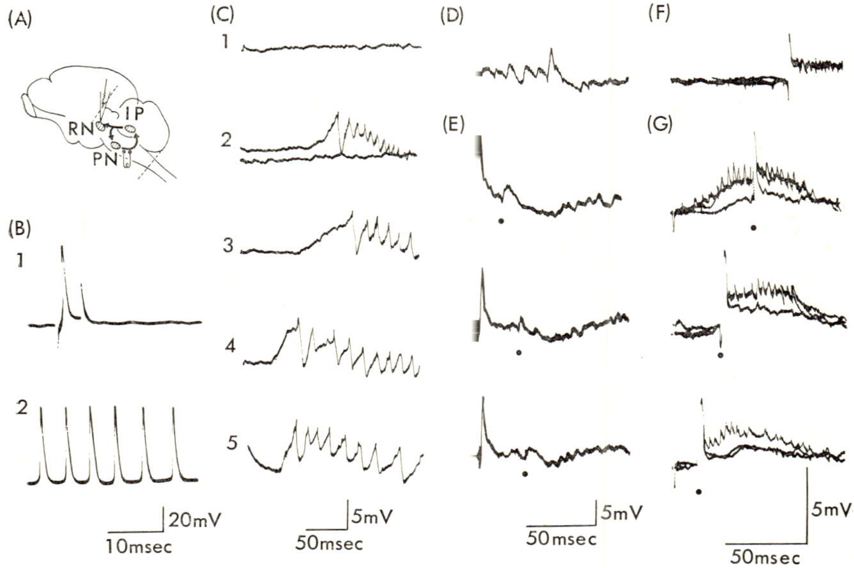

Fig. 6. A, Diagram of the experimental arrangement. Internal capsules of both sides were sectioned at the thalamic level, or cats were decerebrated at the level of A7 of the stereotaxic coordinates. The spinal cord was sectioned at the second cervical segment. In addition to these procedures, the left inferior cerebellar peduncle was sectioned but is not illustrated in the diagram. IP, Nucleus interpositus of left side; RN, red nucleus of right side; and PN, nucleus reticularis tegmenti pontis of right side. B_1, Intracellular spike potential of an RN cell induced by applying a short depolarizing pulse of 2.5×10^{-9}A. B_2, Spontaneous burst of spikes observed in the same cell as in B_1. C, Picrotoxin-induced depolarization of an RN cell evoked by BP stimulation of increasing stimulus intensities of 2, 2.8, 3, 4, 5 V, respectively, from top to bottom. D–G, Excitability test of IP neurons during BP stimulation as tested by the IP-induced EPSP in an RN cell before (D, E) and after (F, G) picrotoxin injection. Testing the IP-induced EPSPs (control in D, F) by preceding BP stimulation. Dots indicate the moment of IP stimulation. Time scale of 10 msec and voltage calibration of 20 mV apply to B_1 and B_2 (from Tsukahara *et al.* 1971).

cerebello-rubro-olivo-cerebellar loop or the cerebello-rubro-reticulo-cerebellar loop are the most likely candidates other than the cerebello-pontine loop under consideration. Since the former two loops pass through the inferior cerebellar peduncle, this possibility was tested by sectioning the inferior cerebellar peduncle.

Figure 6C exemplifies the depolarization of an RN cell by BP stimulation where the internal capsule, cervical spinal cord, and inferior cerebellar peduncle were sectioned. Thus the conclusion

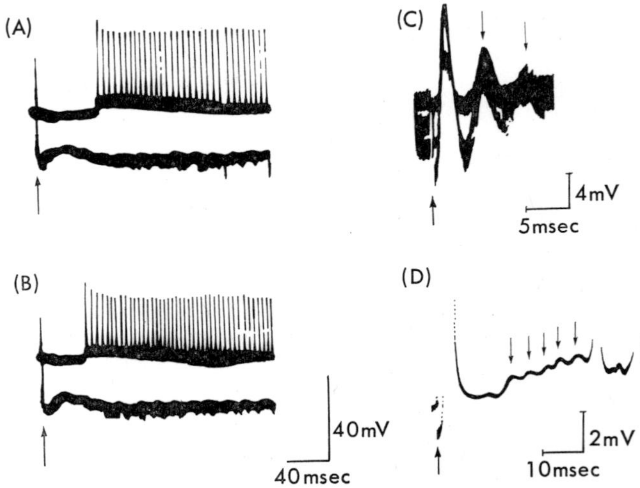

Fig. 7. A, B, Simultaneous recording of BP-induced responses from RN and PN neurons. Upper trace, intracellular response from an RN neuron. Lower trace, extracellular response from PN neurons. C, D, Ripples of the membrane potentials super-imposed on the picrotoxin-induced depolarization or evoked EPSPs. C, PN recording. D, RN recording. C, Intracellular potential of a PN cell induced by VL stimulation. Evoked EPSP is followed by additional depolarization. D, Intracellular potential of an RN cell induced by BP stimulation. The initial EPSP is followed by the picrotoxin-induced depolarization. Small wavelets of depolarization are noticed as indicated by downward arrows. Upward arrows indicate the moment of stimulation.

could be drawn that there is a mechanism in the brain stem and the cerebellum which gives tonic bombardment of EPSPs onto the RN cell through the intact BP and IP. In the light of the present evidence that the IP innervates the PN monosynaptically with excitatory synapses and the PN gives monosynaptic excitation to the IP via the BP, this mechanism must be the impulse reverberation along the excitatory IP-PN loop. This hypothesis was further supported by the fact that the long-lasting depolarization and the repetitive spike initiation were similarly observed in the neurons of the IP and PN.

Picrotoxin-induced depolarization and repetitive firing of the RN cell were recorded simultaneously with the extracellular recording of PN cells during a single shock applied to the BP. Figure 7A and B illustrates sets of records indicating that during repetitive firing of the RN cell, PN neurons are also discharging tonically in parallel with the RN cell discharges. Since the IP is the main input of RN cells, it is suggested that both IP and PN neurons are discharged tonically with a similar time course by a single BP stimulus.

Further evidence that the picrotoxin-induced depolarization is due to the repetitive firing of the IP neurons is obtained in the following way. Figure 6D–G illustrates the excitability tests of IP neurons during BP stimulation before (Fig. 6E) and after (Fig. 6G) picrotoxin administration. As reported previously (Tsukahara, Korn & Stone, 1968), BP stimulation induced in RN cells a sequence of membrane potential changes, depolarization-hyperpolarization-depolarization. Test stimuli were applied at various stimulus intervals after conditioning BP stimulation (Fig. 6E).

The amplitude of the IP-induced EPSPs decreased during the membrane hyperpolarization phase (control in Fig. 6D). Under picrotoxin, BP stimulation initiated a prolonged depolarization in RN cells. IP stimuli were applied at various stimulus intervals after BP stimulation. As can be seen in Fig. 6G, the amplitudes of the IP-EPSPs increased during the picrotoxin-induced depolarization (control in Fig. 6F). These results indicate that during picrotoxin-induced depolarization, the excitability of IP neurons markedly increased.

Finally, the possibility that the picrotoxin-induced depolarization is due to some pacemaker mechanisms either in IP or PN neurons was excluded by one experiment where, in addition to the internal capsule and inferior cerebellar peduncle, the BP was sectioned. In this preparation, picrotoxin-induced depolarization or repetitive spike potentials were not observed when monitored in RN or PN cells. Stimulation of the IP or the cut end of the BP induced only a single spike in RN or PN cells and never a burst of spikes. Therefore, it is unlikely that the picrotoxin-induced depolarization is caused by pacemaker activity in the IP or PN. It is more likely that it is produced by the presence of a nerve connection passing through the BP.

The picrotoxin-induced depolarization had a kind of threshold. Figure 6C shows the records of the depolarization by changing the stimulus intensities. At stimulus intensities of 2.8 V (Fig. 6C-2), the depolarization appeared in an all-or-none manner as noted in two traces with the same stimulus intensity. With further increase of stimulus intensity, the onset of the picrotoxin-induced depolarization shortened. In addition, the rising time course of the slow depolarization became faster as in Fig. 6C-4 and -5 as compared to the traces in Fig. 6C-2 and -3.

It is interesting to note that on the falling phase of the evoked EPSP or on the rising phase of the prolonged depolarization, ripples of EPSPs were frequently superimposed. Figure 7C illustrates a ripple in the intracellular record of a PN cell caused by stimulating the VL. It is noted the peak intervals were 4.3 and 4.5 msec. Another example of these ripples is shown in Fig. 7D. This record represents the BP-induced depolarization of an RN cell. It must be noted that on the rising phase of the prolonged depolarization, there are ripples with intervals of about 4 msec. These intervals of successive wavelets of ripples are in good agreement with the estimated loop time of the cerebello-pontine reverberating circuit.

MODEL

The main features of the dynamic properties of the reverberating circuit in the previous section can be summarized as follows: (1) Membrane depolarization of RN cells induced by picrotoxin injection exhibited a kind of threshold phenomenon. This would mean that the whole IP-PN network has a threshold of average activity below which it becomes quiescent and above which it becomes more active. (2) The rising time course of the picrotoxin-induced depolarization depends on how many IP or PN cells are excited initially by brief stimulus pulses applied either to the BP or VL. (3) The IP-PN reverberating circuit has a loop time of 4.4 msec.

In order to explain these properties, a model is proposed in this section. The idea underlying this model is that the population behavior of the IP or PN neurons is assumed to be described by a set of nonlinear differential equations. By using the experimentally determined parameters in these equations, it is possible to solve the equation and to reproduce the above-mentioned properties of the neural network.

We may assume the behavior of the IP or PN neurons is described by the following equation (McCulloch & Pitts, 1943).

$$x_i(t+\tau) = 1\left[\sum_{j=1}^{n} a_{ij}x_j(t) - \theta_i\right] \tag{1}$$

$$1[x] = \begin{cases} 1 & x \geq 0 \\ 0 & x < 0 \end{cases}$$

where $x_i(t)$ is the state of the i-th neuron at time t, and a_{ij} is the coefficient that determines the strength and polarity of the interaction from the j-th neuron to the i-th neuron. The i-th neuron gives a unit output if it gets an input at least a certain number, θ_i, of its input. It is supposed that the neuron can only change its state at one of a series of discrete equally spaced times. Thus, if one of these possible times is t and the next is $t+\tau$, then subsequent ones are $t+2\tau, \ldots$.

From the neurophysiological point of view, it may be objected that this ideal neuron is too unrealistic, especially in relation to its time dependency, because a network of these formal neurons behaves in a synchronized fashion, t and $t+\tau, \ldots$. However, it must be realized that the great advantage of this formulation is its simplicity, which is necessary for treating the complex network behavior here.

To make theoretical analysis possible for the IP-PN reverberating circuit, the following two points are assumed. (1) IP is composed of m neurons. There is no intranuclear connection between these m IP neurons. Each IP neuron receives excitatory connections from all PN neurons with the same coupling coefficient, k'. (2) Similarly, PN is composed of $n-m$ neurons. There is no intranuclear connection between these PN neurons. Each PN neuron receives excitatory connections from all IP neurons with the same coupling coefficient, k. From these assumptions, it follows,

$$\sum_{i=1}^{m} x_i(t+\tau) = \sum_{i=1}^{m} 1\left[k' \sum_{j=m+1}^{n} x_j(t) - \theta_i\right] \tag{2}$$

$$\sum_{i=m+1}^{n} x_i(t+\tau) = \sum_{i=m+1}^{n} 1\left[k \sum_{j=1}^{m} x_j(t) - \theta_i\right] \tag{3}$$

Here we shall introduce $I(t)$, and $P(t)$ as

$$I(t) \equiv \frac{1}{m} \sum_{i=1}^{m} x_i(t)$$

$$P(t) \equiv \frac{1}{n-m} \sum_{i=m+1}^{n} x_i(t) .$$

Hence, $I(t)$ is a quantity between 0 and 1, and represents a fraction of the excited IP neurons out of the total m IP neurons and similarly $P(t)$ represents a fraction of excited PN neurons out of the total $n-m$ PN neurons. If m or $n-m$ is sufficiently large, $I(t)$ or $P(t)$ would be regarded as a continuous quantity in the range between 0 and 1. By substituting in Eqs. (2) and (3) with $I(t)$ and $P(t)$, we have the following equations.

$$I(t+\tau) = \frac{1}{m} \sum_{i=1}^{m} 1[k'(n-m)\cdot P(t) - \theta_i] \tag{4}$$

$$P(t+\tau) = \frac{1}{n-m} \sum_{i=m+1}^{n} 1[km\cdot I(t) - \theta_i] \tag{5}$$

In order to solve these equations using experimental parameters, we assume the following approximation.

$$I(t+\tau) \simeq I(t)+\tau\cdot\frac{dI(t)}{dt} \tag{6}$$

$$P(t+\tau) \simeq P(t)+\tau\cdot\frac{dP(t)}{dt} \tag{7}$$

The introduction of Eqs. (6) and (7) is highly speculative from the mathematical point of view, but fairly reasonable in view of the continuous nature of time in the real brain. By introducing the non-linear functions $\Psi[P(t)]$ and $\Phi[I(t)]$ which are defined by Eqs. (10) and (11) the simplified expressions obtained are:

$$\frac{dP(t)}{dt} = -\frac{1}{\tau}P(t)+\Phi[I(t)] \tag{8}$$

$$\frac{dI(t)}{dt} = -\frac{1}{\tau}I(t)+\Psi[P(t)] \tag{9}$$

where

$$\Psi[P(t)] = \frac{1}{\tau m}\sum_{i=1}^{m} 1[k'(n-m)P(t)-\theta_i] \tag{10}$$

$$\Phi[I(t)] = \frac{1}{\tau(n-m)}\sum_{i=m+1}^{n} 1[kmI(t)-\theta_i]\,. \tag{11}$$

$\Psi[P(t)]$ is a quantity between $1/\tau$ and 0 and relates the population behavior of the IP nucleus when excited by PN input. Similarly, $\Phi[I(t)]$ is a quantity between $1/\tau$ and 0, and relates the population behavior of PN neurons excited by IP inputs.

It is important to note that $\Psi[P(t)]=0$, as long as the condition that $k'(n-m)P(t)<\theta_i$ is fulfilled, regardless of the value of θ_i. Similarly, it is noted that $\Phi[I(t)]=0$, as long as the condition that $kmI(t)<\theta_i$ is fulfilled. Furthermore, it is noted that $\Psi[P(t)]=1/\tau$ as long as $k'(n-m)P(t)>\theta_i$ is satisfied regardless of the value of θ_i. Thus if $\theta_i\equiv\theta$ for all values of i, $\Psi[P(t)]$ or $\Phi[I(t)]$ is a step function as shown in Fig. 8A. If the frequency distribution of θ_i is given as the inset diagram in Fig. 8B, $\Psi[P(t)]$ or $\Phi[I(t)]$ is given by the solid curve of Fig. 8B.

In applying the solution of this equation to the cerebro-pontine-reverberating circuit, it is necessary to have the following parameters: the functional form of $\Psi[P(t)]$, $\Phi[I(t)]$, and τ and the initial conditions P $(t=0)$ and I $(t=0)$. Since $\Psi[P(t)]$ represents the relation of a fraction of IP neurons excited by PN inputs, it is possible to obtain

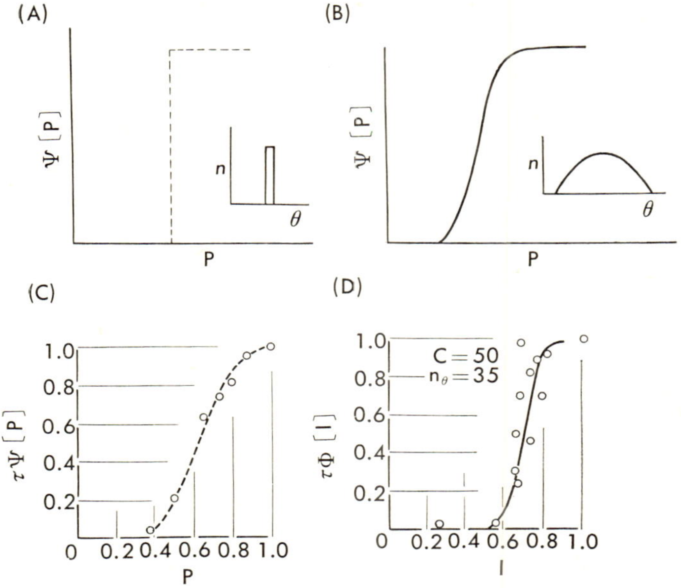

Fig. 8. A, The form of the $\Psi[P]$ versus P relation when all θ_i are equal to θ as indicated in the inset. B, The expected curve of $\Psi[P]$ versus P when the frequency distribution is given as the inset of B. C, Input-output relation of the IP neuronal pool as determined by the field potential in the IP nucleus by stimulating BP. Abscissa; Amplitude of presynaptic component of the field potential (indicated by maximal amplitude at 1). Ordinate; Amplitude of postsynaptic component of the field potential as expressed by the ratio of maximal amplitude. D, Input-output relation of PN neuronal pool. Open circles, experimental data as determined by the field potential in the PN nucleus by stimulating IP. Solid line represents the relation determined by Eq. (13). Abscissa; Amplitude of presynaptic component of the field potential (indicated by the maximal amplitude at 1). Ordinate; Amplitude of the postsynaptic component of the field potential as expressed by the ratio of the maximal amplitude. For the theoretical relation indicated by the solid line, the abscissa represents m/M, and the ordinate plots n/N of Eq. (13).

this relation experimentally. In order to do this, a microelectrode was inserted in the IP region, and the BP was stimulated. The field potential was recorded in this region in response to BP stimulation. The field potentials were composed of two component potentials; peak latencies of these potentials were 0.5 and 1.1 msec.

Several lines of evidence led us to conclude that the initial component is presynaptic in nature while the second component is postsynaptic. Figure 8C demonstrates the relation of amplitudes of both pre- and postsynaptic potentials obtained from a set of records applying stimulus to BP at various stimulus strengths. If it is assumed that each pre- and postsynaptic fiber contributes the same amount of potential changes to the extracellular field potentials, the relation in Fig. 8C may be considered to represent a gross input-output relation, and it can be helpful for deriving the $\Psi[P(t)]$ versus $P(t)$ relation, if this curve does not change with time.

Similarly, $\Phi[I(t)]$ represents the relation of the fraction of PN neurons excited by the IP inputs. By recording the extracellular field potentials in the PN nucleus, it was found that the field potentials consist of two component potentials, 0.5 and 1.3 msec in peak latencies. The initial component potential was assumed to be presynaptic in origin and the second component postsynaptic in nature. Therefore, a similar input-output relation was constructed and is illustrated in Fig. 8D.

In this case, however, we derived the similar input-output relation in a different way, using the parameters of the individual PN neurons determined experimentally in Section II. Suppose that the PN nucleus is composed of N PN neurons and each PN neuron receives C inputs from the IP nucleus. The whole PN nucleus is supplied from the IP nucleus with M excitatory fibers, which contact individual PN neurons at random. Suppose also m fibers are active out of the total M input fibers. For any PN neuron, the probability that x inputs are excited out of the total C inputs to the PN neuron; $P(x:m)$, would be described by the following bionominal expression.

$$P(x:m)=\binom{c}{x}\left(\frac{m}{M}\right)^{x}\left(1-\frac{m}{M}\right)^{c-x} \tag{12}$$

Therefore, the fraction of the excited PN neurons over the total N PN neurons, n/N, can be calculated by the following equation,

$$f\left(c, n_\theta, \frac{m}{M}\right)=\frac{n}{N}=\sum_{x=n_\theta}^{c}\frac{c!}{x!(c-x)!}\left(\frac{m}{M}\right)^{x}\left(\frac{M-m}{M}\right)^{c-x} \tag{13}$$

where n_θ denotes the threshold value of PN neurons, which was estimated as 35 in Section II. C was estimated as 50. Substituting $C=50$, and $n_\theta=35$ in Eq. (13), a theoretical curve was obtained as illustrated by the solid line in Fig. 8D. Comparing this curve with the experimental values derived from field potential analysis as described by open circles, both relations agree quite well. In this article, $\Phi[I(t)]$ or $\Psi[P(t)]$ was derived from these curves.

As for the other parameter in Eqs. (8) and (9), τ, it may be reasonably considered that the PN or IP nucleus receives excitatory

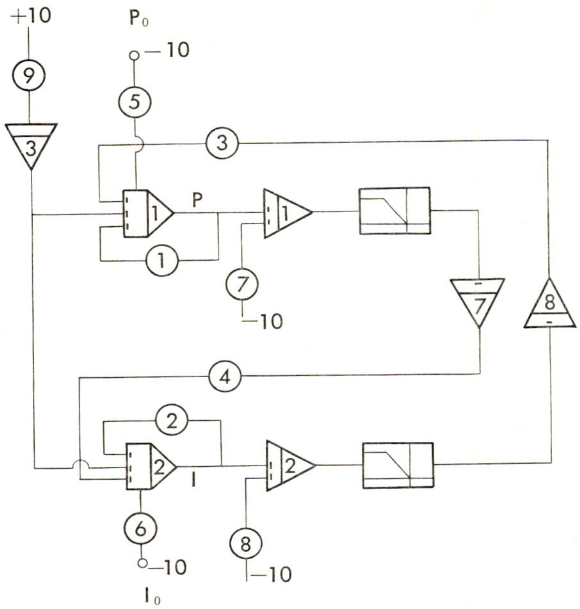

Fig. 9. Analog computer program for solving Eqs. (8) and (9). P_0, I_0 represent the initial condition of P and I at $t=0$, respectively. See text.

impingement every discrete time. In Section II, the loop time of the cerebello-pontine reverberating circuit was estimated as 4.4 msec. Therefore, it was assumed to be $2\tau=4$ msec.

In order to solve Eqs. (8) and (9) at various initial conditions, an analog computer was used. Figure 9 illustrates the program of the computer. The nonlinear function $\Phi[I(t)]$ or $\Psi[P(t)]$ was generated by the function generator of the computer. Figure 10 illustrates the family of curves obtained from these equations by plotting $I(t)$ against time. As can be seen from these curves, the agreement is qualitative rather than quantitative, but the model fairly successfully reproduces the main features of responses. A difference exists in the time course of $I(t)$. Experimentally, a rather slow rise of the membrane depolarization is obtained, but the theoretical curve represents a rather steep increase. This discrepancy may partly be due to the fact that we neglected the contribution of the refractory period of neurons.

In spite of such a difference, the general similarity between the two sets of curves (Figs. 10 and 5H or 6C) is rather striking and suggests that the main characteristics of the picrotoxin-induced depolarization,

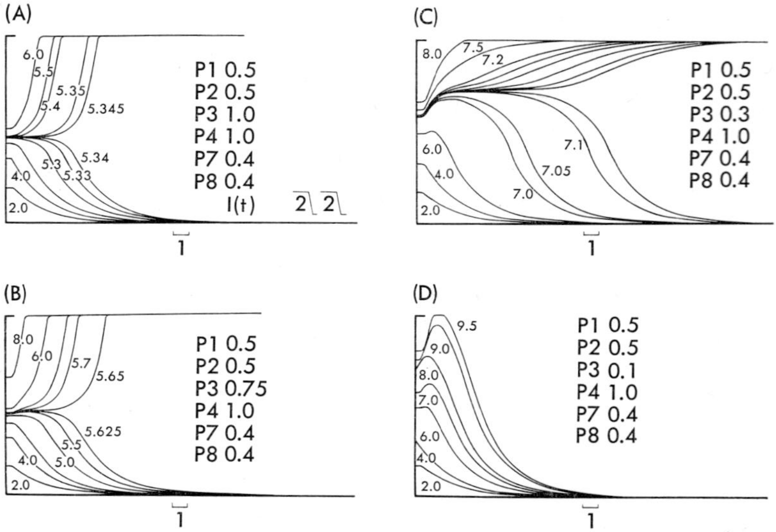

Fig. 10. The relation of $I(t)$ *versus* t as calculated by the analog computer. Initial condition, $P_0 = I_0$, is indicated on each trace in the figure.

threshold phenomenon, and the effects of the initial conditions can be explained by this model. It does not seem worth trying other values for the constants in the hope of obtaining a better fit, because at our present stage of knowledge of the connectivity of IP and PN neurons, it would be a pointless complication.

CONCLUSION

We have reviewed the major physiological and theoretical results that led the authors to conclude that the neuronal network in question is a reverberating circuit composed of neurons of the nucleus interpositus of the cerebellum and those of the nucleus reticularis tegmenti pontis. Individual cell properties and their connections were investigated by applying microelectrode techniques to both IP and PN neurons. In addition, the population behavior of IP neurons was monitored by an intracellular microelectrode as a membrane potential change of the red nucleus cell. The latter stems mainly from the following experimental results: (1) The IP constitutes a main excitatory input onto RN neurons (Tsukahara *et al.* 1967). Approximately 50 IP axons converge onto an RN cell (Toyama *et al.* 1970). (2) Each IP axon produces 0.8 mV depolarization in an RN cell

(Toyama *et al.* 1970). The EPSP decays exponentially with a time constant of 7 msec (Tsukahara *et al.* 1967). (3) The belief that the RN cells serve as monitors of the IP-PN reverberating circuit arises from the collision experiments between IP- and PN-induced EPSP in an RN cell, indicating that more than 80% of IP axons that send synapses to RN cells also make synapses to PN neurons. (4) It is possible to estimate the mean firing frequency of the IP axons from the amount of depolarization of RN cells by the following equation.

$$f = d/s \cdot n \tag{14}$$

where s denotes the time integral of a unitary EPSPs which was estimated as 5.1 mV•msec (Toyama, Tsukahara & Udo, 1968). The convergence number of RN cells from IP neurons is n, and d denotes the mean depolarization as measured in a monitored RN cell. It must be noted therefore that the time course of the membrane depolarization of RN cells serves as a good physiological monitor of the recruitment of IP neurons due to impulse reverberation.

The major characteristic of the properties of the reverberating circuit is its regenerative behavior. Above and below threshold, there are two distinct stable states, the excited state where many constituent neurons become active and the resting state where all neurons are inactive. However, the time course of the transient behavior of the neuronal recruitment depends on how many cells are excited initially. These two basic properties were successfully reproduced by a mathematical model proposed in this article.

However, there is one point that was not elucidated either experimentally or theoretically. For some mechanisms that are obscure at present, the excited state shifts to the resting state automatically after about 500 msec or so. It is of interest in this connection that in the isolated cerebral cortical slab Burns (1958) found that the prolonged electrical activity subsided after several seconds, and the duration of the prolonged activity depended upon the size of the slab that was isolated from the rest of the cerebral cortex.

Ebina succeeded in reproducing the resetting of the excited state to the resting state in his theoretical consideration of the cerebello-pontine reverberating circuit by assuming that there is some accommodative process in the neuronal firing of either IP or PN neurons. It is possible, however, to reproduce this as in Fig. 10 without assuming any of these accommodative processes, but by taking into account the different input-output relations of IP and PN nuclei. However, the authors would not argue that the mechanism discussed above is the sole cause of the resetting of the excited state to the resting state which was monitored in RN neurons as the termination of the prolonged depolarization. If the accommodation or the depression of the synaptic transmission is assumed, the n_θ of Eq. (13) may alter to n_θ' where $n_\theta' < n_\theta$. It is easy to demonstrate $f(c, n_\theta, m/M) > f(c, n_\theta', m/M)$. Therefore, the curve of Fig. 8C and D could be less steep. This would result in a shift from the excited state to the resting state. The question must, however, remain open for future experimental study.

Even confined to a rather simple reverberating circuit, the experiments and the model developed above may be criticized in several respects. From the experimental point of view, it is possible that the reverberating circuit just mentioned is contaminated with other reverberating circuits, the dentato-pontine or interposito-rubro-interposeal loop (Ito, 1969), although other possible loops were excluded by trans-section experiments. Unfortunately, these postulated loop systems are not well understood so it is difficult to evaluate the contribution of these loops, although from the neurophysiological data, it is possible that the latter may not be important.

One may also consider it far from realistic to assume a complex spatial interconnection between IP and PN to be homogeneous as mentioned in Section IV.

The assumption that there is no intranuclear connection in either IP or PN nuclei needs some comment. Recent neuroanatomical study (Matsushita & Iwahori, 1971) revealed that there are some recurrent collaterals in the IP that may terminate in the interneurons in the IP nucleus. However, there is no electrophysiological proof of their existence, nor any detailed physiological study of these postulated recurrent fibers. Therefore, it must be considered that the model was constructed based on the most prominent connections so far examined. This kind of simplification is, however, rather unavoidable for understanding the dynamic aspect of the network.

The use of the McCulloch-Pitts model for the real neuron may also be questioned. At best, it would approximate only the threshold behavior of a neuron. Actually, the McCulloch-Pitts model has been replaced by the more realistic Caianiello model (Caianiello, 1961), which takes into account refractoriness and the time-dependent coupling coefficient. However, in view of the present stage of neurophysiology, it is difficult, or even impossible, to reconcile the many variables of Caianiello's equation to the experimental results. Therefore, in this work an attempt is made to derive simple differential equations starting from the McCulloch-Pitts model, which can be correlated with the experimentally determinable variables. It is to be hoped that in the near future more rigorous mathematical formulation will be developed in linking the properties of a network with those of the constituent neurons.

From the functional point of view, the reverberating circuit may act as a short-term memory storage in the central nervous system, in view of the presence of the two stable states in its dynamic properties. Present experiments do demonstrate that it transforms a millisecond electrical activity to a hundred millisecond neuronal firing. In the neuroendocrine system, on the other hand, we may assume that some intrinsic rhythm must be generated in the central nervous system. Therefore, we must seek the generator either in the cell properties of the endocrine center, such as the pacemaker activity, or in the assembly of neurons, each of which has no definite intrinsic rhythmic activity. The present example may provide some clue for understanding the latter mechanism.

ACKNOWLEDGEMENTS

The authors thank Prof. H. Tamura in our department for his discussion on the analogue computer program. We also thank Miss M. Kakinoki for typing the manuscript. This research was partly supported by the Ministry of Education of Japan. Grant No. 92105 and 92475.

REFERENCES

Burns, B. D. (1958). *The Mammalian Cerebral Cortex.* pp. 24–33, London: Edward Arnold.

Caianiello, E. R. (1961). Outline of a theory of thought processes and thinking machines. *J. theor. Biol.* **1**, 204–235.

Ebina, Y. Generation and disappearance of impulses in excitatory connected two modeled neuron groups (in preparation).

Eccles, J. C. (1967). Circuits in the cerebellar control of movement. *Proc. nat. Acad. Sci. U.S.A.* **58**, 336–343.

Eccles, J. C., Ito, M. & Szentagothai (1967). *The Cerebellum as a Neuronal Machine.* p. 257, Berlin: Springer-Verlag.

Forbes, A. (1922). The interpretation of spinal reflexes in terms of present knowledge of nerve conduction. *Physiol. Rev.* **2**, 361–414.

Ito, M. (1969). Neurons of cerebellar nuclei. In *The Interneuron,* ed. Brazier, A. B., pp. 309–327, UCLA Forum.

Ito, M., Highstein, S. M. & Fukuda, J. (1970). Cerebellar inhibition of the vestibuloocular reflex in rabbit and cat and its blockage by picrotoxin. *Brain Res.* **17**, 524–526.

Ito, M., Yoshida, M., Obata, K., Kawai, N. & Udo, M. (1970). Inhibitory control of intracerebellar nuclei by the Purkinje cell axons. *Exp. Brain Res.* **10**, 64–80.

Lorente de Nó, R. (1938). Analysis of the activity of the chains of internuncial neurons. *J. Neurophysiol.* **1**, 207–244.

Matsushita, M. & Iwahori, N. (1971). Structural organization of the interpositus and the dentate nuclei. *Brain Res.* **35**, 17–36.

McCulloch, W. S. & Pitts, W. H. (1943). A logical calculus of the ideas immanent in neural nets. *Bull. math. Biophys.* **5**, 115–133.

Obata, K., Takeda, K. & Shinozaki, H. (1970). Further study on pharmacological properties of the cerebellar-induced inhibition of Deiters neurons. *Exp. Brain Res.* **11**, 327–342.

Sasaki, S., Kawaguchi, S., Shimono, T. & Yoneda, Y. (1969). Responses evoked in the cerebellar cortex by the pontine stimulation. *Jap. J. Physiol.* **19**, 95–109.

Toyama, K., Tsukahara, N., Kosaka, K. & Matsunami, K. (1970). Synaptic excitation of red nucleus neurones by fibres from interpositus nucleus. *Exp. Brain Res.* **11**, 187–198.

Toyama, K., Tsukahara, N. & Udo, M. (1968). Nature of the cerebellar influences upon the red nucleus neurones. *Exp. Brain Res.* **4**, 292–309.

Tsukahara, N. (1972). The properties of the cerebellopontine reverberating circuit. *Brain Res.* **40**, 67–71.

Tsukahara, N. & Bando, T. (1970). Red nuclear and interpositonuclear excitation of pontine nuclear cells. *Brain Res.* **19**, 295–298.

Tsukahara, N., Bando, T., Kitai, S. T. & Kiyohara, T. (1971). Cerebello-pontine reverberating circuit. *Brain Res.* **33**, 233–237.

Tsukahara, N., Fuller, D. R. G. & Brooks, V. B. (1968). Collateral pyramidal influences on the corticorubral system. *J. Neurophysiol.* **31**, 467–484.

Tsukahara, N., Korn, H. & Stone, J. (1968). Pontine relay from cerebral cortex to cerebellar cortex and nucleus interpositus. *Brain Res.* **10**, 448–453.

Tsukahara, N. & Kosaka, K. (1968). The mode of cerebral excitation of red nucleus neurons. *Exp. Brain Res.* **5**, 102–117.

Tsukahara, N., Toyama, K. & Kosaka, K. (1967). Electrical activity of the red nucleus neurons investigated with intracellular microelectrodes. *Exp. Brain Res.* **4**, 18–33.

SEXUAL DIFFERENTIATION AND DEVELOPMENT OF THE HYPOTHALAMUS AND STEROID-INDUCED STERILITY

YASUMASA ARAI

Regardless of genetic sex, rats are born with a sexually undifferentiated hypothalamus which is of the female pattern. It is only in the first 7–10 postnatal days that this pattern becomes established in the female. But in the male the normal masculine pattern of GTH regulation is organized by the internal secretion of the neonatal testes during the first few days of life. This process involves the differentiation of the preoptic-anterior hypothalamic regulation of GTH secretion, which determines whether this region can have the functional cyclicity or not. From recent experimental data, it is likely that exposure of the developing hypothalamus to androgen alters the maturation and the normal development of steroid-receptors permanently and perhaps changes the integrate mechanisms of the preoptic-anterior hypothalamus to the acyclic masculine pattern.

INTRODUCTION

It is generally accepted that the neural control of reproductive activity of the male is different from that of the female. The mechanisms that regulate the cyclic release of the pituitary gonadotropins (GTH) are apparently absent in the male. An ovary transplanted to any region outside the hepatic circulation of the male rat contains large vesicular follicles but no corpora lutea (CL). This indicates that the male rat does not display cyclic fluctuations in GTH secretion. However, ovaries transplanted to adult male rats that have been castrated within a few days of the day of birth show normal follicular growth and formation of CL (Pfeiffer, 1936; Yazaki, 1960;

Kawashima, 1960; Takewaki, 1962; Harris, 1964; Gorski & Wagner, 1965). Ovarian transplants of this structure reflect the female pattern of cyclic release of GTH by the host. Thus, it can be assumed that, regardless of genetic sex, the rat brain has the ability to maintain the cyclic female pattern of GTH regulation at birth. The undifferentiated sexual centers must be female. For the development of the masculine (acyclic) pattern of GTH regulation in the rat, the presence of endogenous testicular activity during the neonatal period seems to be the critical factor in the male. This conjecture is additionally supported by the fact that testicular transplantation to newborn female rats induces the differentiation of GTH regulation in the male direction, *i.e.*, they lose the ability to maintain the cyclic release of GTH (Pfeiffer, 1936; Gorski *et al.* 1965). Similarly, the existence of a neonatal steroid-sensitive period during which exogenous androgen or estrogen may cause disturbances of the differentiation of the cyclic female pattern of GTH regulation has been demonstrated in the female rat. It is only in the first 7–10 postnatal days that single or multiple injections of sex steroids to female rats can induce acyclic anovulatory sterility (Takewaki, 1962; Harris, 1964; Jacobsohn, 1965; Barraclough, 1966, 1967; Gorski, 1966, 1971; Duncan, Wyngarden & Cornette, 1967; Flerkó, 1967; Saunders, 1968; Arai, 1968, 1972*a*). As a matter of fact, in normal female rats the development of the neural substrates for the female pattern of GTH regulation is not disturbed and does continue free from the steroidal influence during that steroid-sensitive period, because neonatal ovaries are thought to be incapable of synthesizing enough sex steroids (Price & Ortiz, 1944).

The influence of neonatal endogenous or exogenous sex steroids is now thought to be exerted presumably upon the hypothalamus (Barraclough & Gorski, 1961; Wagner, Erwin & Critchlow, 1966), but the mode of action is still obscure. The term hypothalamic masculinization would be acceptable for the phenomenon of losing the cyclicity of GTH secretion induced by neonatal steroidal influence. However, the process of hypothlamic masculinization is not always a physiological phenomenon if induced experimentally, because of its dependence on the dosage of hormone injection and the length of the injection period. The functional alterations in the hypothalamic sexual centers may vary in degree according to these treatment schedules (Gorski & Barraclough, 1963; Arai, 1964*a*). In this chapter, the current understanding of the concept of sexual differentiation of the brain is reviewed briefly, problems surrounding the mechanisms of sex steroid-induced hypothalamic sterility are discussed from the physiological level to the pharmacological level.

DEVELOPMENT OF THE ACYCLICITY OF GTH REGULATION IN THE MALE

As mentioned above, Pfeiffer (1936) was the first to report the basic findings that, following castration on the day of birth (day 1), the male rat can form CL in an ovarian transplant. Gorski (1966) has proposed the term *fale* (feminine male) to describe this type of male rat which is capable of secreting GTH in a cyclic or female pattern. Yazaki (1960) demonstrated that castration as late as day 3 is still compatible with CL formation, and the presence of neonatal testes for more than 3 days of life is necessary to secure proper development of the normal masculine pattern of GTH secretion. Since then, similar attempts have been made by several investigators to examine the critical period during which castration can stop the development of the masculine pattern of GTH regulation. Summaries of the results are presented in Table 1. The data from albino strains (T. Wistar & Sprague-Dawley) mostly confirm the findings of Yazaki (1960). Interestingly enough, the results from pigmented strains (Lister, Wayne & Long-Evans) differ from those of albino rats. The critical period is apparently shorter in the pigmented rats than

TABLE 1. Summary of the effects of neonatal castration on development of the hypothalamic GTH regulation in the male rat

Strain	Critical period for production of *fales**							Reference
	(1 = day of birth)							
	1	2	3	4	5	6	7	
Albino								
T	+	+	+	±	—	—	—	Yazaki (1960)
Sprague-Dawley	+	(+)**	+	(—)	—	—	(—)	Gorski & Wagner (1965)
Wistar	+	+	+	±***	—	—	—	Arai, unpublished
Pigmented								
Lister	+	—	—	—	—	—	—	Harris (1964); Harris & Levine (1965)
Wayne	+	—	—	(—)	—	(—)	(—)	Inoué (1966)
Long-Evans	+	—	—	—	(—)	(—)	(—)	Arai, unpublished

* Feminine male, see text; ** not examined; *** small number of animals became *fales*.

in albino rats. The cause of this strain difference is not clear at present. Recently we found that a single injection of testosterone propionate (TP, 50 μg) to 5-day-old female rats or to 5-day-old male rats which had been orchidectomized on day 1 (potential *fales*) was equally effective in masculinizing the pattern of GTH secretion among rats of the strains examined (Wayne, Long-Evans, Wistar & Sprague-Dawley; Arai, unpublished). This would indicate that

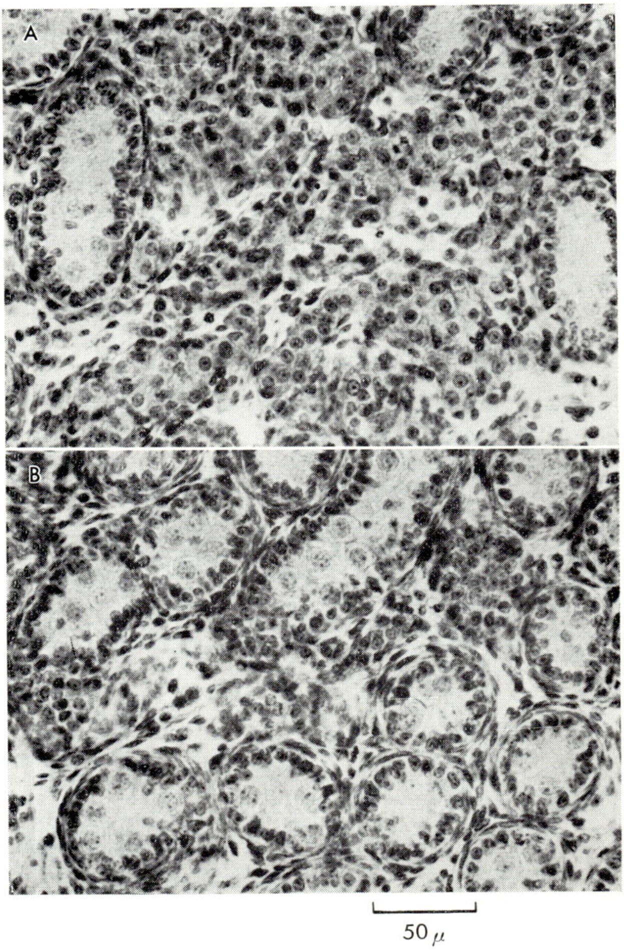

Fig. 1. Sections of testes of 3-day-old rats. A, A section of a testis of a rat treated with 40 IU HCG for the first 2 days of life. Note that intersuitial tissue (Leydig cells) is markedly increased. B, A section of a testis of a control rat treated with saline. H-E stain.

the competence of the developing neurons of the hypothalamus to respond to exogenous androgen is not markedly different among these strains of rats. Presumably the difference exists in the testicular activity. The testicular activity in pigmented rats may reach the functional level for masculinizing their brain earlier than that in albino rats. If that is the case, what brings about the difference in testicular activity? Since the testes of the neonatal rats are possibly under pituitary control (Yaginuma, Matsuda, Murasawa, Kobayashi & Kobayashi, 1969; Goldman, Grazia, Kamberi & Porter, 1970; Goldman & Gorski, 1971), it is of interest to speculate that there is a strain difference in the degree of development of the basal GTH secretory mechanisms even before the differentiation of GTH regulatory pattern occurs.

In unpublished observations from the author's laboratory, Arai and Serisawa have been able to accelerate the sexual differentiation of the brain (hypothalamic masculinization) by neonatal administration of HCG. In our colony (Wistar), castration on day 3 is still capable of producing *fales* (see Table 1), but in male rats receiving daily injections of 40–10 IU HCG for the first 2 days of life, castration on day 3 was no longer capable of producing *fales*. Histologic examination of their neonatal testes showed marked stimulation by HCG, Leydig cells exhibiting hyperplasia and hypertrophy (Fig. 1A), compared with nontreated neonatal testes (Fig. 1B). The results suggest that neonatal testes of these rats are able to secrete a physiological level of androgen if enough GTH (presumably LH) is available at that time. Sutherland (personal communication) found that a sharp increase in serum LH occurred on day 3 in the neonatal male rat, although FSH levels were rather constant during neonatal days.

Although the results of neonatal castration implicate testicular androgen, Resko, Feder & Goy (1968) gave biochemical proof that neonatal testes are actually secreting testosterone in the blood. The concept that testicular testosterone is the differentiating factor in the rat is now acceptable. Neumann & Kramer (1967) reported that antiandrogen, cyproterone acetate (1,α-methylen-6-chloro-$\Delta^{4,6}$-pregnadiene-17α-ol-3,20dione-17α-acetate), was effective in preventing normal differentiation of the male hypothalamus. Recently we obtained results indicating that administration of a biochemical inhibitor of androgen synthesis, Su-10603 (7-Cl-3,4-dihydro-2-(3-pyridyl)-1(2H)-naphthalenone), also somehow prevented the hypothalamic masculinization (Arai, unpublished). Su-10603 has been reported to inhibit 17β-hydroxylase and C_{17-20}-lyase selectively in rat fetal

TABLE 2. Effects of various steroids on the hypothalamic masculinization of potential *fales**

Steroid injected on day 4 or 5	Percent masculinized	References
10 μg TP (day 4)	100	Gorski & Wagner (1965)
50 μg TP (day 5)	100	Arai (1972 *b*)
500 μg DHT* (day 5)**	0	Arai (1972 *b*)
50 μg TP+500 μg DHT (day 5)	100	Arai (1972 *b*)
5 μg EB*** (day 5)	100	Gorski & Wagner (1965)
50 μg diethylstilbestrol	100	Arai, unpublished
50 μg hexestrol	90	Arai, unpublished

* All test animals were castrated on day 1; ** 5α-dihydrotestosterone; *** estradiol benzoate.

testis (Goldman, 1971). In our preliminary experiments, daily injections of Su-10603 (500 μg) were given to the male for the first 4 days of life, and castration was performed on day 5. CL formation was found in 4 out of 11 rats treated with Su-10603 neonatally, but CL formation was never found in any of the castrated controls (0/10). This suggests that biochemical inhibition of the neonatal testicular activity (presumably inhibition of testosteronogenesis) may disturb the sexual differentiation of the male hypothalamus.

The influence of exogenous sex steroids other than testosterone on the hypothalamic masculinization of potential *fales* has also been examined by several investigators. The results are summarized in Table 2. Ten μg of TP is the minimal dosage of exogenous steroids that will masculinize potential *fales* (Gorski *et al.* 1965). In contrast, it is of interest that 500 μg of 5α-dihydrotestosterone (DHT, 5α-androstan-17β-ol-3-one) was not effective in inducing hypothalamic masculinization (Arai, 1972*b*). Recently it has been suggested that one of the most active forms of androgen may be this steroid in peripheral target organs such as prostates and seminal vesicles (Anderson & Liao, 1968; Bruchovsky & Wilson, 1968). In these tissues, the conversion of testosterone to DHT appears to be a prerequisite for the initiation of protein synthesis. The presence of an enzyme system that converts testosterone to DHT has also been shown in the hypothalamus (Jaffe, 1969; Martini, 1971). Further, McDonald (1971) and Beyer, Morali & Cruz (1971) reported that DHT at relatively low dosages effectively inhibits ovarian compensatory hypertrophy in the rat. Beyer *et al.* (1971) demonstrated that this inhibitory action of DHT was exerted on the hypothalamus.

Therefore, failure of DHT to masculinize the *fale* brain is in strong contrast to these cases. The fact that DHT given in combination with TP could not attenuate the masculinizing effect of TP (see Table 2) would indicate that the binding of TP (presumably testosterone) to the possible receptors in the hypothalamus involved in sexual differentiation cannot be affected by the simultaneous presence of DHT, and the receptor site of DHT might not be the same as TP if DHT is taken up by the neonatal hypothalamic tissue. Therefore, DHT may not be normally involved in the hypothalamic masculinization. The nature of testosterone receptors involved in hypothalamic masculinization may differ from that of receptors in the peripheral target organs, in which DHT has been reported to be the most active in masculinizing activity when given prenatally (Goldman & Baker, 1971).

Hypothalamic masculinization may not be androgen specific. Other than androgen, natural and synthetic estrogen have been also found to be effective. Five μg estradiol benzoate (EB) is the minimal dosage of this steroid for hypothalamic masculinization, but this seems to be pharmacologically high (Gorski *et al.* 1965). Ramirez & McCann (1963) stated that 5 μg EB would be hunderds of times above the physiological level of estrogen in immature rats. Furthermore, Flerkó, Petrusz & Tima (1967) reported that 0.1 μg EB administered repeatedly to the male rat for the first 10 days of life could prevent the masculinizing effect of testicular androgen. Although the mechanisms of this action of low dosages of EB are not certain yet, they considered two possibilities: antagonism between two steroids, and negative feedback action of EB, which might inhibit GTH secretion by the neonatal pituitary. However, the presence of an aromatizing enzyme system has been demonstrated in the developing human male hypothalamus (Naftolin, Ryan & Petro, 1967). Thus, whether the hypothalamic aromatization of androgen to estrogen is normally involved in the processes of sexual differentiation of the rat brain is now an interesting problem to be solved.

MASCULINIZATION OF THE FEMALE HYPOTHALAMUS BY ANDROGEN

Testicular transplantation or administration of androgen to newborn female rats is capable of masculinizing their brains in a manner comparable to the *fales*. These female rats do not ovulate and show persistent vaginal estrus. Their ovaries contain numerous large

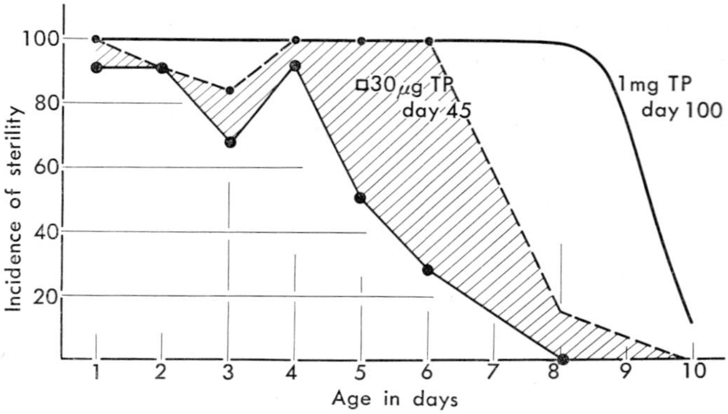

Fig. 2. Influence of age at injection of 10 μg TP (30 μg, 1 mg TP) and at observation, on the incidence of sterility (No. anovulatory/No. injected). The shaded area reflects those rats that exhibited the DAS (modified from Gorski, 1968). ●——●, day 45 (TP 10 μg); ●----●, day 100 (TP 10 μg).

vesicular follicles and exhibit interstitial hypertrophy, but CL are absent. The histologic features are almost identical to those of ovarian transplants placed in males. Because of lack of cyclic GTH regulation, associated with anovulatory conditions, these masculinized females are sterile. This type of female rat (androgenized rats) has some advantage as a model for study of the mechanisms of hypothalamic masculinization. As shown in Table 1, it is clear that there is a critical period (day 1–3) for the production of the *fale*. However, it is rather difficult in the male to determine whether the limiting factor exists in the sensitivity of the hypothalamic neurons to androgen or in the testicular activity. Potential *fales* (castrated on day 1) can be masculinized following TP injection later than this critical period (see Tables 1 and 2). In females the amount of androgen and the time of initial exposure of the hypothalamic competent neurons to androgen can be controlled because neonatal ovaries are hormonally inactive.

The relative importance of the dose and time of injection is demonstrated well by Gorski (1968) (Fig. 2). The sensitivity of the hypothalamus to androgen apparently decreases with age, and eventually the neural substrates become refractory to the organizing action of androgen by day 10. Half of the rats injected with 10 μg TP on day 5 ovulated during their postnatal period (Gorski, 1968). However, when they were autopsied around day 100, all were anovulatory. In contrast, 30 μg TP on day 5 resulted in as high an incidence of anovulatory conditions in early postpubertal days as 10 μg TP on days 1–4 did. The threshold apparently increased during days 4–5.

TABLE 3. Effects of ovariectomy on the hypothalamic pituitary system in the DAS rats (Arai, 1971 *a*)

Treatment	Vaginal smear before ovariectomy	Ovarian graft* at autopsy (day 140)
10 μg TP on day 5, ovariectomized on day 30	—	CL 7/8
10 μg TP on day 5, ovariectomized on day 90	PrE** or PE***	CL 0/6
1 mg TP on day 5, ovariectomized on day 30	—	CL 0/5
1 mg TP on day 5, ovariectomized on day 90	PE	CL 0/5
Oil on day 5, ovariectomized on day 90	cyclic	CL 5/5

* Transplantation of ovarian grafts was performed on day 100, and autopsy was performed on day 140. No. rats with CL/No. all rats. ** Prolonged estrus; *** persistent estrus.

This delayed development of the anovulatory condition following the neonatal injection of low doses of TP is called the delayed anovulation syndrome (DAS) (Gorski, 1968).

Since 30 μg TP on day 5 did not cause the DAS, it is thought to be the result of incomplete or partial differentiation following exposure to subthreshold doses.[1)] The onset of the DAS depends on the postpubertal ovarian activity (Arai, 1971*a*). Prepubertal ovariectomy (day 30) could in some way prevent rats injected with 10 μg TP from losing the capacity to secrete enough LH for ovulation at the time of the expected onset of anovulatory persistent estrus (Table 3). When ovariectomy was delayed until day 90 (most of the rats injected with 10μg TP on day 5 had developed persistent vaginal estrus by that time), CL formation was no longer found at autopsy in the ovarian transplants (grafted on day 100). However, CL formation still occurred in the ovarian grafts similarly treated but ovariectomized on day 30. This would suggest that the postpubertal influence of ovarian hormonal feedback by estrogen and/or androgen participates in the development of the DAS, as previously suggested by Kikuyama & Kawashima (1966). The gradual loss of the functional capacity for ovulation in the DAS is thought to be due to an independent modification of a neonatally partially differentiated system under postpubertal ovarian influence. The onset of the DAS is unlikely to be ascribable to continued development of the masculinizing process initiated by androgen during the period of neuronal competence.

1) The threshold dose of exogenous androgen should mimic the effect induced by the endogenous testicular activity. The neonatal female rats which receive a transplant of a single littermate testis show anovulatory persistent estrus since the onset of puberty.

LOCALIZATION AND SEXUAL DIFFERENTIATION OF THE NEURAL SUBSTRATES FOR REGULATION OF OVULATION

According to the current concept, the neural regulation of GTH secretion is divided into at least two fundamental functional levels (Fig. 3). The first level involves the continuous discharge of GTH necessary for follicular development and estrogen secretion, but cannot independently provoke ovulation (Barraclough, 1966, 1967; Gorski, 1966, 1971). The center for the tonic discharge of GTH is in the medial basal hypothalamus (MBH), including the arcuate nucleus and the median eminence region. This area corresponds to the hypophysiotropic area of Halász, Pupp & Uhlark (1962), which can maintain relatively normal adenohypophysial function following implantation of a pituitary graft in this area in hypophysectomized rats. Local implantation of estrogen implicates the MBH as the

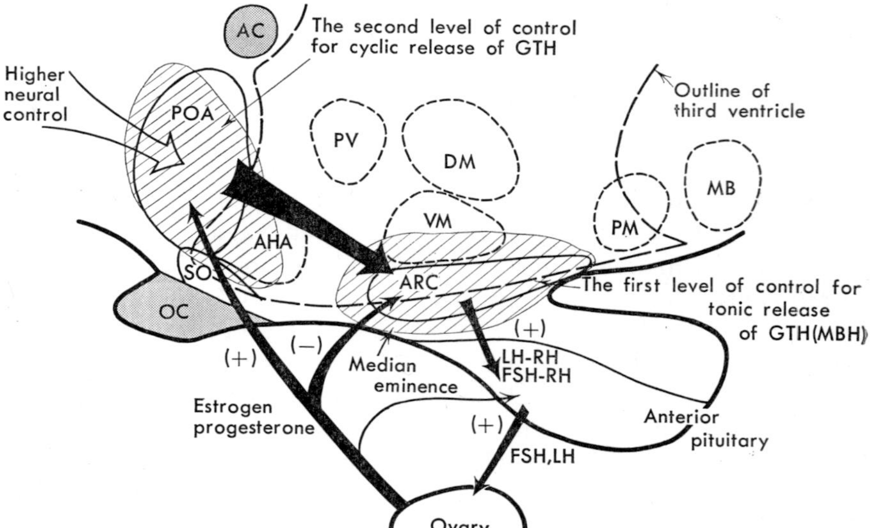

Fig. 3. Schematic model of neuroendocrine regulation of GTH secretion in the normal female rat. Abbreviations: AC, anterior commissure; AHA, anterior hypothalamic area; ARC, arcuate nucleus; DM, dorsomedial nucleus; MB, mamillary body; OC, optic chiasma; PM, premammillary nucleus; POA, preoptic area; PV, paraventricular nucleus; SC, suprachiasmatic nucleus; VM, ventromedial nucleus.

main site for negative feedback inhibition of GTH secretion (Lisk, 1960; Davidson & Sawyer, 1961). This mechanism appears to be preserved in androgenized rats, because of the appearance of castration cells in the anterior pituitary following ovariectomy (Arai, 1964*b*; Harris & Levine, 1965) and apparent inhibition by estrogen of ovarian compensatory hypertrophy in androgenized rats (Petrusz & Nagy, 1967). The second level is responsible for the control of cyclic GTH (especially LH) release; it may be located in the medial preoptic area (POA) and the anterior hypothalamus (AH). Electrolytic lesions within these areas or the surgical interruption of all inputs from the POA-AH to the MBH prevents spontaneous ovulation but does not induce gonadal atrophy (Flerkó, 1968). Conversely, electrical stimulation of the POA-AH-MBH in barbiturate-blocked normal proestrous rats readily induces ovulation (Critchlow, 1958; Terasawa & Sawyer, 1969). In androgenized rats, electrical stimulation of only the arcuate region of the MBH can induce ovulation, but the POA is refractory to the stimulus (Barraclough *et al.* 1961). Based on this observation and the identical occurrence of anovulatory persistent vaginal estrus both in androgenized and POA-lesioned rats, Barraclough & Gorski (1961) concluded that the center for cyclic LH release in the POA is affected by the neonatal androgen treatment which results in dysfunction.

Subsequently, several studies have confirmed this conclusion. Implantation of minute androgen crystals to the POA-AH of neonatal female rats has been shown to be able to induce anovulatory persistent estrus (Wagner *et al.* 1966; Nadler, 1968). Other evidence indicating that the POA of androgenized rats is functionally depressed is the lack of positive feedback regulation of GTH by estrogen or progesterone. In normal female rats, the POA has been demonstrated to be the focus of the positive feedback action of estrogen and/or progesterone to facilitate GTH secretion (Quinn & Zarrow, 1964; Döcke & Dörner, 1965; Davidson, 1969; Taleisnik, Velasco & Astrada, 1970; Kobayashi & Miyake, 1971). However, the POA of androgenized rats and male rats seems to be incapable of responding to stimulatory steroidal influence (Brown-Grant, Quinn & Zarrow, 1964; Wagner, 1968; Taleisnik, Caligaris & Astrada, 1971).

Recently, other than electrical stimulation, electrochemical (EC) or electrolytic stimulation has been shown to be effective in demonstrating the localization of the neural substrates for GTH regulation (Everett, 1964; Terasawa *et al.* 1969). Terasawa, Kawakami & Sawyer (1969) compared the effectiveness of these two types of stimula-

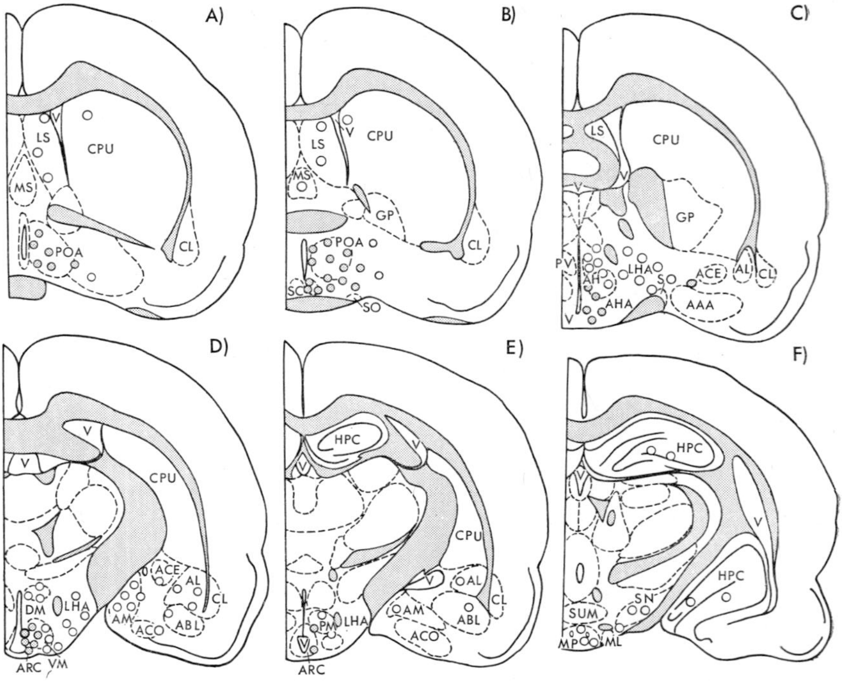

Fig. 4. A-F. Diagrammatic representations of the frontal sections of rat forbrain indicating sites of EC stimulation. ● indicates positive site of stimulation. ○ indicates negative site. When about 2 months old, the animals were orchidectomized and at the same time received a subcutaneous ovarian graft. The brain stimulation (100 μA of DC for 10 sec) was performed 4–6 weeks after the transplantation. Abbreviations: AAA, anterior amygdaloid area; ABL, lateral basal amygdaloid nucleus; ACE, central amygdaloid nucleus; ACO, cortical amygdaloid nucleus; AH, anterior hypothalamic nucleus; AHA, anterior hypothalamic area; AL, lateral amygdaloid nucleus; AM, medial amygdaloid nucleus; ARC, arcuate nucleus; CL, claustrum; CPU, caudate nucleus and putamen; DM, dorsomedial nucleus; GP, globus pallidus; HPC, hippocampus; LHA, lateral hypothalamic area; LS, lateral septum; ML, lateral mammillary nucleus; MP, posterior mammillary nucleus; POA, preoptic area; PV, paraventricular nucleus; SC, suprachiasmatic nucleus; SN, subthalamic nucleus; SO, supraoptic nucleus; V, ventricle; VM, ventromedial nucleus.

tion when applied to the POA in androgenized rats. They confirmed the previous findings of Barraclough *et al.* (1961) that the POA of androgenized rats is insensitive to electrical stimulation. However, they found that EC stimulation of the POA was capable of inducing ovulation in androgenized rats. This suggests that the POA of androgenized rats still retains the potential to stimulate the MBH to release pituitary LH. The question remaining is how should the lack of spontaneous ovulation in androgenized rats be interpreted.

Although the significance of responsiveness to EC stimulation is not yet conclusive, EC stimulation appears to be effective for examining possible components participating in GTH regulation not only in the hypothalamus but also in some extrahypothalamic areas. Figure 4A-F illustrates the results of EC stimulation on male castrated rats with an ovarian transplant (Arai, unpublished). The distribution of sites responsive to EC stimulation is spread over the POA and the rostral part of the AH, but becomes restricted to the basal areas at the caudal end of the optic chiasma and eventually converges with the MBH (arcuate-median eminence region). This distribution, as far as the POA-AH-MBH is concerned, is almost the same as that obtained in normal females receiving EC stimulation (Everett, 1964; Clemens, Shaar, Kleber & Tandy, 1971; Kalara, Ajika, Krulich, Fawcett, Quijada & McCann, 1971). However, EC stimulation other than to the POA-AH-MBH was totally negative.

Recently participation of higher neural mechanisms (higher than the second level of regulation by the POA) in GTH regulation has been suggested. Afferent inputs to the hypothalamus from the amygdala and hippocampus appear to be involved in modulating

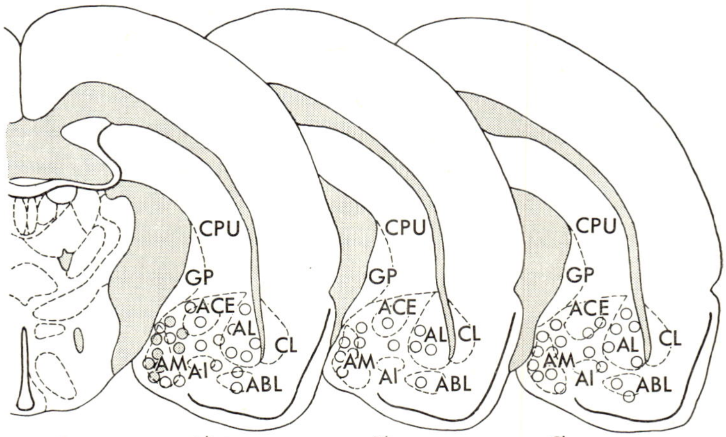

Fig. 5. Diagrammatic representations of stimulation sites in the amygdala. Anodal dc of 200 μA for 20 sec was applied. A, Light-induced persistent estrous rats. B, Androgenized rats (1 mg TP on day 5). C, Castrated male rats bearing an ovarian graft. For abbreviations, see Fig. 5 (Arai, 1971*b*).

hypothalamic GTH regulation (Kawakami *et al.* in this monograph). EC stimulation of the medial amygdaloid complex has been shown to induce ovulation in light-induced persistent estrous rats (Velasco & Taleisnik, 1969; Arai, 1971*b*) and in normal female rats (Kawakami *et al.* 1970). However, as shown in Fig. 5, the EC stimulation of the amygdaloid complex did not induce ovulation and CL formation in the male rats bearing an ovarian graft even if the stimulation was applied to the medial amygdaloid nucleus, the stimulation of this nucleus being successful in inducing ovulation in light-induced persistent estrous rats. Similar results were also obtained in our recent comparative studies of the responsivity of the amygdala to EC stimulation among nonandrogenized light-induced persistent estrous rats, androgenized rats and castrated male rats with an ovarian transplant (Arai, 1971*b*). In our studies, the amygdala of androgenized rats was highy resistent to EC stimulation, no ovulatory response being recognized following the stimulation (Fig. 5). Terasawa (1971) also reported the failure of EC stimulation of the amygdala to induce ovulation in androgenized rats.

Velasco *et al.* (1969) have suggested the existence of a functional sexual difference in the amygdala, based on the results that high levels of plasma LH following EC stimulation of the medial amygdaloid complex and subsequent ovulation were observed in female rats, but there was no increase in plasma LH after stimulation in males. The apparent difference in the responsivity of the amygdala to EC stimulation between nonandrogenized (light-induced) persistent estrous rats and androgenized persistent estrous rats or male rats indicates that the functional role or capacity of the amygdala in modulating the hypothalamic GTH regulation is different, because EC stimulation of the POA-AH-MBH did induce ovulation equally in both types of persistent estrous rats and also in male rats (Quinn, 1966; Moll & Zeilmaker, 1966; Everett, Holsinger, Zeilmaker, Redmond & Quinn, 1970; Arai, 1972*a*; Arai, unpublished, Table 4). Furthermore the medial amygdaloid complex of the *fales* did respond to EC stimulation, and subsequent CL formation occurred in their ovarian transplants, while potential *fales* injected with 50 μg TP on day 5 failed to respond to EC stimulation of the amygdala (Arai, unpublished, Table 4). Therefore, it is suggested that the neonatal influence of androgen is focused not only on the POA-AH, but also on the amygdala. The refractoriness of the amygdala to EC stimulation in male and androgenized rats represents a possible masculine character of this structure. However, it is still questionable whether the primary cause of the lack of cyclic

TABLE 4. Responsivity of the amygdala of the *fale* to EC stimulation (Arai, unpublished)

Treatment	Response of ovarian transplant to EC stimulation				
	No. stim.	ARCN*	POA	MAMG**	LAMG***
	(No. of rats with CL/No. of all rats)				
Castration on day 1	0/5****	2/2	5/5	4/7	0/5
Castration on day 1 +50 μg TP on day 5	0/4	2/2	5/5	0/5	0/3
Castration on day 30	0/5	4/4	5/5	0/8	0/4

All animals were kept under constant illumination to block spontaneous CL formation in their ovarian grafts. The stimulation was performed more than 60 days after exposure to constant light. * Arcuate nucleus; ** medial amygdaloid complex; *** lateral amygdaloid complex; **** no sign of recent CL formation was observed, although old luteal tissues were occasionally found.

activity of the POA in androgenized or male rats is due to this functional difference of the amygdaloid structures. Raisman & Field (1971) found sexual differences in the mode of termination of the afferent fibers to the POA. These results suggest that functional differences in the POA between female and male (or androgenized) rats may be due to some difference in the neural connection with afferent fibers from outside the POA rather than some intrinsic property of the POA neurons.

POSSIBLE MECHANISMS OF HYPOTHALAMIC MASCULINIZATION

Since Kikuyama (1961, 1962) found that simultaneous administration of the tranquilizing agents reserpine and chlorpromazine with gonadal steroids prevents the development of anovulatory persistent estrus, a number of attempts to identify chemical agents that can protect the developing hypothalamus from androgen have been made. Among them, the findings of Kincl & Maqueo (1965) and Cagnoni, Fantini, Morace & Ghetti (1965) that progesterone can protect 5-day-old female rats from the masculinizing effect of androgen are of special interest, because the hormonal environment of pregnant animals is progesterone-dominant and this condition may protect the male fetal hypothalamus from the masculinizing effect of testicular androgen before birth. In this connection, the findings of Flerkó *et al.* (1967) that low doses of estrogen alone or concurrently with progesterone inhibit hypothalamic masculinization of neonatal males are also important.

TABLE 5. Inhibition of the masculinizing action of TP by tranquilizers, steroids and barbiturates (modified from Arai & Gorski, 1968 *a*)

Injection(s) on day 5	Incidence of sterility No. anovulatory/No. injected			
	Day 45–60	%	Day 90–120	%
30 μ TP	32/37	87	15/16	94
30 μg TP+10 μg reserpine	6/16	37	11/16	69
30 μg TP+500 μg chlorpromazine	5/12	42	9/12	75
30 μg TP+2.5 mg progesterone	2/10	20	7/10	70
30 μg TP+1 mg DCA	3/9	33	5/9	55
30 μg TP+1.5 mg pregnanedione	3/13	23	3/5	60
30 μg TP+0.3 mg pentobarbital*	0/15	0	4/15	27
30 μg TP+0.5 mg phenobarbital**	2/17	12	4/17	23
Inhibitory agent only***	1/37	3	1/37	3

* Two injections given 4–5 hr apart; ** single injection simultaneous with TP; *** reserpine, chlorpromazine, progesterone, or barbiturates.

Table 5 summarizes our recent data on the protective effects of steroids, tranquilizers, and barbiturates (Arai & Gorski, 1968*a*). These data confirm the observations that certain tranquilizers and progesterone can protect neonatal rats from androgen (Kikuyama, 1961, 1962; Kincl *et al.* 1965; Cagnoni *et al.* 1965). However, this protection was apparent only in rats of early postpubertal age. Most of the rats showed the DAS, eventually resulting in persistent vaginal estrus by days 90–120. In these cases, the protective action of these agents would be insufficient to counteract the effect of 30 μg TP given neonatally. More marked protection against androgen as judged both by laparotomy on day 45 and autopsy on day 90 was obtained by simultaneous injection of pentobarbital or phenobarbital[1]. The mechanisms by which tranquilizers, steroids, and barbiturates inhibit or delay the development of the sterility syndrome are not clear. Antagonism between androgen and progesterone or DCA is already recognized at the level of the reproductive target organs (Roberts & Szego, 1953; Arai, 1960). Recently it has been reported that uptake of radioactive gonadal steroids by hypothalamic tissues may be competitively inhibited by concurrent administration of a hormonal antagonist (Eisenfeld & Axelord, 1967; Kato, Kobayashi & Villee, 1968; Stern & Eisenfeld, 1971). It is possible that this kind of interaction between TP and the other steroids occurs in the

[1] This inhibitory action of barbiturates was mostry confirmed by Sutherland (see Gorski, 1971), but recently Brown-Grant *et al.* (1971) could not do it.

neonatal hypothalamus. The results that the antiandrogen cyproterone acetate effectively counteracts the masculinizing action of TP on the developing hypothalamus in female rats give good support to this conjecture (Wollman & Hamilton, 1967; Arai & Gorski, 1968*c*).

On the other hand, progesterone, pregnanedione, and DCA can act as anesthetics (P'an & Laubach, 1962; Heuser, 1967), In fact, one common feature of all the drugs used (Table 5) is that they have CNS-depressing properties. Protection by progesterone and other steroids might be comparable to protection by CNS depressants such as tranquilizers and barbiturates. It is of more interest to speculate that barbiturates interact with androgen at the neuronal level and perhaps prevent some action of androgen on the cell membrane and/or intarcellular biochemical processes. However, other than neuronal factors, general factors, blood flow, hypothermia and liver metabolism, *etc*., which might be affected by barbiturates are still unknown.

In another series of experiments, we have attempted to determine the length of time that neural elements must be exposed to the inductive stimulation of androgen for the initiation of the masculinizing process (Arai *et al.* 1968*b*, *c*). As the basic procedure, 30 μg TP

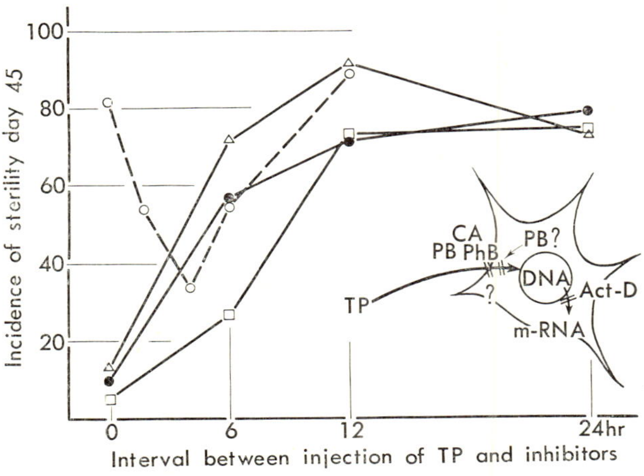

Fig. 6. Incidence of sterility (No. anovulatory/No. injected $\times$ 100) at 45 days of age following injection of 30 μg TP to 5-day-old female rats simultaneously with or at various intervals prior to cyproterone acetate (CA, 0.6 mg), pentobarbital (PB, 0.3 mg $\times$2), phenobarbital (PhB, 0.5 mg), or actinomycin-D (Act-D, 1 μg). The incidence of sterility following 30 μg TP alone on day 5 was 83.3 (modified from Arai & Gorski, 1968*b*, *c*; Kobayashi & Gorski, 1970). ●, CA; □, PB; △, PhB; ○, Act-D.

was injected into 5-day-old female rats. Following TP, cyproterone acetate, pentobarbital or phenobarbital was injected at different intervals to inhibit the masculinizing action of TP. The results are presented in Fig. 6. The results from the three different agents are roughly comparable. Since cyproterone acetate or barbiturates could inhibit the masculinizing action of TP only when given concurrently with TP or shortly after TP, androgen, if enough TP is taken up by the hypothalamic neurons, may exert its inductive action on the developing hypothalamus in a relatively short period, during which time unknown but presumably intracellular biochemical processes are initiated. This hypotheses is supported by the recent findings by Kobayashi & Gorski (1970) that antibiotics such as actinomycin-D or puromycin can inhibit the masculinizing action of TP if given shortly before or after TP injection. Since these antibiotics have been shown to inhibit DNA-dependent RNA and protein synthesis, respectively, these intracellular biochemical processes may be involved in hypothalamic masculinization and may be triggered within a short time (less than 12 hr). In this regard, Shimada & Gorbman (1970) found that by the use of DNA-RNA hybridization techniques, neonatal TP treatment causes a permanent reduction of the synthetic rate of RNAs that are common in the normal female brain. At the same time, they also found that new RNA species that are not characteristic if the normal female brain are produced by neonatal TP treatment. Concerning this hypothesis in the brief exposure time for hypothalamic masculinization, however, the dosage of the hormone, the time of injection during the neuronal competent period, and strain difference must be considered as influencing factors. Recently Alklint & Norgren (1971) stated that if nonesterified androgens were used the minimal exposure time would be longer than 12 hr, as we predicted.

As mentioned earlier, androgenized or male rats are not sensitive to the stimulatory or positive feedback action of estrogen or progesterone (Brown-Grant *et al.* 1964; Wagner, 1968; Taleisnik *et al.* 1971). Thus, reduction of sensitivity to sex steroids in androgenized rats has been suggested (Petrusz *et al.* 1967). Based on the hypothesis that relative insensitivity to steroids might be accompanied by a decrease in their uptake in presumed target tissues, the uptake of ^{3}H-estradiol by the POA or the AH of androgenized rats has been studied by several investigators. Although experimental procedures were different for the results reported, most of them showed a decreased estradiol-binding capacity of the hypothalamus in androgenized rats (Flerkó,

Mess & Illei-Donhoffer, 1969; McGuire & Lisk, 1969; Flerkó, Illei-Donhoffer & Mess, 1971; Maurer & Woolley, 1971; Tuohimaa & Johansson, 1971). Green, Luttge & Whalen (1969), however, did not find any difference between the sexes. These results suggest that neonatal androgen action may interfere with the normal development of estrogen receptor proteins in the hypothalamus and may result in alterations of postpubertal hypothalamic functions.

Significant changes in neuronal nuclear size of the POA-AH that correlate with the sexual differentiation of GTH regulation have been reported, but the changes produced by neonatal castration were opposite in the results from two different laboratories (Pfaff, 1966; Dörner & Staudt, 1968). Another morphological sexual difference, although limited to the monkey, has been reported in the ependymal linings of the third ventricle of the hypothalamus (Kumar, 1968). The significance of this sexual dimorphism in the ependyma is not clear but is apparently dependent on the reproductive activity of the animals. Raisman *et al.* (1971) (*vide ante*) found that the mode of synaptic connection of the POA neurons is sexually differentiated. In female rats, afferent fibers to the neurons of the POA terminate more frequently on the dendritic spines than on the dendritic shafts, compared with those in males. It is not clear at present whether the sexual difference in the integrity of the POA correlates with the structural difference in the POA between the sexes, but it is possible.

PERSISTENT DIESTRUS SYNDROME—A PHARMACOLOGICAL EFFECT OF NEONATAL TREATMENT WITH SEX STEROIDS

The role of sex steroids as inductive substances for the differentiation of the hypothalamic sexual centers has been discussed so far mostly at the physiological level. As mentioned earlier, neonatal estrogen treatment has been found to modify hypothalamic GTH regulation permanently (Takewaki, 1962; Barraclough, 1966, 1967; Gorski, 1966). Whether the nature of the modification by estrogen is just the same as that by androgen is not yet certain. Gorski (1966, 1971) is of the opinion that the masculinizing effect of estrogen is pharmacologic. This concept seems reasonable because neonatal ovaries are hormonally inactive and because the only available sex steroid during the neonatal period is testicular androgen. However, the direct application of EB to the POA-AH of newborn female rats has been demonstrated to induce persistent vaginal estrus (Sutherland

& Gorski, 1970). Therefore, this steroid seems to be able to act directly on the neurons of the POA-AH to induce sterility. In adult female rats, the hypothalamic sexual centers are normally regulated by the stimulatory or inhibitory influences of estrogen and progesterone, but not by androgen. The hypothalamic centers, especially in prepubertal rats, have been reported to be highly sensitive to estrogen (Ramirez *et al.* 1963). Thus it is possible to speculate that the development of prospective estrogen-sensitive components of the hypothalamus can be permanently disturbed at the very labile stage of neonatal days by unexpected exposure to estrogen. Based on this interpretation, the processes of hypothalamic masculinization might be considered to be disturbances of the normal development of the female GTH regulation caused by sex steroids administered neonatally. In this sense, hypothalamic masculinization is not androgen specific. Recently the possibility that the hypothalamic aromatization of androgen to estrogen may participate in the process of sexual differentiation was discussed by Naftolin *et al.* (1967). In addition, the conversion of testosterone to DHT, which is a prerequisite for the chromosomal activation of the nuclear apparatus by androgen in peripheral tissues, does not appear to be normally involved in the processes of hypothalamic masculinization (Brown-Grant, Munck, Naftolin & Sherwood, 1971; McDonald, 1971; Whalen & Luttge, 1971; Arai, 1972*b*). Therefore, in order to justify these possibilities, it must be determined whether testosterone itself, estrogen transformed from androgen, or both are responsible for initiation of hypothalamic masculinization.

Finally, hypothalamic sterility can be discussed at the pharmacological level. As previously reported, the functional alterations in the hypothalamic sexual centers caused by the neonatal treatment with sex steroids vary in degree according to the treatment schedules (Gorski *et al.* 1963; Arai, 1964*a*). Neonatal females that receive a transplant of a single litter-mate testis never ovulate even in early postnatal days (Gorski, 1966). In these rats the degree of hypothalamic masculinization appears to be much closer to the normal male than in females receiving exogenous androgen. Since male rats castrated later than the neuronal competent period or females with a testicular graft do not show the DAS, the hypothalamic sexual centers of the DAS rats seem to be less masculinized. Among rats showing the persistent vaginal estrus syndrome, functional alterations or disturbances in the hypothalamic sexual centers are variable depending on the dosage, the time of injection and the length of the injection period. Gorski (1966) has suggested that subcutaneous injection of 30 μg TP on day 5 is the most appropriate dose for the physiological threshold. Therefore, it is not correct to consider that the acylicity of GTH secretion is the only characteristic of the masculine pattern of GTH regulation. The persistent vaginal diestrus syndrome can be produced by injections of large doses of androgen or estrogen to female rats for the first 30 days of life (Takasugi, 1956; Takewaki, 1968). In such rats ovaries are quite atrophic and similar in structure to those of weanling or hypophysectomized rats, showing neither follicular growth nor CL formation; vaginal cornification does not occur permanently because of the lack of ovarian estrogen. The secretion and synthesis of GTH, not only LH but also FSH, and of the hypothalamic GTH-releasing hormones are permanently suppressed (Arai, 1963, 1964*a*, *b*; Arai & Kusama, 1968). In this case, it is apparent that the degree of hypothalamic masculization (or sterilization) goes far beyond physiological levels. The disturbance of the hypothalamic sexual centers must exist not only in the second level of GTH regulation by the POA, but also in the first level—the tonic regulation of GTH by the MBH. In these persistent diestrous (PD) rats, EC stimulation of medial POA failed to induce ovulation (Arai & Masuda, 1970). The production of LRF (LRH) in PD rats has been reported to be decidedly lower than that of normal females (Arai, 1963). The observation that the pituitary castration reaction, such as the appearance of castration cells in the anterior pituitary, fails to occur in this type of PD rats following ovariectomy (Arai, 1964*b*, 1965) suggests that possible disturbance exist also in the hypophysiotrophic mechanisms of the MBH. This contrasts with the case of androgenized or estrogenized persistent estrous rats, which invariably show a marked increase in gonadotropes in the anterior pituitary, and these cells are eventually transformed into signet ring cells (Arai, 1964*b*; Harris *et al.* 1965) Such rats show a postcastration increase in pituitary LH content (Van Rees & Gans, 1966) because the first level of regulation by the MBH is still intact. Furthermore, ovarian follicular growth is completely inhibited in PD rats. This indicates that the regulation of the tonic release of GTH is disturbed, and at the same time the FSH control mechanisms, which has been suggested to be located in the AH (Flerkó, 1963, 1968), would also be impaired. Therefore, it can be assumed that the derangement of the hypothalamic sexual centers in PD rats is much wider than that of androgenized or estrogenized persistent estrous rats. Two steroid-induced syndromes, persistent estrus and persistent diestrus, are seemingly independent phenomena. However, we must emphasize again that the degree of the disturbances in these two syndromes are highly dependent on the dose, the time of injection and the length of the injection period.

ACKNOWLEDGEMENTS

The author expresses his thanks to Dr. F. Neumann, Schering A. G., Berlin, Germany, for cyproterone acetate, to Dr. J. J. Chart, CIBA Pharmaceutical Co., Summit, U.S.A., for Su-10603, and to Dr. M. Sawai, Teikoku Hormone Mfg. Co, Ltd., Tokyo, for sex steroids and HCG. The author's research is mostly supported by grants from the Ministry of Education of Japan and partly by a grant from the Ford Foundation. The author has received bibliographic aid from UCLA Brain Information Service, which is supported by the NINDB grant.

REFERENCES

Alklint, T. & Norgren, A. (1971). Effects of neonatally injected non-estrified testosterone on reproductive functions in female rats. *Acta endocr., Copenh.* **66**, 720–726.

Anderson, K. M. & Liao, S. (1968). Selective retention of dihydrotestosterone by prostatic nuclei. *Nature, Lond.* **219**, 277–279.

Arai, Y. (1960). Effects of adrenocorticoid on response of reproductive accessories to androgen in castrated male rats. *J. Fac. Sci. Univ. Tokyo* IV, **9**, 111–115.

Arai, Y. (1963). The content of luteinizing hormone of the anterior pituitary and of luteinizing hormone releasing factor of the hypothalamus in estrogen-induced persistent-diestrous rats. *Proc. Japan Acad.* **39**, 605–609.

Arai, Y. (1964*a*). Persistent-estrous and -diestrous conditions induced by early postnatal administration of estrogen in female rats. *Endocr. Jap.* **11**, 204–208.

Arai, Y. (1964*b*). Changes in the hypothalamic-pituitary system after ovariectomy in persistent-estrous and -diestrous rats. *J. Fac. Sci. Univ. Tokyo* IV, **10**, 369–379.

Arai, Y. (1965). Failure of ovariectomy to induce pituitary castration reaction in estrogen-induced persistent-diestrous rats. *Proc. Japan Acad.* **41**, 163–168.

Arai, Y. (1968). Sexual differentiation of the hypothalamus and steroid-induced sterility. *Juntendo Med. J.* **14**, 145–157.

Arai, Y. (1971*a*). Possible process of the secondary sterilization : Delayed anovulation syndrome. *Experientia* **27**, 463–464.

Arai, Y. (1971*b*). Effect of electrochemical stimulation of the amygdala on induction of ovulation in different types of persistent estrous rats and castrated male rats with an ovarian transplant. *Endocr. Jap.* **18**, 211–214.

Arai, Y. (1972*a*). Neonatal influence of sex hormones on the developing hypothalamus. *Advances in Neurol. Sci.* **16**, 364–368.

Arai, Y. (1972*b*). Effect of 5α-dihydrotestosterone on differentiation of masculine pattern of the hypothalamus in the rat. *Endocr. Jap.* **19**, 389-393.

Arai, Y. & Gorski, R. A. (1968*a*). Protection against the neural organizing effect of exogenous androgen in the neonatal female rat. *Endocrinology* **82**, 1005–1009.

Arai, Y. & Gorski, R. A. (1968*b*). Critical exposure time for androgenization of the developing hypothalamus in the female rat. *Endocrinology* **82**, 1010–1014.

Arai, Y. & Gorski, R. A. (1968*c*). Critical exposure time for androgenization of the rat hypothalamus determined by anti-androgen injection. *Proc. Soc. exp. Biol. Med.* **127**, 590–593.

Arai, Y. & Kusama, T. (1968). Effect of neonatal treatment with estrone on hypothalamic neurons and regulation of gonadotrophin secretion. *Neuroendocrinology* **3**, 107–1014.

Arai, Y. & Masuda, S. (1970). Effect of electrochemical stimulation of the preoptic area on induction of ovulation in estrogenized rats. *Endocr. Jap.* **17**, 237–239.

Barraclough, C. A. (1966). Modification in the CNS regulation of reproduction after exposure of prepubertal rats to steroid hormones. *Recent Prog. Horm. Res.* **22**, 503–539.

Barraclough, C. A. (1967). Modifications in reproductive function after exposure to hormones during the prenatal and early postnatal period. In Martini, L. & Ganong, W. F., *Neuroendocrinology* II. New York: Academic Press.

Barraclough, C. A. & Gorski, R. A. (1961). Evidence that the hypothalamus is responsible for androgen-induced sterility in the female rat. *Endocrinology* **68**, 68–79.

Beyer, C., Morali, G. & Cruz, M. L. (1971). Effect of 5α-dihydrotestosterone on gonadotropin secretion and estrous behavior in the female Wistar rat. *Endocrinology* **89**, 1158–1161.

Brown-Grant, K., Quinn, D. L. & Zarrow, M. X. (1964). Superovulation in the androgen-treated immature rat. *Endocrinology* **74**, 811–813.

Brown-Grant, K., Munck, A., Naftolin, F. & Sherwood, M. R. (1971). The effect of the administration of testosterone propionate alone or with phenobarbitone and of testosterone metabolites to neonatal female rats. *Horm. Behav.* **2**, 173–182.

Bruchovsky, N. & Wilson, J. D. (1968). The intranuclear binding of testosterone and 5α-androstan-17β-ol-3-one by rat prostate. *J. biol. Chem.* **243**, 5953–5960.

Cagnoni, M., Fantini, F., Morace, G. & Ghetti, A. (1965). Failure of testosterone propionate to induce the " early-androgen " syndrome in rats previously injected with progesterone. *J. Endocr.* **33**, 527–573.

Clemens, J. A., Shaar, C. J., Kleber, J. W. & Tandy, W. A. (1971). Areas of the brain stimulatory to LH and FSH secretion. *Endocrinology* **88**, 180–184.

Critchlow, V. (1958). Ovulation induced by hypothalamic stimulation in the anesthetized rat. *Amer. J. Physiol.* **195**, 171–174.

Davidson, J. M. (1969). Feedback control of gonadotropin secretion. In *Frontiers in Neuroendocrinology*, ed. Ganong, W. F. & Martini, L., New York: Oxford Univ. Press.

Davidson, J. M. & Sawyer, C. H. (1961). Effects of localized intracerebral implantation of oestrogen on reproductive function in the female rabbit. *Acta endocr., Copenh.* **37**, 385–393.

Döcke, F. & Dörner, G. (1965). The mechanisms of the induction of ovulation by estrogen. *J. Endocr.* **33**, 491–499.

Dörner, G. & Staudt, J. (1968). Structural changes in the preoptic anterior hypothalamic area of the male rat, following neonatal castration and androgen substitution. *Neuroendocrinology* **3**, 136–140.

Duncan, G. W., Wyngarden, L. J. & Cornette, J. C. (1967). Ante- and postnatal steroid environment as it affects development of mammalian reproductive mechanisms. *Rev. Can. Biol.* **26**, 237–258.

Eisenfeld, A. J. & Axelord, J. (1967). Evidence for estradiol binding sites in the hypothalamus-effect of drugs. *Biochem. Pharmac.* **16**, 1781–1785.

Everett, J. W. (1964). Preoptic stimulative lesions and ovulation in the rat: " Threshold " and LH-release time in late diestrus. In *Major Problems in Neuroendocrinology*, ed. Bajusz, E. & Jasmin, G., New York / Basel: Karger.

Everett, J. W., Holsinger, J. W., Zeilmaker, G. H., Redmond, W. C. & Quinn, D. L. (1970). Strain differences for preoptic stimulation of ovulation in cyclic, spontaneous persistent-estrous, and androgen-sterilized rats. *Neuroendocrinology* **6**, 98–108.

Flerkó, B. (1963). The central nervous system and the secretion and release of luteinizing hormone and follicle stimulating hormone. In Nalbandov, A. V., *Advances in Neuroendocrinology*. Urbana: Univ. Illinois Press.

Flerkó, B. (1967). Brain mechanisms controlling gonadotrophin secretion and their sexual differentiation. In *Symposium on Reproduction. Congress of the Hungarian Society for Endocrinology and Metabolism*, ed. Lissak, K., Budapest: Akademiai Kiado.

Flerkó, B. (1968). The experimental polyfollicular ovary. In *Endocrinology and Human Behavior*. ed. Michael, R. P., London: Oxford Univ. Press.

Flerkó, B., Illei-Donhoffer, A. & Mess, B. (1971). Oestradiol-binding capacity in neural and non-neural target tissues of neonatally androgenized female rats. *Acta Biol. Acad. Sci. Hung.* **22**, 125–130.

Flerkó, B. Mess, B. & Illei-Donhoffer, A. (1969). On the mechanism of androgen sterilization. *Neuroendocrinology* **4**, 164–169.

Flerkó, B., Petrusz, P. & Tima, L. (1967). On the mechanism of sexual differentiation of the hypothalamus: Factors influencing the "critical period" of the rat. *Acta Biol. Acad. Sci. Hung.* **18**, 27–36.

Goldman, A. S. (1971). Production of hypospadias in the rat by selective inhibition of fetal testicular 17β-hydroxylase and C_{17-20}-lyase. *Endocrinology* **88**, 527–531.

Goldman, A. S. & Baker, M. K. (1971). Androgenicity in the rat fetus of metabolites of testosterone and antagonism by cyproterone acetate. *Endocrinology* **89**, 276–280.

Goldman, B. D. & Gorski, R. A. (1971). Effects of gonadal steroids of LH and FSH in neonatal rats. *Endocrinology* **89**, 112–115.

Goldman, B. D., Grazia, Y. R., Kamberi, I. A. & Porter, J. C. (1970). Serum gonadotropin concentrations in intact and castrated neonatal rats. *Endocrinology* **88**, 771–776.

Gorski, R. A. (1966). Localization and sexual differentiation of the nervous structures which regulate ovulation. *J. Reprod. Fert.*, Suppl. **1**, 67–88.

Gorski, R. A. (1968). Influence of age on the response to paranatal administration of a low dose of androgen. *Endocrinology* **82**, 1001–1004.

Gorski, R. A. (1971). Gonadal hormones and the perinatal development of neuroendocrine function. In *Frontiers in Neuroendocrinology*, ed. Martini, L. & Ganong, W. F., New York: Oxford Univ. Press.

Gorski, R. A. & Barraclough, C. A. (1963). Effects of low dosages of androgen on the differentiation of hypothalamic regulatory control of ovulation. *Endocrinology* **73**, 210–216.

Gorski, R. A. & Wagner, J. W. (1965). Gonadal activity and sexual differentiation of the hypothalamus. *Endocrinology* **76**, 226–239.

Green, R., Luttge, E. G. & Whalen, R. E. (1969). Uptake and retention of tritiated estradiol in brain and peripheral tissues male, female and neonatally androgenized female rats. *Endocrinology* **85**, 373–378.

Halász, B., Pupp, L. & Uhlark, S. (1962). Hypophysiotrophic area in the hypothalamus. *J. Endocr.* **25**, 147–154.

Harris, G. W. (1964). Sex hormones, brain development and brain function. *Endocrinology* **75**, 627–648.

Harris, G. W. & Levine, S. (1965). Sexual development of the brain and its experimental control. *J. Physiol.* **181**, 379–400.

Heuser, G. (1967). Induction of anesthesis, seizures and sleep by steroid hormones. *Anesthesiology* **28**, 173–183.

Inoué, S. (1966). Strain difference in the time of the brain masculinization in male rats. *Bull. Res. Inst. Dent. Mater.* **2**, 793–803.

Jacobsohn, D. (1965). Some effects of sex hormones on the fetus and newborn in relation to later reproductive functions. *Acta Univ. Lund.* II, **17**, 1–19.

Jaffe, R. B. (1969). Testosterone metabolism in target tissues: Hypothalamic and pituitary tissues of the adult rat and human fetus, and the immature rat epiphysis. *Steroids* **14**, 483–498.

Kalara, S. P., Ajika, K., Krulich, L., Fawcett, C. P., Quijada, M. & McCann, S. M. (1971). Effects of hypothalamic and preoptic electrochemical stimulation on gonadotropin and prolactin release in proestrous rats. *Endocrinology* **88**, 1150–1158.

Kato, J., Kobayashi, T. & Villee, C. A. (1968). Effect of clomiphene on the uptake of estradiol by the anterior hypothalamus and hypophysis. *Endocrinology* **82**, 1049–1052.

Kawakami, M., Terasawa, E. & Ibuki, T. (1970). Changes in multiple unit activity of the brain during the estrous cycle. *Neuroendocrinology* **6**, 30–48.

Kawashima, S. (1960). Secretion of androgen by ovarian grafts in male rats castrated at birth. *J. Fac. Univ. Tokyo* IV, **9**, 117–125.

Kikuyama, S. (1961). Inhibitory effect of reserpine on the induction of persistent estrus by sex steroids in the rat. *Annot. Zool. Japon.* **34**, 111–116.

Kikuyama, S. (1962). Inhibition of induction of persistent estrus by chlorpromazine in the rat. *Annot. Zool. Japon.* **35**, 6–11.

Kikuyama, S. & Kawashima, S. (1966). Formation of corpora lutea in ovarian grafts in ovariectomized adult rats subjected to early postnatal treatment with androgen. *Sci. Papers. Coll. Gen. Educ. Univ. Tokyo* **16**, 69–74.

Kincl, F. A. & Maqueo, M. (1965). Prevention by progesterone of steroid-induced sterility in neonatal male and female rats. *Endocrinology* **77**, 859–862.

Kobayashi, F. & Gorski, R. A. (1970). Effects of antibiotics on androgenization of the neonatal female rat. *Endocrinology* **86**, 285–289.

Kobayashi, F. & Miyake, T. (1971). Acute effect of hypothalamic defferentation of progesterone induced ovulating hormone release in the rats. *Endocr. Jap.* **18**, 395–402.

Kumar, T. C. A. (1968). Sexual differences in the ependyma lining the third ventricle in the area of the anterior hypothalamus of adult rhesus monkeys. *Zeit. Zellforsch.* **90**, 28–36.

Lisk, R. D. (1960). Estrogen-sensitive centers in the hypothalamus of the rat. *J. exp. Zool.* **145**, 197–208.

Martini, L. (1971). Feedback mechanisms and the control of the hypothalamo-hypophysial complex. In *Steroid Hormones and Brain Functions,* ed. Sawyer, C. H. & Gorski, R. A., Los Angeles: Univ. Calif. Press.

Maurer, R. & Woolley, D. (1971). Distribution of ^{3}H-estradiol in clomiphene-treated and neonatally androgenized rats. *Endocrinology* **88**, 1281–1287.

McDonald, P. G. (1971). Some biological actions of dihydrotestosterone (DHT). *J. Reprod. Fert.* **25**, 309–310.

McGuire, J. L. & Lisk R. D. (1969). Oestrogen receptors in androgen or oestrogen sterilized female rats. *Nature, Lond.* **221**, 1068–1069.

Moll, J. & Zeilmaker, G. H. (1966). Ovulatory discharge of gonadotrophins induced by hypothalamic stimulation in castrated male rats bearing a transplanted ovary. *Acta endocr., Copenh.* **51**, 281–289.

Nadler, R. D. (1968). Masculinization of female rats by intracranial implantation of androgen in infancy. *J. comp. Physiol. Psychol.* **66**, 157–166.

Naftolin, F., Ryan, K. J. & Petro, Z. (1967). *J. clin. Endocr. Metab.* **33**, 377–379.

Neumann, F. & Kramer, M. (1967). Female brain differentiation of male rats as a result of early treatment with an androgen antagonist. In *Hormonal Steroids*, ed. Martini, L., Fraschini, F. & Motta, M., Amsterdam: Excepta Medica.

P'an, S. Y. & Laubach, G. D. (1962). Steroid central depessants. In *Methods in Hormone Research* III, ed. Dorfman, R. I., New York: Academic Press.

Petrusz, P. & Nagy, E. (1967). On the mechanism of sexual differentiation of the hypothalamus: Decreased hypothalamic oestrogen sensitivity in androgen-sterilized rats. *Acta Biol. Acad. Sci. Hung.* **18**, 21–26.

Pfaff, D. W. (1966). Morphological changes in the brains of adult male rats after neonatal castration. *J. Endocr.* **36**, 415–416.

Pfeiffer, C. A. (1936). Sexual differences of the hypophysis and their determination by the gonads. *Amer. J. Anat.* **58**, 195–226.

Price, D. & Ortiz, E. (1944). The relation of age to reactivity in the reproductive system of the rat. *Endocrinology* **34**, 215.

Quinn, D. L. (1966). Luteinizing hormone release following preoptic stimulation in the male rat. *Nature, Lond.* **209**, 891–892.

Quinn, D. L. & Zarrow, M. X. (1964). Inhibition of pregnant mare's serum-induced ovulation in the immature rat. *Endocrinology* **74**, 309–313.

Raisman, G. & Field, P. M. (1971). Sexual dimorphism in the preoptic area of the rat. *Science, N. Y.* **173**, 731–733.

Ramirez, V. D. & McCann, S. M. (1963). Comparison of the regulation of luteinizing hormone (LH) secretion in immature and adult rats. *Endocrinology* **72**, 452–464.

Resko, J. A., Feder, H. H. & Goy, R. W. (1968). Androgen concentrations in plasma and testis of developing rats. *J. Endocr.* **40**, 485–491.

Roberts, S. & Szego, C. M. (1953). Steroid interaction in the metabolism of reproductive target organs. *Physiol. Rev.* **33**, 593–629.

Saunders, F. J. (1968). Effects of sex steroids and related compounds on pregnancy and on development of the young. *Physiol. Rev.* **48**, 601–643.

Shimada, H. & Gorbman A. (1970). Long lasting changes in RNA synthesis in the forebrains of female rats treated with testosterone soon after birth. *Biochem. biophys. Res. Commun.* **38**, 423–430.

Stern, J. M. & Eisenfeld, A. J. (1971). Distribution and metabolism of ^{3}H-testosterone in castrated male rats: Effects of cyproterone, progesterone and unlabeled testosterone. *Endocrinology* **88**, 1117–1125.

Sutherland, S. D. & Gorski, R. A. (1970). Intrahypothalamic interaction of steroids and pentobarbital in the neonatal rat. *Anat. Rec.* **166**, 386.

Takasugi, N. (1956). Veränderungen der hypophysären, gonadotropen Aktivität der daueranoestrishen und der ovariektomierten Ratten unter Stree-Situationen. *J. Fac. Sci. Univ. Tokyo* IV, **7**, 625–639.

Takewaki, K. (1962). Some aspects of hormonal mechanism involved in persistent estrus in the rat. *Experientia* **18**, 1–6.

Takewaki, K. (1968). Reproductive organs and anterior hypophysis of neonatally androgenized female rats. *Sci. Rep. Tokyo Woman's Christian Coll.* **1** (3), 31–41.

Taleisnik, S., Caligaris, L. & Astrada, J. J. (1971). Sex difference in hypothalamo-hypophysial function. In *Steroid Hormones and Brain Function*, ed. Sawyer, C. H. & Gorski, R. A., Los Angeles: Univ. Calif. Press.

Taleisnik, S., Velasco, M. E. & Astrada, J. J. (1970). Effect hypothalamic deafferentation on the control of luteinizing hormone secretion. *J. Endocr.* **46**, 1–7.

Terasawa, E. (1971). Changes in electrical activity in the androgenized and the anterior deafferented rat. *Folia Endocrinol. Japon.* **46**, 1057–1096.

Terasawa, E., Kawakami, M. & Sawyer, C. H. (1969). Induction of ovulation by electrochemical stimulation in androgenized and spontaneously constant-estrous rats. *Proc. Soc. Exp. Biol. Med.* **132**, 497–501.

Terasawa, E. & Sawyer, C. H. (1969). Electrical and electrochemical stimulation of the hypothalamo-adenenohypophysial system with stainless steel electrodes. *Endocrinology* **84**, 918–925.

Tuohimaa, P. & Johansson, R. (1971). Decreased estradiol binding in the uterus and anterior hypothalamus of androgenized female rats. *Endocrinology* **88**, 1159–1164.

Van Rees, G. P. & Gans, E. (1966). Effect of gonadectomy and estrogen on pituitary LH-content and organ weights in androgen-sterilized female rats. *Acta Endocr., Copenh.* **52**, 471–477.

Velasco, M. E. & Taleisnik, S. (1969). Release of gonadotropins induced by amygdaloid stimulation in the rat. *Endocrinology* **84**, 132–139.

Wagner, J. W. (1968). Luteinization of ovarian transplants in gonadectomized, pregnant mare's serum-primed immature male rats. *Endocrinology* **83**, 479–484.

Wagner, J. W., Erwin, W. & Critchlow, V. (1966). Androgen sterilization produced by intracerebral implants of testosterone in neonatal female rats. *Endocrinology* **79**, 1135–1142.

Whalen, R. E. & Luttge, W. G. (1971). Perinatal administration of dihydrotestosterone to female rats and the development of reproductive function. *Endocrinology* **89**, 1320–1322.

Wollman, A. L. & Hamilton, J. B. (1967). Prevention by cyproterone acetate of androgenic, but not gonadotrophic, elicitation of persistent estrus in rats. *Endocrinology* **81**, 350–356.

Yaginuma, T., Matsuda, A., Murasawa, Y., Kobayashi, T. & Kobayashi, T. (1969). Presence of hypothalamo-pituitary testicular axis in the early postnatal period. *Endocr. Jap.* **16**, 5–10.

Yazaki, I. (1960). Further studies on endocrine activity of subcutaneous ovarian grafts in male rats by daily examination of smears from vaginal grafts. *Annot. Zool. Japon.* **33**, 217–225.

CIRCADIAN RHYTHM OF GLUCOCORTICOID SECRETION

Takashi Moroji, Kosuke Yamazaki, Fumihiko Okada, Yoshiro Saito, Yutaka Asano, and Itaru Yamashita

The circadian rhythm of adrenocortical activity, usually enhanced in the morning and decreased at night in man, was found to be frequently disturbed in subjects with CNS disorders such as head injury or general paresis.

In relation to this rhythm, circadian variations in contents of CNS transmitters in various developmental stages were investigated in rats. Serotonin showed a high level in the light and a decrease in the dark periods, while norepinephrine had a peak in the dark and low values in the light phases. This pattern became manifest on 35–37 postnatal days and continued constant thereafter. Brain acetylcholine also indicated a circadian change with a rise in the morning and a downward trend toward evening. Methamphetamine abolished an elevation of hypothalamic norepinephrine content during the dark phase, whereas it had no effect on circadian rhythm of plasma corticosterone.

The relationship between the circadian periodicity of adrenocortical function and that of CNS contents of transmitters still awaits further clarification.

INTRODUCTION

The presence of a rhythmic variation in adrenocortical function was suggested by Pincus (1943). He found that the rates of urinary excretion of 17-ketosteroids (17-KS) were significantly lower at night than in the day, and the maximum rate occurred during the morning. Halberg & Visscher (1950) demonstrated that the circulating eosinophil counts that reflect adrenocortical function were lower at night than in the morning in the tail blood of mice and rats.

Following the development of highly specific quantitative methods for measuring plasma corticosteroids, Bliss, Sandberg, Nelson & Eik-

Nes (1953) demonstrated a morning fall in plasma corticosteroid levels which continued during the course of the day. Migeon, Tyler, Mahoney, Florentin, Castle, Bliss & Samuels (1956) first disclosed the circadian variation of plasma 17-hydrocorticosteroid (17-OHCS) concentrations, which are lowest at 22.00 and 02.00 and highest at 06.00. Perkoff, Eik-Nes, Nugent, Fred, Nimer, Rush, Samuels & Tyler (1959) also indicated, in their study of plasma samples obtained at 4 hr intervals, that the level of 17-OHCS was maximum at 08.00 and tended to decrease during the rest of the day to a minimum level between 20.00 and 04.00. Doe, Vennes & Flink (1960) demonstrated that the urinary 17-OHCS excretion lags behind the plasma concentration by about 3 hr.

The circadian variation in adrenocortical function has been studied not only in control conditions, but also under a variety of extraordinary situations, such as sleep reversal (Perkoff *et al.* 1959), rapid transition across a time zone (Flink & Doe, 1959; Conroy & Hall, 1969), imposition of abnormal time schedules, *i.e.*, 12, 19, or 32 hr rest-activity cycles (Orth, Island & Liddle, 1967), and conditions with all known Zeitgebers excluded (Mills, 1964). These studies reveal that it takes at least about 1 week for the adrenocortical function to adapt to any newly imposed living schedule.

With regard to possible control mechanism for the circadian rhythm, there are some postulations, including (1) variations in steroid removal from the blood, (2) variations in adrenocortical sensitivity to ACTH, and (3) changes in the rate of ACTH secretion. According to Perkoff *et al.* (1959), variations in both steroid removal from the blood and adrenocortical sensitivity to ACTH are of minor importance. DeMoor, Heirwegh, Heremans & Declerck-Raskin (1962) suggested that the circadian rhythm of the plasma corticoid level was not attributable to any change in binding capacity of transcortin.

It appears very likely, on the other hand, that the circadian rhythm of plasma corticosteroid concentration is due to changes in the rate of ACTH secretion. Actually a rhythmic variation in plasma ACTH concentration, higher levels at 06.00 than at 18.00, was shown by Ney and his associates, Schimizu, Nicholson, Island & Liddle (1963). They also demonstrated that the variation in plasma ACTH levels closely correlated with changes in plasma 17-OHCS. The cortisol-deficient patients with a controlled level of plasma cortisol or recovering from long-term supression of pituitary function exhibited the circadian variation in plasma ACTH level (Graber, Givens, Nicholson, Island & Liddle, 1965; Graber, Ney, Nicholson, Island & Liddle, 1965). Demura, West, Nugent, Nakagawa & Tyler (1966), using a sensitive radio-immunoassay technique, estimated ACTH levels in plasma at 2 hr intervals and found the maximum values in the morning.

It is well known that the secretion of ACTH from the pituitary is controlled by means of a peptide, the corticotropin-releasing factor (CRF) released from the hypothalamus (Schally, Anderson, Lipbscomb, Long & Guillman, 1960). In *in vitro* studies, Ungar (1964) suggested the existence of a circadian rhythm in the CRF activity in the mouse hypothalamus. Lately, Hiroshige, Sakakura & Itoh (1969) and Hiroshige & Sakakura (1971) verified that the hypothalamic content of CRF shows a marked circadian variation in correlation with changes in plasma corticosteroid level.

Furthermore, evidence of the control of the hypothalamus-pituitary-adrenocortical function by certain higher centers in the central nervous system (CNS) have been provided by a number of observations in human subjects as well as in experimental animals. Alterations in consciousness and sleep-wakefulness pattern closely associated with changes in the plasma corticosteroid rhythm suggest that the CNS is involved in the mechanism for the development of the circadian rhythm of corticosteroids. Hökfelt & Luft (1959) demonstrated the disturbed circadian rhythm in plasma corticosteroids in 4 out of 7 cases of suprasellar tumor with the displacement of the hypothalamus. According to another report, loss of the circadian rhythm in plasma corticosteroid levels was also observed in patients, in alert conditions, with focal diseases of the CNS, particularly in those with CNS diseases affecting the pretectum, temporal lobe or hypothalamus (Krieger, 1961; Krieger & Krieger, 1966; Krieger, Silverberg, Rizzo & Krieger, 1968*b*).

There have been other experimental studies concerning the control mechanism for the circadian rhythm. Extensive brain ablation including the thalamus and hypothalamus prevented the circadian rhythm of pituitary ACTH contents in mice (Galicich, Halberg, French & Ungar, 1965). Electrolytic lesions of the anterior periventricular zone and the arcuate nucleus of the rat brain disrupted the circadian rhythm of plasma corticosteroids (Slusher, 1964) and the surgical interruption of neural afferent from the anterior to the medial basal hypothalami in rats disturbed the circadian ACTH rhythm (Halász, Vernicos-Danellis & Gorski 1967). The bilateral fornix section in the monkey altered the phase of urinary 17-OHCS excretion (Mason, Mangan, Conrad & Brady, 1957). A chronic implant of cortisol in the median eminence reduced the evening rise in plasma corticosterone in the rat (Slusher, 1966). The implants in the ventral hippocampus and the midbrain reticular formation elevated the morning level of the plasma corticosterone and inhibited the normal circadian rise at 17.00, resulting in disappearance of the plasma corticosterone variation.

We have performed a series of studies for several years on the circadian rhythm of urinary 17-OHCS or plasma 17-OHCS levels in patients with various mental disorders (Suwa & Yamashita, 1972). Marked impairment of the rhythm was observed in patients with some disorders in the CNS, such as brain lesions with disturbed consciousness (Moroji, Yamazaki, Kato & Onodera, 1966), general paresis (Yamazaki, 1970), and hard-controlled epilepsy (Sakashita, 1972). The disturbance of the rhythm was also observed in many patients with endogenous psychosis, particularly those in acutely disturbed states of schizophrenia (Moroji, 1966) or manic-depressive psychosis (Kato, 1970) and also subjects with emotionally blunted and deteriorated schizophrenia. The first part of the present article describes the circadian rhythm of adrenocortical activity in patients with mental disorders, especially in cases of brain lesions and general paresis.

More recently, it was postulated by some investigators (Krieger & Krieger, 1967, 1968; Krieger, Silverberg, Rizzo & Krieger, 1968; Krieger & Rizzo, 1969; Scapagnini, Moberg, Van Loon & De Groot, 1970; Scapagnini, Van Loon, Moberg & Ganong, 1970; Scapagnini Moberg, Van Loon, De Groot & Ganong, 1971) that transmitters in the CNS, such as serotonin (5-HT) and acetylcholine (Ach), play an important role in development of the circadian rhythm of corticosteroids. Actually, it was demonstrated that the contents of these neurotransmitters in the brain show distinct circadian rhythm (Albrecht, Visscher, Bittner & Halberg, 1956; Asano, 1971; Asano & Moroji, 1972; Dixit & Buckley, 1967; Matussek & Patschke, 1963; Okada, 1971; Quay, 1964, 1965, 1968; Reis, Corvelli & Conners, 1969; Scheving, Harrison, Gordon & Pauly 1968; Hanin, Massarelli & Costa, 1970). We have studied the circadian rhythm of the contents of the CNS of these synaptically active agents, their alterations under various circumstances, and the development of the rhythms of 5-HT and noradrenaline (NE). The experimental results are summarized in the second part of this paper.

CIRCADIAN RHYTHM OF PITUITARY—ADRENOCORTICAL FUNCTION IN PATIENTS WITH CENTRAL NERVOUS SYSTEM DISEASES

Brain lesions associated with disturbed consciousness

Eik-Nes & Clark (1958) first reported disturbance of the circadian rhythm of plasma 17-OHCS levels in subjects with severe brain damage accompanied by impaired consciousness. Later, Perkoff *et al.* (1959) observed marked derangement of the rhythm with alterations in consciousness or sleep-pattern in only 19 out of 31 patients examined. These findings led us to consider that the loss of the circadian corticosteroid rhythm may be attributable to either disturbance of consciousness itself or focal CNS lesions interfering with consciousness. We, therefore, have carried out investigations in the circadian pattern of adrenocortical function in patients with various CNS diseases associated with altered consciousness.

In normal subjects, plasma 17-OHCS levels reached the maximum value at 08.00, tended to decrease to a minimum level at 02.00, and returned rapidly to the peak value during sleep the following morning (Fig. 1). The circadian rhythm in urinary excretion of 17-OHCS followed the same variation as the plasma levels with a phase-delay of a few hours. A circadian variation in urinary 17-KS excretion was similar to that of urinary 17-OHCS (Fig. 1). The circadian rhythm of urinary excretion of 17-OHCS was found to be reproducible in the measurements carried out on 3 successive days or 3 times in a month for each subject. The circadian rhythms of the plasma 17-OHCS level, urinary excretion of 17-OHCS, and 17-KS observed in normal subjects are consistent with the previous reports cited above.

In all patients who were in various states of disturbed consciousness (Table 1), the loss of normal circadian rhythm in urinary 17-OHCS and 17-KS excretion was remarkable (Fig. 2). Three different types of relationships between the circadian rhythm of corticosteroids and the states of consciousness were discriminated as follows. (a) Continuous loss of the circadian rhythm with a persistent derangement of consciousness (Cases 1, 2, and 6 in Table 1). (b) Irregular rhythm patterns even after complete recovery from disturbance of consciousness (Cases 3, 4, 7, and 8 in Table 1). (c) Restoration of a normal rhythm as the patients recovered from disturbed consciousness (Cases 5 and 8 in Table 1).

It is noteworthy that either irregular patterns (Type b) or no

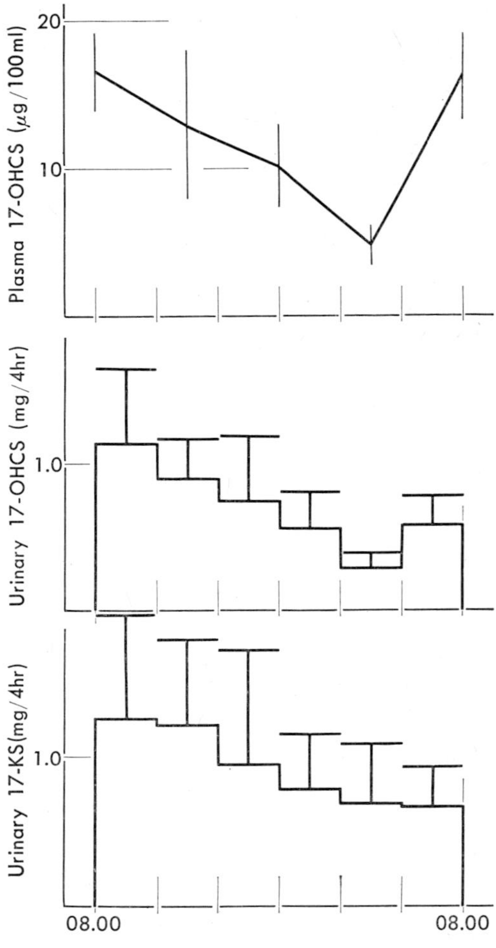

Fig. 1. Mean values of plasma 17-OHCS levels and urinary 17-OHCS and 17-KS output for 24 hr in normal subjects, illustrating the close parallel between them (Plasma 17-OHCS levels: 5 males, aged 27–34. Urinary 17-OHCS and 17-KS output: 6 males and 4 females, aged 21–43). Plasma 17-OHCS was measured by Peterson's modification, urinary 17-OHCS by Gleen-Nelson's modification (Kanbegawa, 1963), and urinary 17-KS by the method of Shida, Shimazaki & Takei, 1963. Most of the subjects were asleep from midnight until about 07.00 (from Moroji *et al.* 1966).

abnormalities (Type c) of the rhythm were observed in the patients who had recovered from disturbed consciousness. Further studies disclosed that dilatation of the ventricular system including the Sylvius, aqueduct, and third ventricle was observed by pneumo-

TABLE 1. Clinical data on 9 patients with brain lesions (from Moroji *et al.* 1966)

Case No.	Age	Sex	Diagnosis	Symptoms	Duration disturb. conscious	EEG	PEG	Circad. rhythm of plasma 17-OHCS
1	23	M	brain contusion	semi-coma decerebr. rigidity	9 month	irregul. flat	marked dilat.	irregular
2	50	M	IIIrd. ventr. tumor.	semi-coma decerebr. rigidity	9 month	diffuse slow	—	irregular flat
3	23	M	brain contusion	dementia abulia	2 week	low volt. irreg. α	marked dilat.	irregular
4	73	F	lt. front. intracereb. hematoma	dementia	1 week	—	—	irregular
5	36	F	ruptur. lt. mid-cereb. art.	rt. leg paresis aphasia	9 week	diffus. α sporad. slow w.	nearly normal	irregular ↓ normal
6	36	M	anoxic encephalo-pathy	semi-coma decerebr. rigidity	1 year 8 month	flat	marked dilat.	irregular
7	29	M	cerebro-vascular accident	rt. hemi plegia aphasia	1 week	diffus. α	slight dilat.	irregular
8	29	M	rt. subdr. hematoma	normal	1 week	normal	normal	irregular ↓ normal
9	28	M	aneurysm lt. int. carotid	coma rt. hemi plegia	1 month	low volt irreg. slow	—	irregular

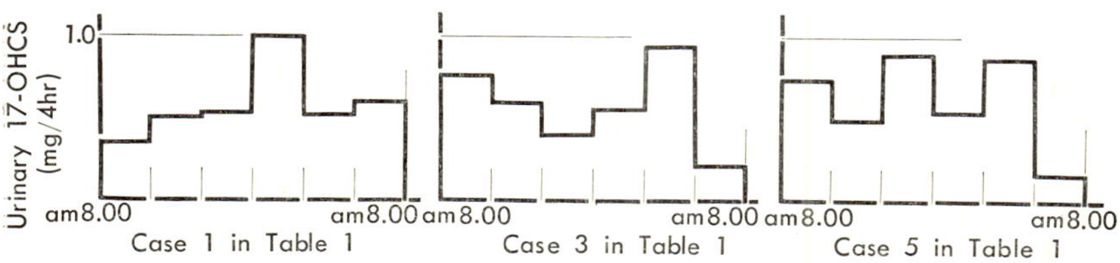

Fig. 2. Diurnal variations of urinary 17-OHCS and 17-KS excretion in patients with disturbed consciousness due to brain damage (from Moroji *et al.* 1966).

encephalogram in patients of Type b. No enlargement of the ventricles, however, was noted in patients of Type c. These results suggest that some other mechanism and/or structure in the CNS besides that concerned with the maintenance of consciousness is also implicated in the development of the circadian rhythm of 17-OHCS.

TABLE 2. Circadian variations of plasma 17-OHCS levels and pneumoencephalography (PEG) data in 13 patients with general paresis. The plasma 17-OHCS was measured by Peterson's modification (1957) of Silber-Porter procedure (from Yamazaki, 1970)

Case No.	Age	Sex	Plasma 17-OHCS μg/100 ml					PEG findings		
			8.00	14.00	20.00	2.00	8.00	IIIrd ventr. width (mm)	IIIrd ventr. area (cm²)	Lat. ventr. area (cm²) rt : lt
1	36	M	9.5	4.1	2.7	30.1	52.0	9	1.2	4.7:3.1
2	43	M	8.0	4.1	19.1	32.8	23.0			
3	43	M	30.1	39.1	1.3	26.0	16.7			
4	54	M	23.6	7.5	24.7	18.2	26.8	15	1.9	4.3:2.2
5	44	M	11.3	5.1	1.5	0.3	0.3	14	2.0	4.9:8.0
6	50	M	7.2	0.6	1.9	0.9	13.5	9	1.0	5.5:6.4
7	53	M	26.2	26.2	19.4	6.4	18.9	18	3.8	7.5:8.4
8	46	M	15.4	11.6	5.7	10.1	14.5	3	0.7	2.5:2.0
9	59	M	24.5	11.8	20.3	20.3	10.1	8	0.9	3.2:3.2
10	47	M	15.0	10.3	24.7	7.4	12.5			
11	36	M	28.8	17.7	24.5	33.0	23.7	11	2.3	6.2:6.0
12	43	M	18.7	8.2	13.7	8.6	21.0			
13	38	M	13.2	19.8	5.0	18.5	10.5			
Mean			17.8	12.8	12.6	16.3	18.6			
S.D.			8.2	11.0	10.1	12.0	12.6			

General paresis

A more extensive area of the brain than that related to consciousness seems to be responsible for the development of the circadian rhythm of corticosteroids. Therefore, we investigated the circadian rhythm of plasma 17-OHCS in patients with general paresis, which is a chronic spirochetal meningoencephalitis and seldom accompanies disturbance of consciousness.

Thirteen male patients with general paresis were examined (Table 2). All of them had been treated by antibiotics and fever therapy and were already in the chronic stage of the disease. None had overt anxiety, sleep disturbance, or impairment of consciousness. Their symptoms ranged from relatively slight to extreme dissolution of mental functions. Their sleep-wakefulness patterns conformed to the usual hospital regimen.

The plasma 17-OHCS levels in the patients are shown in Table 2. One of 13 patients exhibited a normal circadian rhythm of plasma 17-OHCS concentrations. It is of interest that this patient is the only one who had a normal ventricular system. This observation is consistent with the results from patients with disturbed consciousness and normal PEG. The present results in the patients with general paresis and with other CNS diseases suggest that alteration in the corticosteroid rhythm may occur in patients with CNS diseases in the absence of altered consciousness or sleep disturbance. Seven of the patients exhibited an enlargement of the lateral as well as third ventricles ranging from moderate to marked degrees on the PEG. Absence of the rhythm in these cases may be ascribable to degeneration of the areas of the CNS concerned with the development of the corticoid rhythm, possibly due to chronic meningoencephalitic process. A loss of circadian variation in plasma 17-OHCS levels may be regarded as an unspecific manifestation of general paresis. Thus, it is concluded that the brain structures indispensable for the development of the circadian rhythm in adrenocortical activity are distinct from those involved in the maintenance of consciousness.

CORRELATION BETWEEN CIRCADIAN RHYTHM OF PLASMA CORTICOSTEROID CONCENTRATIONS AND DAILY VARIATION IN CONTENTS OF TRANSMITTERS IN THE BRAIN

Serotonin (5-hydroxytryptamine) contents in mouse brain were found by Albrecht *et al.* (1956) to be higher before and around noon than at or after 16.00. From the observations that daily adrenal activation in mice occurs during roughly the same period as a small but significant decrease in brain 5-HT concentration, they presumed that circadian variations in the brain contents might be related to the plasma corticosteroid rhythm. However, Dixit *et al.* (1967) failed to demonstrate such a relationship, although they observed a typical daily change in brain 5-HT contents as well as the circadian rhythm of plasma corticosteroid levels in rats.

Implants of 5-HT in selected areas of the hypothalamus of cats were capable of eliciting an acute rise in plasma 17-OHCS levels (Krieger *et al.* 1968). This observation suggested that 5-HT facilitates the release of the corticotropin-releasing factor from the hypothalamic secretory neuron, which in turn enhances the adenohypophysial-adrenal activity. Atropine administered just prior to the time of an expected circadian rise in plasma 17-OHCS levels blocked the elevation of the plasma 17-OHCS level in cats (Krieger, Glick, Silverberg & Krieger, 1968*a*). A cholinergic as well as monoaminergic mechanism was suggested for the control of ACTH release of the initiation of the circadian rise in adrenocortical function.

The circadian rise of plasma 17-OHCS levels in the cat was abolished by four different types of drugs affecting serotonin mechanisms (Krieger *et al.* 1969). (i) 3-(2-aminobutyl) indole acetate, which increases the content of 5-HT in CNS, (ii) *p*-chloramphetamine, which depletes CNS 5-HT contents, (iii) 2-(3-dimethyl-aminopropylthio-) cinnamanilide, which acts as a competitive inhibitor of 5-HT at the receptor site, and (iv) cyproheptadine, which is a serotonin antagonist. Therefore, it appears plausible that serotonin plays some role in the neural mechanism to cause the circadian periodicity of the plasma 17-OHCS levels.

As to the role of norepinephrine (NE) in ACTH secretion, an increased turnover rate of NE in the CNS was observed after a variety of stress procedures (Mangili, Motta & Martini, 1966), although Krieger *et al.* (1969) found that the agents known to deplete CNS norepinephrine levels (reserpine, L-α-methyl-*p*-tyrosine) are ineffective in blocking the circadian rise of plasma 17-OHCS levels.

Norepinephrine and serotonin are found in appreciable amounts in the brain, in higher concentrations particularly in the hypothalamus, thalamus, midbrain and medulla oblongata (Dahlström & Fuxe, 1964; Andén, Dahlström, Fuxe, Larsson, Olsen & Ungerstedt, 1966; Hillarp, Fuxe & Dahlström, 1966). The contents of these biogenic amines fluctuate in step with the circadian rhythm in different regions of the CNS (Friedman & Walker, 1968; Growinski, Axelrod & Iversen, 1966; Manshardt & Wurtman, 1968; Quay, 1965, 1968; Reis, 1968; Reis *et al.* 1969; Reis & Wurtman, 1968) as well as in the whole brain (Albrecht *et al.* 1956; Dixit *et al.* 1967; Matussek *et al.* 1963; Quay, 1964; Scheving *et al.* 1968). Therefore, we have performed a series of studies on the relationship between the daily rhythm of the CNS concentrations of neurotransmitters and the circadian rhythm in plasma corticosteroid levels.

Maturation of the circadian rhythm of brain serotonin contents in rats

Allen & Kendall (1967) demonstrated that no definite circadian rhythm of corticosteroids is present until days 30–32 in unhandled rats. Manual handling or electric shock during the early postnatal period accelerated the maturation of plasma corticosteroid rhythm (Ader, 1969). The circadian rhythm in corticotropin-releasing activity as well as in plasma corticosteroid concentration developed

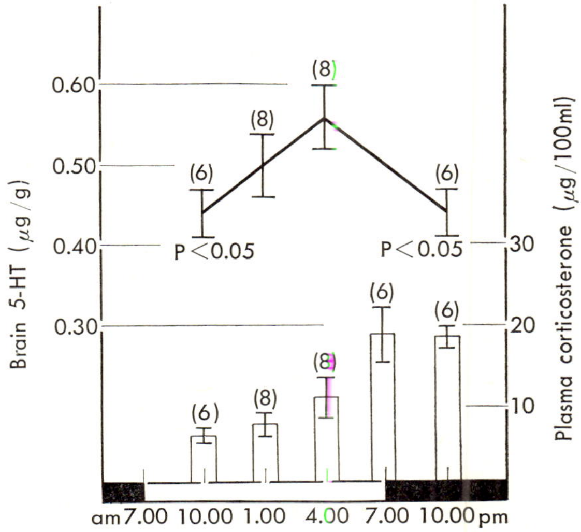

Fig. 3. Circadian patterns for brain 5-HT levels and plasma corticosterone on days 35–37 in male rats. Solid line, 5-HT. Column, Corticosterone. Bar, Standard error of mean value. P, Difference between the 5-HT levels in the 16.00 and 10.00 groups, and the 16.00 and 22.00 groups. The 5-HT assay was carried out by the method of Snyder, Axelrod & Zweig (1965), and plasma corticosterone was measured by the method of Mattingly (1962), as modified for microdetermination (from Okada, 1971).

during the third week of life in handled rats (Hiroshige *et al.* 1969). The maturation of the circadian rhythm in brain 5-HT contents and its correlation to the circadian rhythm of plasma corticosterone levels in rats are discussed below.

The 5-HT contents in the brain were measured in Wistar-King A strain rats at 17–20, 27–30, 35–37, and 49–52 days of age. Circadian variations of the 5-HT contents were not present until the 35th–37th postnatal day (Fig. 3). The maximum value was obtained at 16.00 and the minimum value at 22.00 in the dark period. The circadian rhythm of the brain 5-HT contents in rats of 35–37 days of age is very similar to that reported so far in the adult rat (Dixit *et al.* 1967; Quay, 1964; Scheving *et al.* 1968). The present results of the onset of rhythm are quite consistent with those of Baker & Quay (1969) that the 5-HT level in the rat brain at birth is about 40% that of an adult and increases gradually until maturity, and the circadian variation of brain 5-HT levels becomes prominent on approximately the 35–37th day, when the brain 5-HT is about 75% that of the adult.

On the other hand, the circadian rhythm of plasma corticosterone concentrations in the rat, which is characterized by an increase in the evening, was first observed on days 27–30. This rhythm became more conspicuous thereafter. The onset of the circadian rhythm of brain 5-HT is somewhat slower than the development of the circadian rhythm of plasma corticosterone. Consequently, it seems difficult at present to postulate any physiological correlation between these two circadian rhythms.

Maturation of the circadian rhythms of brain serotonin and norepinephrine in rats

The previous study created a need for further studies to clarify the relationship between the circadian rhythms of brain 5-HT contents and plasma corticosterone levels. We conducted a series of experiments in an attempt to disclose the maturation process of the circadian rhythm not only of serotonin but also of norepinephrine levels in the rat brain.

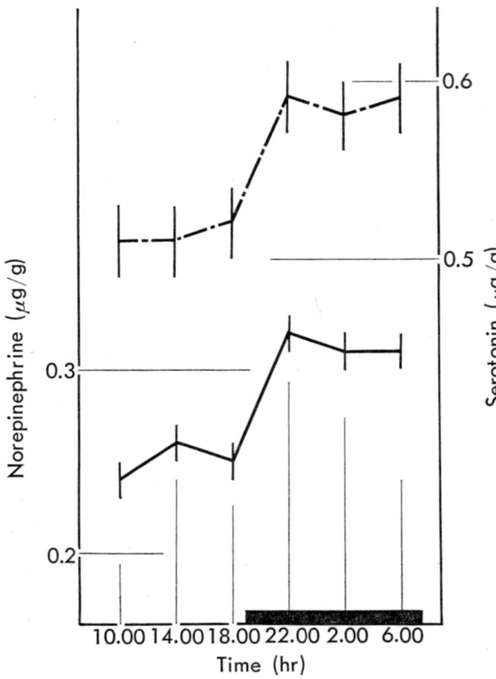

Fig. 4. Circadian rhythms of brain NE and 5-HT on days 15–17; each value from 9 rats. For the determination of NE and 5-HT, the method of Chang (1964) modified by Maickel, Cox, Saillant & Miller (1968) was used (from Asano, 1971; Asano *et al.* 1972).

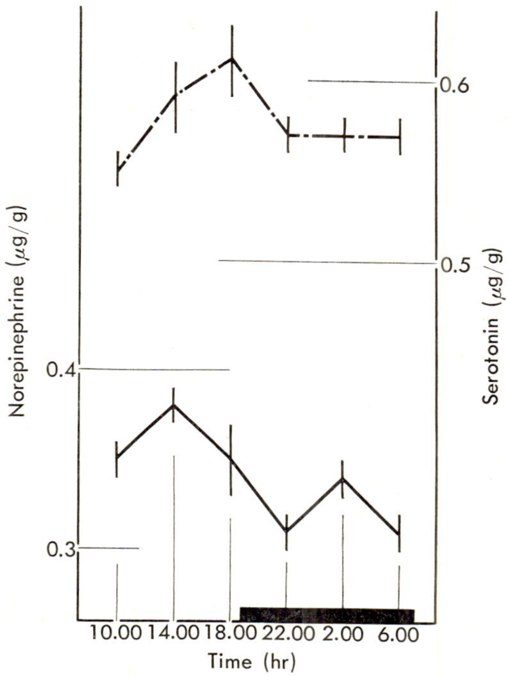

Fig. 5. Circadian rhythms of brain NE and 5-HT on day 21; each value from 10 rats (from Asano, 1971; Asano *et al.* 1972).

The circadian variations of norepinephrine and serotonin contents in the rat brain were investigated at 8–9, 15–17, 21, 27–28, 35–37, and 56–66 days after birth. Both serotonin and norepinephrine showed a common pattern of circadian changes from postnatal days 8–9 to days 15–17, exhibiting a peak in the dark and a reduction in the light periods (Fig. 4). This pattern was reversed on days 21 and 27–28, when both amine levels increased in the light and decreased in the dark periods (Fig. 5). On days 35–37 and after this stage, however, a discrepancy began to appear between the circadian variations of NE and 5-HT contents in the brain (Fig. 6). The level of norepinephrine was low during the light period and then increased rapidly to reach a peak at the beginning of the dark phase followed by a decrease for the remainder of the dark period. On the other hand, the serotonin content remained at a high level in the light phase and tended to decrease during the dark period. This tendency became more manifest and concrete on days 55–66 than on days 35–37.

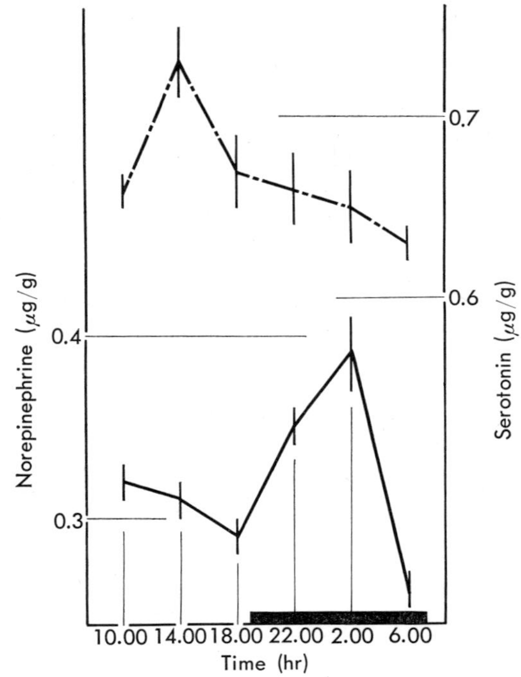

Fig. 6. Circadian rhythms of brain NE and 5-HT on days 56–66; each value from 10 rats (from Asano, 1971; Asano *et al.* 1972).

The present observations are consistent with the histochemical findings. Loizou (1969) and Loizou & Salt (1970) demonstrated that the norepinephrine-containing cell bodies and nerve terminals developed slowly over the first 2 weeks, and thereafter their norepinephrine content increased rapidly to an adult level by the 4–5th week of life, whereas the adult pattern of serotonin-containing neurons appeared by the 3rd week of postnatal life. Furthermore, the present results indicate that the onset of an adult pattern of serotonin is nearly coincident with that of the circadian rhythm of plasma 17-OHCS levels (Allen *et al.* 1967; Ader, 1969; Hiroshige *et al.* 1969; Okada, 1971).

A diurnal variation in the serotonin contents of the hippocampus and amygdala was generally parallel to the circadian fluctuation of plasma corticosterone, which is composed of a progressive increase in 2 phases during the light period and a sharp fall during the dark period (Scapagnini, Moberg, Van Loon & De Groot, 1970). The treatment with *p*-chlorphenylalanine, which blocks 5-HT synthesis, abolished the diurnal variation of plasma corticosterone (Scapagnini, Moberg, Van Loon & De Groot, 1970; Scapagnini, Van Loon, Moberg & Ganong, 1971).

Effects of methamphetamine on the circadian rhythms of norepinephrine contents in the hypothalamus and plasma corticosterone

The highest concentration of NE in brain structures was demonstrated in the hypothalamus by the use of fluorescent histochemical as well as biochemical methods (Andén *et al.* 1966; Dahlström *et al.* 1964). The level of NE content in the hypothalamus exhibits a daily rhythm in rats. The circadian rhythm is also established for the hypothalamic corticotropin-releasing activity (Hiroshige *et al.* 1969; Hiroshige *et al.* 1971). These findings caused us to assume that hypothalamic NE functions as a neurotransmitter to produce the circadian rhythm of pituitary-adrenocortical functions. Thus, we studied the effects of methamphetamine, known to reduce the hypothalamic NE content, on the circadian rhythm of hypothalamic NE content and plasma corticoid level in rats.

A distinct daily rhythm was observed in hypothalamic NE contents in the controls (Fig. 7). The maximum value was found at 22.00 and the minimum at 07.00. The levels during the dark period were higher than those in the light phase. Methamphetamine abolished an elevation of hypothalamic NE content during the dark

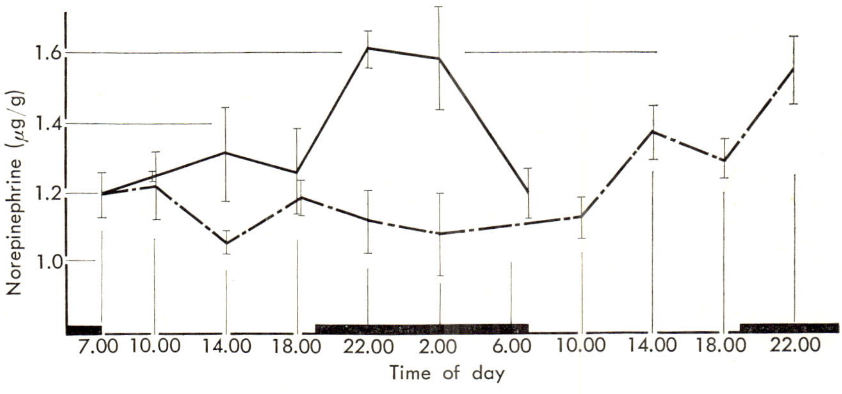

Fig. 7. Effects of methamphetamine on the circadian rhythm of the hypothalamic NE content in rats. The individual plotted values are the means with standard errors. Solid line, Saline 0.5 ml i.p. at 07.00. Broken line, Methamphetamine 5 mg/kg i.p. at 07.00. Norepinephrine was measured by the method of Chang (1964) modified by Maickel *et al.* (1968) (from Asano & Moroji, unpublished).

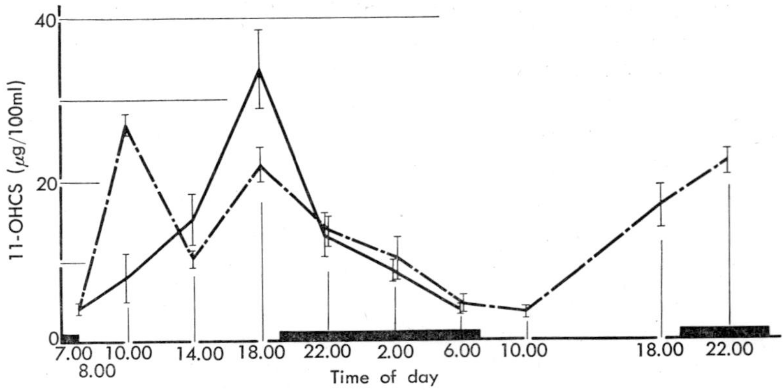

Fig. 8. Effects of methamphetamine on the circadian rhythm of plasma corticosterone in rats. Solid line, Saline 0.5 ml i.p. at 07.00. Broken line, Methamphetamine 5 mg/kg i.p. at 07.00. Bar, Standard error (from Asano & Moroji, unpublished).

phase for 24 hr after the injection. On the following day, however, a rhythm similar to that of the controls was restored (Fig. 7).

Plasma corticosterone level exhibited a difinite circadian variation, as shown in Fig. 8. The injection of methamphetamine produced a peak at 10.00. The circadian rise early in the evening, however, was observed in the treated animals as well as in the controls. On the following day, the time of the maximal level shifted to 22.00 (Fig. 8). Behavioral changes due to methamphetamine were observed for about 4 hr after the injection.

An inverse relationship between brain NE and plasma corticosterone rhythms seems to be consistent with the conception that the central adrenergic neuronal system inhibits ACTH secretion (Bhattacharya & Marks, 1969; Van Loon, Hilger, Cohen & Ganong, 1969; Van Loon & Ganong, 1969; Van Loon, Scapagnini, Moberg & Ganong, 1970; Van Loon, Nicholson & Liddle, 1971). The circadian rhythm of plasma corticosterone levels was maintained in rats treated with methamphetamine, although hypothalamic NE was depleted for approximately 24 hr after the injection. Methamphetamine at a dose of 5 mg/kg is known to cause a marked depletion of NE, particularly in the hypothalamus, without altering the hypothalamic serotonin content (Fuxe & Ungerstedt, 1970; Growinski *et al.* 1966; Ott, Schmitt, Pohle & Matthies, 1971).

The present results are in agreement with the findings that drugs known to deplete NE in the CNS, such as reserpine and L-alpha-methyl-*p*-tyrosine (Krieger *et al.* 1970), and also an alpha-adrenergic blocker, phenoxybenzamine (Krieger *et al.* 1968), were ineffective in blocking the circadian rise of plasma corticoid level. Accordingly, it is concluded that the daily rhythm in hypothalamic NE content is not responsible for the development of the circadian rhythm of pituitary-adrenocortical function.

Circadian rhythm of brain acetylcholine level

The level of acetylcholine (Ach) in the brain was shown to vary with convulsion (Richter & Crossland, 1949; Crossland, 1953; Takahashi & Aprison, 1964), excitation (Richter & Crossland, 1949; Anand, 1952; Aprison, Kariya, Hingtgen & Toru, 1968), anesthesia (Richer *et al.* 1949; Crossland & Merrick, 1954; Elliot, Swank & Henderson, 1950; Tobias, Lipton & Lepinat, 1946), and sleep (Yamaguchi, Ling & Marcznski, 1963). Acetylcholine was effective in initiating the circadian rise of plasma 17-OHCS only for a relatively short period of time in the 24 hr cycle (Krieger *et al.* 1967; Krieger *et al.* 1968). Experiments to provide evidence for the presence of a circadian rhythm of brain Ach concentrations in rats are described below.

Total Ach content in the rat brain began to increase after the light

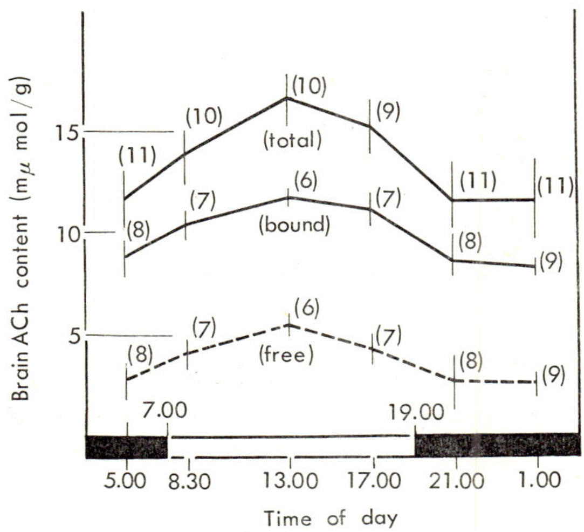

Fig. 9. The free, bound, and total Ach contents in the rat brain; each value represents the mean with standard error of 6–11 animals. For estimation of free and bound Ach contents in the rat brain, the chemical method of Fellman (1969) was employed (from Saito, 1971).

was turned on in the morning, reached a peak at 13.00, and took a downward trend toward evening when the light was turned off (Fig. 9). The minimum value was found at 21.00, but the levels during the dark period were generally as low as this value. The free and bound Ach exhibited very similar circadian patterns to that of total levels (Fig. 9).

The present results are essentially compatible with those reported by Hanin *et al.* (1970). However, they observed peak concentration that continued from 2 hr after the light was turned on to 6 hr after the light was turned off. This difference is difficult to explain at present. The rhythmic changes in the Ach contents are quite similar to the circadian rhythm of plasma corticosteroids.

CONCLUSION

The circadian periodicity of adrenocortical functions was estimated in 9 patients with brain lesions and 13 with general paresis. It was shown that the subjects having extensive CNS damage or lesions at the brain stem frequently demonstrated disturbed patterns of circadian variation in plasma and urinary 17-OHCS levels.

The relationships between the circadian rhythm of plasma corticosteroids and the daily rhythms of 5-HT, NE, and Ach contents in the rat brain were investigated in the hope that the circadian rhythm of plasma corticosteroids may be attributable to these neurotransmitters in the brain. The results indicated that the daily rhythm of brain serotonin plays a certain role in the development of the corticosteroid rhythm, whereas that of brain NE, particularly hypothalamic NE content, has no relation to the control mechanism for the circadian rhythm of plasma corticosteroids. The function of brain Ach remains a subject for further study.

ACKNOWLEDGEMENTS

We are greatly indebted to Professor N. Suwa for his advice and encouragement. This work was supported in part by a grant from the Ministry of Education of Japan.

REFERENCES

Ader, R. (1969). Early experiences accelerate maturation of the 24-hour adrenocortical rhythm. *Science, N.Y.* **163**, 1225–1226.

Albrecht, P., Visscher, M. B., Bittner, J. J. & Halberg, R. (1956). Daily changes in 5-hydroxytryptamine concentration in the mouse brain. *Proc. Soc. exp. Biol Med.* **92**, 703–706.

Allen, C. & Kendall, J. W. (1967). Maturation of the circadian rhythm of plasma corticosterone in the rat. *Endocrinology* **80**, 926–930.

Anand, B. K. (1952). Influence of temperature on vagal inhibition and liberation of acetylcholine in frog hearts. *Am. J. Physiol.* **168**, 218–225.

Andén, N. E., Dahlström, A., Fuxe, K., Larsson, K., Olsen, L. & Ungerstedt, U. (1966). Ascending monoamine neurones to the telencephalon and diencephalon. *Acta Physiol. Rev.* **67**, 313–326.

Aprison, M. H., Kariya, T., Hingtgen, J. N. & Toru, M. (1968). Neurochemical correlates of behaviour changes in acetylcholine, norepinephrine and 5-hydroxytryptamine concentrations in several discrete brain areas of the rat during behavioural excitation. *J. Neurochem.* **15**, 1131–1139.

Asano, Y. (1971). The maturation of the circadian rhythm of brain norepinephrine and serotonin in the rat. *Life Sci. Oxford* **10**, Part 1. 883–894.

Asano, Y. & Moroji, T. (1972). The maturation of the circadian rhythm of brain norepinephrine and serotonin in the rat. *Brain and Nerve.* (Tokyo) **24**, 1445–1551.

Baker, P. C. & Quay, W. B. (1969). 5-Hydroxytryptamine metabolism in early embryogenesis, and the development of brain and retinal tissues. A review. *Brain Res.* **12**, 273–295.

Bhattacharya, A. N. & Marks, B. H. (1969). Reserpine-and-chlorpromazine-induced changes in hypothalamo-hypophyseal-adrenal system in rats in the presence and absence of hypothermia. *J. Pharmac. exp. Ther.* **165**, 108–116.

Bliss, E. L., Sandberg, A. A., Nelson, D. H. & Eik-Nes, K. (1953). The normal levels of 17-hydroxycorticosteroids in the peripheral blood of man. *J. clin. Invest.* **32**, 818–823.

Chang, C. C. (1964). A sensitive method for spectrophotofluorometric assay of catecholamines. *Int. J. Neuropharmac.* **3**, 643–649.

Conroy, R. T. W. L. & Hall, M. D. (1969). Adrenal cortical function and body temperature rhythms after a transatlantic flight. *J. Physiol.* **200**, 123.

Crossland, J. (1953). The significance of brain acetylcholine. *J. Ment. Sci.* **99**, 247–251.

Crossland, J. & Merrick, A. J. (1954). The effect of anesthesia on the acetylcholine content of the brain. *J. Physiol.* **125**, 56–66.

Dahlström, A. & Fuxe, K. (1964). Evidence for the existence of monoamine-containing neurons in the central nervous system. *Acta physiol. scand.* **62**, Suppl. 232, 1–55.

DeMoor, P., Heirwegh, K., Heremans, J. F. & Declerck-Raskin M.

(1962). Protein binding of corticoids studied by gel-filtration. *J. clin. Invest.* **41**, 816–827.

Demura, H., West, C. D., Nugent, C. A., Nakagawa, K. & Tyler, F. (1966). A sensitive radioimmunoassay for plasma ACTH levels. *J. clin. Endocr. Metab.* **26**, 1297–1303.

Dixit, B. N. & Buckley, J. P. (1967). Circadian changes in brain 5-hydroxytryptamine and plasma corticosterone in the rat. *Life Sci. Oxford* **6**, 755–758.

Doe, R. P., Vennes, J. A. & Flink, E. B. (1960). Diurnal variation of 17-hydroxycorticosteroids, sodium, potassium, magnesium and creatinine in normal subjects and in cases of treated adrenal insufficiency and Cushing's syndrome. *J. clin. Endocr. Metab.* **20**, 253–265.

Eik-Nes, K. B. & Clark, L. D. (1958) Diurnal variation of plasma 17-hydroxycorticosteroids in subjects suffering from severe brain damage. *J. clin. Endocr. Metab.* **18**, 764–768.

Elliott, K. A. C., Swank, R. L. & Henderson, N. (1950). Effects of anesthetics and convulsants on acetylcholine content of brain. *Amer. J. Physiol.* **162**, 469–474.

Fellman, J. H. (1969). A chemical method for the determination of acetylcholine: Its application in a study of presynaptic release and a choline acetyltransferase assay. *J. Nurochem.* **16**, 135–143.

Flink, E. B. & Doe, R. F. (1959). Effect of sudden time displacement by air travel on synchronization of adrenal function. *Proc. Soc. exp. Biol. Med.* **100**, 498–501.

Friedman, A. H. & Walker, C. A. (1968). Circadian rhythm in rat mid-brain and chaudate nucleus biogenic amine levels. *J. Physiol.* **197**, 77–85.

Fuxe, K. & Ungerstedt, U. (1970). Histochemical, biological and functional studies on central monoamine neurons after acute and chronic amphetamine administration. International symposium on amphetamine and related compounds, ed. Costa, E. & Garattini, S. pp. 257–288. New York: Raven Press.

Galicich, J. H., Halberg, F., French, L. A. & Ungar, F. (1965). Effect of cerebral ablation on a circadian pituitary adrenocorticotropic rhythm in C mice. *Endocrinology* **76**, 895.

Graber, A. L., Givens, J. R., Nicholson, W. E., Island, D. P. & Liddle, G. W. (1965). Persistence of diurnal rhythmicity in plasma ACTH concentrations in cortisol-deficient patients. *J. clin. Endocr. Metab.* **25**, 804–807.

Graber, A. L., Ney, R. L., Nicholson, W. E., Island, D. P. & Liddle, G. W. (1965). Natural history of pituitary-adrenal recovery following long-term suppression with corticosteroids. *J. clin. Endocr. Metab.* **25**, 11–16.

Growinski, J., Axerlord, J. & Iversen, L. L. (1966). Regional studies of catecholamine in the rat brain. *J. Pharmac. exp. Ther.* **153**, 30–41.

Halász, B., Vernikos-Danellis, J. & Gorski, R. A. (1967). Pituitary ACTH content in rats after partial or total interruption of neuronal afferents to the medial basal hypothalamus. *Endocrinology* **81**, 921–934.

Halberg, F. & Visscher, M. D. (1950). Regular diurnal physiological variation in eosinophil leveles in five stocks of mice. *Proc. Soc. exp. Biol. Med.* **75**, 846–847.

Hanin, I., Massarelli, R. & Costa, E. (1970). Acetylcholine concentrations in rat brain: Diurnal oscillation. *Science, N.Y.* **170**, 341–342.

Hillarp. A., Fuxe, K. & Dahlström, A. (1966). Demonstration and mapping of central neurons containing dopamine, noradrenaline and 5-hydroxytryptamine and their reactions to psychopharmaca. *Pharmac. Rev.* **18**, 727–741.

Hiroshige, T., Sakakura, M. & Itoh, S. (1969). Diurnal variation of corticotropin-releasing activity in the rat hypothalamus. *Endocr. Jap.* **16**, 465–469.

Hiroshige, T. & Sakakura, M. (1971). Circadian rhythm of corticotropin-releasing activity in the hypothalamus of normal and adrenalectomized rats. *Neuroendocrinology* **7**, 25–36.

Hökfelt, B. & Luft, R. (1959). The effects of suprasellar tumors on the regulation of adrenocortical function. *Acta endocr., Copenh.* **32**, 177–186.

Kanbegawa, A. (1963). A method for the determination of urinary 17-hydroxycorticosteroid by the use of phenylhydrazine. *Hormone-to-Rinsho* **11**, 191–203

Kato, H. (1970). Studies on diencephalo-pituitary-adrenocortical function in endogenous depression: A psychophysiological study of emotion. *Psychiat. Neurol. Jap.* **72**, 579–595.

Krieger, D. T. (1961). Diurnal pattern of plasma 17-hydroxycorticosteroids in pretectal and temporal lobe disease. *J. clin. Endocr. Metab.* **21**, 695–698.

Krieger, D. T., Glick, S., Silverberg, A. & Krieger, H. P. (1968). A A comparative study of endocrine tests in hypothalamic disease. *J. clin. Endocr. Metab.* **28**, 1589–1598.

Krieger, D. T. & Krieger, H. P. (1966). Circadian variation of the plasma 17-hydroxycorticosteroids in central nervous system disease. *J. clin. Endocr. Metab.* **26**, 929–940.

Krieger, D. T. & Krieger, H. P. (1967). Circadian pattern of plasma 17-hydroxycorticosteroid: Alteration of anticholinergic agents. *Science, N.Y.* **155**, 1421–1422.

Krieger, H. P. & Krieger, D. T. (1968). Effect of central nervous system injection of synaptically active agents on ACTH release. *Fedn. Proc.* **27**, 217.

Krieger, D. T. & Rizzo, F. (1969). Serotonin mediation of circadian periodicity of plasma 17-hydroxycorticosteroids. *Am. J. Physiol.* **217**, 1703–1707.

Krieger, D. T., Silverberg, A. T., Rizzo, F. & Krieger, H. P. (1968). Abolition of plasma 17-OHCS levels in the cat. *Am. J. Physiol.* **215**, 959–967.

Loizou, L. A. (1969). The development of monoamine-containing neurons in the brain of the albino rat. *J. Anat.* **104**, 588.

Loizou, L. A. & Salt, P. (1970). Regional changes in monoamines of the rat brain during postnatal development. *Brain Res.* **20**, 467–470.

Maickel, R. P., Cox, R. H., Jr., Saillant, J. & Miller, F. (1968). A method for the determination of serotonin and norepinephrine in discrete areas of rat brain. *Int. J. Neuropharmac.* **7**, 275–281.

Mangili, G., Motta, M. & Martini, L. (1966). Control of adrenocorticotropic hormone secretion. In *Neuroendocrinology*, Band I, ed. Martini, L. & Ganong, W. pp. 297–370. New York, London: Academic Press.

Manshardt, J. & Wurtman, R. J. (1968). Daily rhythm in the noradrenaline content of rat hypothalamus. *Nature, Lond.* **212**, 574–575.

Mason, J. W., Mangan, F., Conrad, D. G. & Brady, J. V. (1957) (abstract). In *Prog. 39th Meeting Endocr. Soc. New York.* p. 20.

Mattingly, D. (1962). A simple fluorimetric method for the estimation of free 11-hydroxycorticosteroids in human plasma. *J. clin. Path.* **15**, 375–379.

Matussek, N. & Patschke, U. (1963). Beziehung des Schlaf-und Wachsrhythmus zum Noradrenalin- und Serotonin-gehalt im Zentralnervensystem von Hamuster. *Med. Exp.* **11**, 81–87.

Migeon, C. J., Tyler, F. H., Mahoney, J. P., Florentin, A. A., Castle, H., Bliss, E. L. & Samuels, L. T. (1956). The individual variation of plasma 17-hydroxycorticosteroids in normal subjects, night workers and blind subjects. *J. clin. Endocr. Metab.* **16**, 622–633.

Mills, J. N. (1964). Circadian rhythms during and after three months in solitude underground. *J. Physiol.* **174**, 217–231.

Moroji, T. (1966). Psychic states and diurnal rhythm of plasma 17-hydroxycorticosteroids: A psychophysiological study of emotion. *Psychiat. Neurol. Jap.* **68**, 1443–1462.

Moroji, T., Yamazaki, K., Kato, H. & Onodera, I. (1966). Diurnal variation of the plasma levels and the urinary excretion of corticosteroids in normal and disturbed states of consciousness. *Psychiat. Neurol. Jap.* **68**, 1921–1933.

Ney, R. L., Schimizu, N., Nicholson, W. E., Island, D. P. & Liddle, G. W. (1963). Correlation of plasma ACTH concentration with adrenocortical response in normal human subjects, surgical patients, and patients with Cushing's disease. *J. clin. Invest.* **42**, 1669–1677.

Okada, F. (1971). The maturation of the circadian rhythm of brain serotonin in the rat. *Life Sci. Oxford* **10**, Part I, 77–86.

Orth, D. N., Island, D. P. & Liddle, G. W. (1967). Experimental alteration of the circadian rhythm in plasma cotisol (17-OHCS) concentration in man. *J. clin. Endocr. Metab.* **27**, 549–555.

Ott, T., Schmitt, M., Pohle, W. & Matthies, H. (1971). The effect of methamphetamine and 5-hydroxytryptamine contents in eleven rat brain regions. *Brain Res.* **25**, 171–178.

Perkoff, G. T., Eik-Nes, K., Nugent, C. A., Fred, H. L., Nimer, R. A., Rush, L., Samuels, L. T. & Tyler, F. H. (1959). Studies of the diurnal variation of plasma 17-hydroxycorticosteroids in man. *J. clin. Endocr. Metab.* **19**, 432–443.

Peterson, R. E., Kaner, A. & Guena, S. L. (1957). Evaluation of Silber-Porter procedure for determination of plasma hydrocortisone. *Analyt. Chem.* **39**, 144–149.

Pincus, G. (1943). A diurnal rhythm in the excretion of urinary ketosteroids by young men. *J. Clin. Endocrinol.* **3**, 195–199.

Quay, W. B. (1964). Circadian and estrous rhythms in pineal and brain serotonin. *Prog. Brain Res.* **5**, 61–63.

Quay, W. B. (1965). Regional and circadian differences in cerebral cortical serotonin concentrations. *Life Sci. Oxford* **4**, 379–384.

Quay, W. B. (1968). Differences in circadian rhythms in 5-hydroxytryptamine according to brain regions. *Am. J. Physiol.* **215**, 1448–1453.

Reis, D. J. (1968). A circadian rhythm of norepinephrine regionally in cat brain : Its relationship to environmental lighting and to regional diurnal variations in brain serotonin. *J. Pharmac. exp. Ther.* **164**, 135–145.

Reis, D. J., Corvelli, A. & Conners, J. (1969). Circadian and ultradian rhythms of serotonin regionally in cat brain. *J. Pharmac. exp. Ther.* **167**, 328–333.

Reis, D. J. & Wurtman, R. J. (1968). Diurnal changes in brain noradrenaline. *Life Sci. Oxford* **7**, 91–98.

Richter, D. & Crossland, J. (1949). Variation in acetylcholine content of the brain with physiological state. *Am. J. Physiol.* **129**, 31–44.

Saito, Y. (1971). The circadian rhythm of brain acetylcholine levels and motor activity in the rat. *Life Sci. Oxford* **10**, Part I, 735–744.

Sakashita, A. (1972). Diurnal rhythm of plasma cortisol in epileptic patients. *Psychiat. Neurol. Jap.* **74**, 599–621.

Scapagnini, U., Moberg, G. P., Van Loon, G. R. & De Groot, J. (1970). Relation of brain 5-hydroxytryptamine (5-HT) to the diurnal variation in plasma corticosterone (B) in the rat. *Fed. Proc.* **29**, 831.

Scapagnini, U., Moberg, G. P., Van Loon, G. R., De Groot, J. & Ganong, W. F. (1971). Relation of brain 5-hydroxytryptamine content to the diurnalvariation in plasma corticosterone in the rat. *Neuroendocrinology* **7**, 90–96.

Scapagnini, U., Van Loon, G. R., Moberg, G. P. & Ganong, W. F. (1970). Effect of α-methyl-p-tyrosine on the circadian variation of plasma corticosterone in rats. *Europ. J. Pharmacol.* **11**, 266–268.

Schally, A. V., Anderson, R. N., Lipbscomb, H. S., Long, J. M. & Guillman, R. (1960). Evidence for the existence of two corticotrophine-releasing factors, α and β. *Nature, Lond.* **188**, 1192–1193.

Scheving L. E., Harrison, W. H., Gordon, P. & Pauly, J. E. (1968). Daily fluctuation (circadian and ultradian) in biogenic amines of rat brain. *Am. J. Physiol.* **214**, 166–173.

Shida, K., Shimazaki, J. & Takei, K. (1963). A method for the chemical determination of urinary 17-ketosteroid. *Hormone-to Rinsho* **11**, 13–20.

Slusher, M. A. (1964). Effects of chronic hypothalamic lesions on diurnal and stress corticosteroid levels. *Am. J. Physiol.* **206**, 1161–1164.

Slusher, M. A. (1966). Effect of cortisol implants in the brain stem and ventral hippocampus on diurnal corticosteroid levels. *Exp. Brain Res.* **1**, 184–194.

Snyder, S. H., Axelrod, J. & Zweig, M. (1965). A sensitive and specific fluorescence assay for tissue serotonin. *Biochem. Pharmac.* **14**, 831–835.

Suwa, N. & Yamashita, I. (1972). Psychophysiological Studies of Emotion and Mental Disorders. Hokkaido University Medical Library Series Vol. 5. Hokkaido University School of Medicine, Sapporo, Japan.

Takahashi, R. & Aprison, M. H. (1964). Acetylcholine content of discrete areas of the brain obtained by a nearfreezing method. *J. Neurochem.* **11**, 887–898.

Tobias, J., Lipton, M. A. & Lepinat, A. A. (1946). Effect of anaesthetics and convulsants on brain-acetylcholine content. *Proc. Soc. exp. Biol. Med.* **61**, 51–54.

Ungar, F. (1964). In vitro studies of adrenal-pituitary circadian rhythms in the mouse. *Ann. N.Y. Acad. Sci.* **117**, 374–385.

Van Loon, G. R. & Ganong, W. F. (1969). Effect of drugs which alter catecholamine metabolism on the inhibition of stressed-induced ACTH secretion produced by L-dopa. *Physiologist* **12**, 381.

Van Loon, G. R., Hilger, L., Cohen, R. R. & Ganong, W. F. (1969). Evidence for a hypothalamic adrenergic system that inhibits ACTH secretion in the dog. *Fedn. Proc.* **28**, 438.

Van Loon, G. R., Nicholson, W. E. & Liddle, G. W. (1971). Effect of an inhibitor of catecholamine synthesis on plasma and pituitary ACTH in rats. *Fedn. Proc.* **30**, 533.

Van Loon, G. R., Scapagnini, U., Moberg, G. P. & Ganong, W. F. (1970). Evidence for adrenergic inhibition of ACTH secretion in the rat. *Fedn. Proc.* **29**, 311.

Yamaguchi, N., Ling, G. M. & Marcznski, T. J. (1963). The effects of chemical stimulation of the preoptic region, nucleus centralis medialis, or brain stem reticular formation with regard to sleep and wakefulness. *Rec. Advan. Biol. Psychiat.* **6**, 9–20.

Yamazaki, K. (1970). Neuroendocrinological investigations in mental disorders, particularly in terms of pyrogen test. Part I. Standardization of pyrogen test and study of general paresis. *Psychiat. Neurol. Jap.* **72**, 443–466.

NEUROSECRETION

ELECTRICAL CHARACTERISTICS OF THE MEMBRANE IN NEUROSECRETORY NEURONS

SHIZUKO IWASAKI and YOUKO SATOW

The membrane characteristics of the secretory neuron were investigated electrophysiologically with microelectrodes. The secretory neuron of crayfish X-organs generated action potentials in the soma membrane when it was depolarized by a current passed through the membrane. The action potential developed in the soma is somewhat unique in respect to the ionic mechanisms. Na and Ca were the principal ions to develop soma action potentials. The Na component was eliminated by TTX, and the Ca component was eliminated by Mn. In the axon that comes out of the soma only the Na ion contributed to the development of the action potential.

The other unique feature of the secretory neuron soma of crayfish was the automatic rhythmic activity of the membrane. The membrane discharged intermittently without any synaptic bombardment. The interval of the grouped discharges was the function of the membrane potential. It was decreased by a depolarizing current passed through the membrane and was increased by a hyperpolarizing current. Occurrence of the grouped discharge was reset by change of the membrane potential. From these results, it was concluded that the grouped discharge in the neuron soma of the X-organ is due to the endogenous automatic activities in the membrane.

Inhibitory innervation of the secretory neuron of the X-organ of crayfish was demonstrated.

INTRODUCTION

It seems certain that the neurosecretory cell performs an important role as a neuron in the control of the neuroendocrine system of living bodies. In this article, consideration is given to the electrical characteristics of the membrane in the X-organ, which is well known as a typical neurosecretory neuron of crayfish. The findings on this neurosecretory neuron may throw some light on the nature of other neurosecretory neurons not only in invertebrates but also in vertebrates. Before considering the electrical nature of the membrane, some histological pictures of the X-organ of crayfish are presented in the first section. The electronmicroscope photographs are from unpublished data by Uchizono and Iwasaki.

ULTRASTRUCTURE OF THE X-ORGAN—SINUS GLAND SYSTEM OF CRAYFISH

The structure of the sinus gland and the ganglionic X-organ of crayfish is illustrated in Fig. 1. Possible neural connections that have been confirmed electrophysiologically are also shown in the

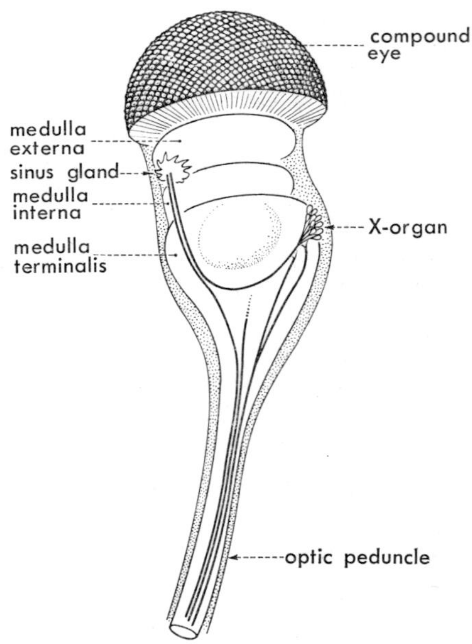

Fig. 1. Diagram of optic ganglia in crayfish. Possible nerve connections of the X-organ and the sinus gland are illustrated. Direct innervation of the optic nerve to the sinus gland was confirmed electrophysiologically (Iwasaki & Satow, unpublished). Innervation by the medullae to the sinus gland is uncertain except by the X-organ.

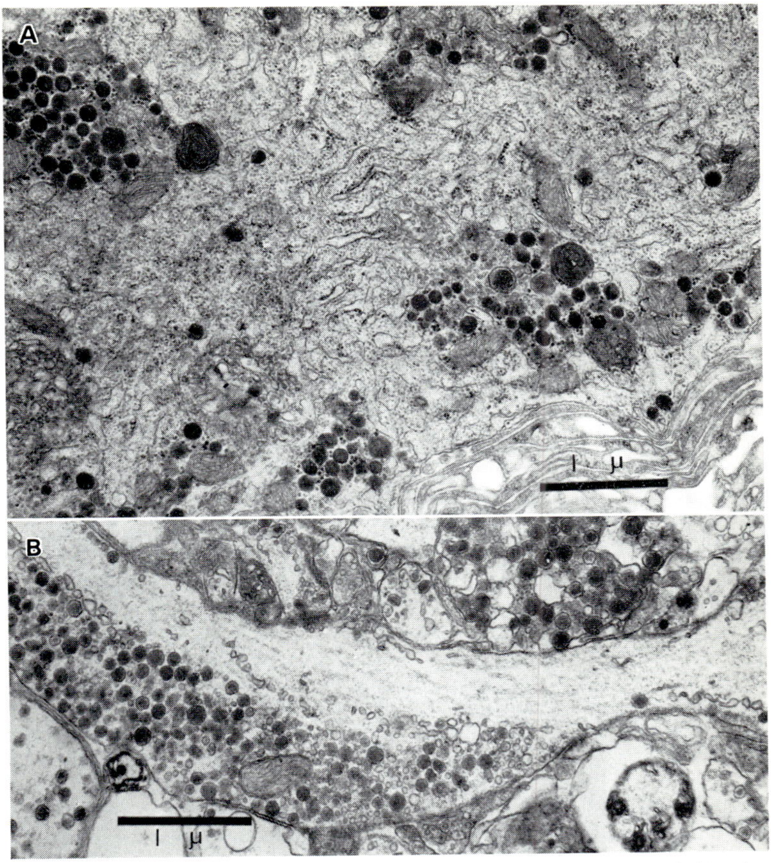

Fig. 2. Ultrastructure of the X-organ of the crayfish *Procambalus clarki*. A, Neuron soma. Presumed secretory granules are abundant in the soma. B, Axon which has originated from the soma. The secretory granules are seen with the varying size of small vesicles along the axon.

figure. Secretory granules are seen in the neuron soma of the X-organ of crayfish in association with the well developed endoplasmic reticulum (Fig. 2A). The diameter and the density of the granules are close to those in the sinus gland. The surface of the soma appears to be naked as illustrated in Fig. 1. In other words, no satellite cells like glia cover the surface of the neuron somata. Furthermore, the soma seems to be in contact with the neighboring somata surrounding it like a cluster of well ripened grapes (not shown in the figure). It might be deducible that the synchronous activation of the neuron in some seasons results from the functional connection between somata.

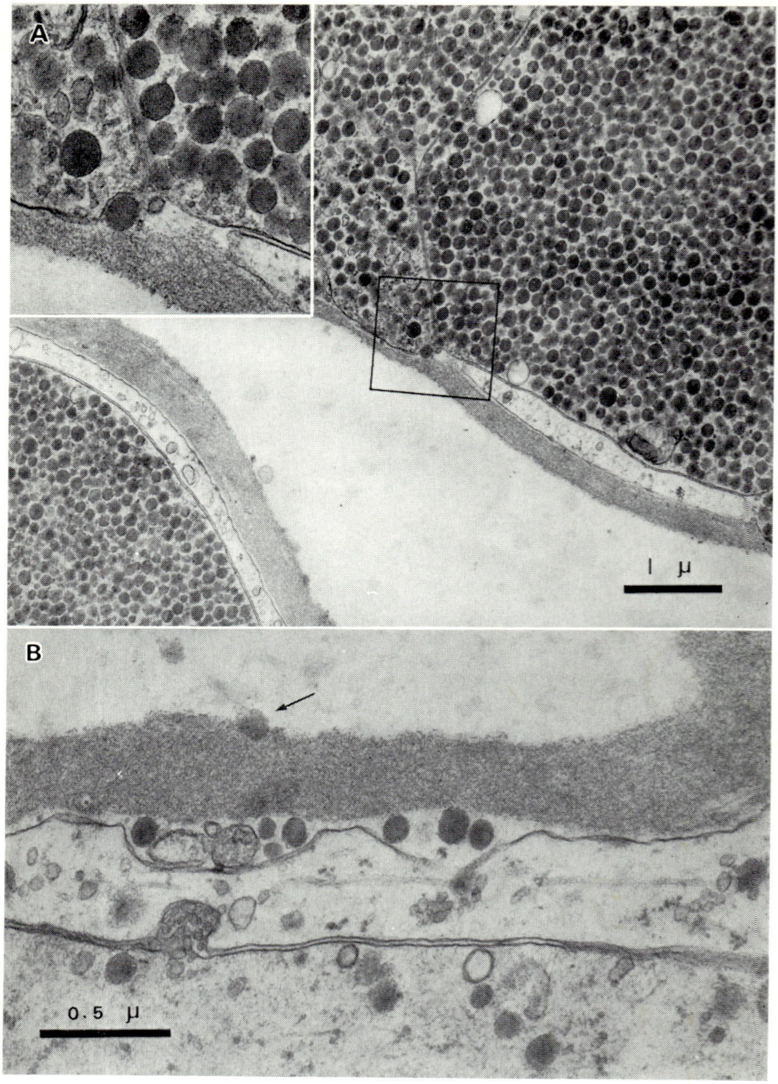

Fig. 3. A, Nerve terminals containing abundant granules are facing the hemocoel. The region indicated by a square is enlarged 4 times in the inset. A granule without the limiting membrane is seen outside of the membrane of the nerve terminal. B, Several granules are seen outside of the cell membrane. They may be released from the nerve terminal, a part of which is seen in the lower part of the plate. One granule is in the border of the basement material (indicated by an arrow).

There is some evidence to indicate a functional coupling between them (unpublished data). The granules can be observed along the axon, which has come out of the soma (Fig. 2B).

The secretory neurons that derive from neurosecretory cells lying in the X-organ, the brain and possibly the thoracic ganglion terminate in the sinus gland (Bliss, Durand & Welsh, 1954) (Fig. 3A). The granules in the sinus gland have varying diameters and electron densities, as reported by Bunt & Ashby (1967), who classified the granules into five types. In conjunction with the mode of release of the hormone, some pictures of the terminals are shown in Fig. 3.

ELECTRICAL CHARACTERISTICS OF THE MEMBRANE IN THE SECRETORY NEURON

The electrical characteristics of the neurosecretory neuron are presented below for the cell body, axon, and neuron terminal. Concerning the findings on the X-organ of crayfish, the apparatus and method were the same as those described previously (Iwasaki & Satow, 1971).

Membrane of the cell body (perikaryon)

Membrane constants

In the hypothalamic neurosecretory cells of fish, it is reported that the specific resistance and the capacitance are 600 $\Omega\cdot cm^2$ and 70 $\mu F/cm^2$, respectively (Kandel, 1964). Considering the existence of dendrites in these neuron somata, the value of resistance has probably been underestimated, and the value of capacitance has probably been overestimated. In the X-organ neuron soma of crayfish, the data of electrical membrane constants reveal an average resistance of $20 \times 10^6 \Omega$ per cell (Iwasaki *et al.* 1971). Taking 30–40 μ as the diameter of a soma that has no dendrites, the specific membrane resistance can be estimated to be about 600–1,000 $\Omega\cdot cm^2$. These figures are comparable with the specific resistance of neuron somata of vertebrates and invertebrates (summarized by Bullock & Horridge, 1965, p. 135). It is unlikely that the neurosecretory neuron soma has specific electrical constants of the resting membrane distinguishing it from membranes of ordinary neurons.

Action potentials

Conclusive evidence to show the excitability of the neurosecretory neuron soma was obtained by microelectrodes inserted into the neuron soma in the supraoptic neuron of the hypothalamus of cats (Yamashita, Koizumi & Brooks, 1970), in the hypothalamic neuron soma of fish (Kandel, 1964; Bennett, Gimenez & Ravitz, 1968), in the caudal neurosecretory neuron soma of fish (Bennett & Fox, 1962; Ishibashi, 1962) and in the X-organ neuron soma of crayfish (Iwasaki & Satow, 1969, 1970, 1971).

With either direct stimulation or anti-or orthodromic stimulation, the soma membrane of the neurosecretory neuron develops action potentials that have an overshoot and a rather long duration relative to that of the ordinary neuron. In Fig. 4A, the typical appearance of the action potentials elicited by direct stimulation is shown. During the injection of the current through the membrane (lower trace in each record), repetitive firing (upper trace), the rate of which depended on the strength of the current, took place. The overshoot

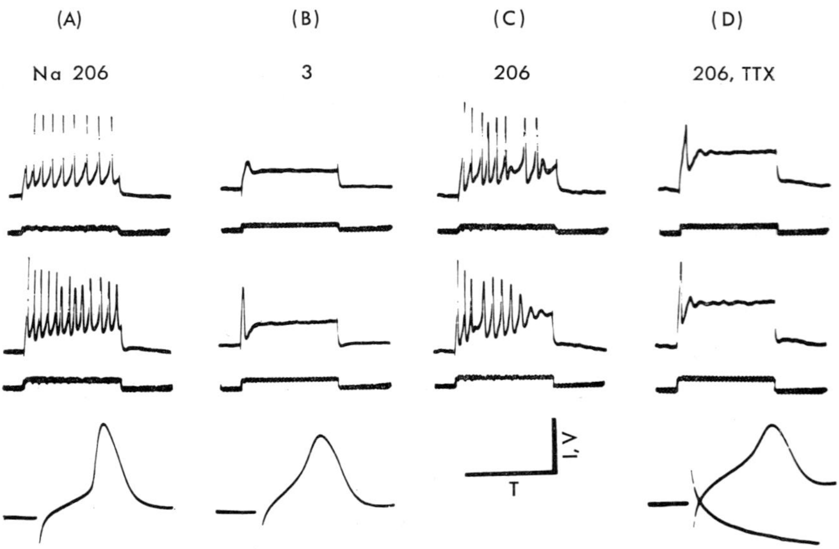

Fig. 4. Intracellular recordings of the action potentials of the X-organ under various conditions. The neuron soma was excited by a direct current through the membrane. The current intensity was monitored under each record. The lowest trace is a record with 20 times faster speed. A, The medium contained the standard amount of ions—NaCl, 205; $CaCl_2$, 13.5; KCl, 5.0; $MgCl_2$, 2.6 in mM. B, NaCl was reduced to 3 mM and Choline Cl was added instead. C, Same medium as A. D, TTX 10^{-7} g/ml was added to A. Development of the action potentials in the Na-deficient and TTX medium was evident. Increase in the threshold and decrease in the amplitude of the action potential and rate of rise should be noted. Calibration I, V and T indicate 4.9 nA, 50 mV, 500 or 25 msec respectively.

of the action potentials was 10–20 mV, and the maximum rate of rise of the potentials was 32±13 V/sec. The duration of the action potentials at the level of 50% spike amplitude was 3.0–8.8 msec. The long duration relative to other excitable membranes might be due to the late development of the Ca component as described below.

One unique response of this neuron soma is the spike initiation without Na ions in the medium. As seen in Fig. 4B, regenerative spike potentials can be evoked by the injected current after removal of 99% of the sodium ions from the solution. The repetitive firing has rarely appeared in Na-deficient solutions. The threshold current in the normal medium failed to evoke the spike potential in the Na-deficient medium (Fig. 4B, upper record). Threshold depolarization for the spike initiation increases invariably by 10–20 mV. Spike initiation without Na activation was confirmed in a medium containing tetrodotoxin (TTX), which has been known as an agent to block the Na activation system without any effect on other systems, such as the K, Ca, and anion activation systems (Fig. 4D). Changes in the Na concentration around the neuron soma have little effect on the spike initiation in the TTX medium.

Which ion or ions are responsible for the initiation of this spike potential? The Na-deficient medium contains K, Ca, Mg, and Cl ions. A permeability increase for K or Cl ions must keep the membrane near the resting potential or hyperpolarize the membrane in some cases. Possible ions responsible for spike initiation in the Na-deficient or TTX-medium must be Ca or Mg or both. The possibility that the Mg ions are responsible was ruled out from the fact that the increase in the Mg concentration diminished the spike potential in the TTX medium and the spike initiation was blocked by the substitution of Mg ions for Ca ions. Ca concentration in the medium modified the amplitude of the action potentials (Fig. 5), and the peak potential of the action potential had a linear relationship

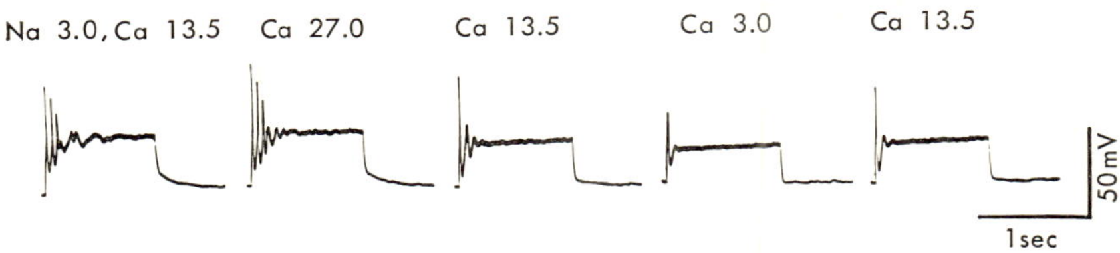

Fig. 5. Effect of external Ca concentration on the action potential in the Na-deficient medium. Ca concentration was changed successively from left to right in the figure (from Iwasaki *et al.* 1971).

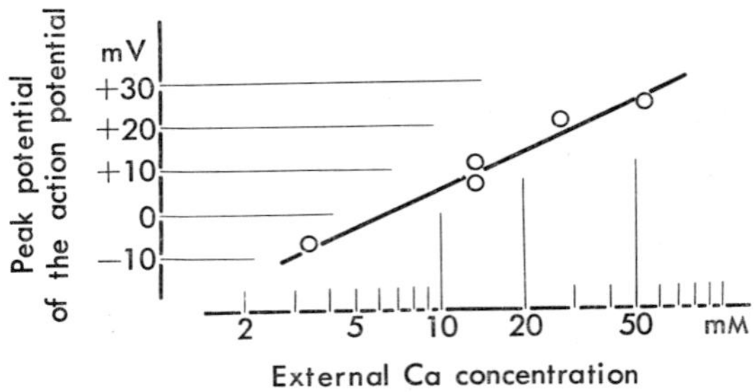

Fig. 6. Ca concentration and the spike amplitude in the TTX medium. Peak potential level of the action potentials was plotted against the external Ca concentration. The relation seems to be linear with the slope of 28 mV per ten fold change in the Ca concentration (from Iwasaki *et al.* 1971).

with the log concentration of Ca ions. The slope was 27–30 mV for a ten fold change in the Ca concentration (Fig. 6). This result leads us to the conclusion that the ions carrying the charge during the spike potential in the Na-deficient or TTX medium are Ca ions in the neuron soma of the X-organ of crayfish.

In a Ca-deficient medium, on the other hand, the Na ion concentration modified the amplitude of the action potential as expected from the Nernst equation. Consequently, it can be deduced that there are two independent activating systems in the neuron soma of the X-organ of crayfish, the Ca and Na systems.

The question now arises whether the neuron soma really has two activating systems. One of them might be assigned to the role of axon. An electrode inserted into the soma is able to record the membrane potential change not only in the vicinity of the electrode but also at some distance from the electrode. Assuming that the resistance of the membrane and the axoplasm are of the same order as those of ordinary neurons and considering the soma as the end of the cable, the potential change at a considerable distance may well spread to the soma where the electrode is inserted. In the present experiment, the potential change of the axon cannot be definitely distinguished from that of the cell body. When the soma membrane was stimulated by a direct current passed through the membrane, the potential rose smoothly from the critical depolarization (Fig. 7). When the soma membrane was activated antidromically, a notch appeared at the rising phase of the action potential. The potential level of the notch coincided with the critical depolarization by direct stimulation. This may suggest the onset of activation in the soma

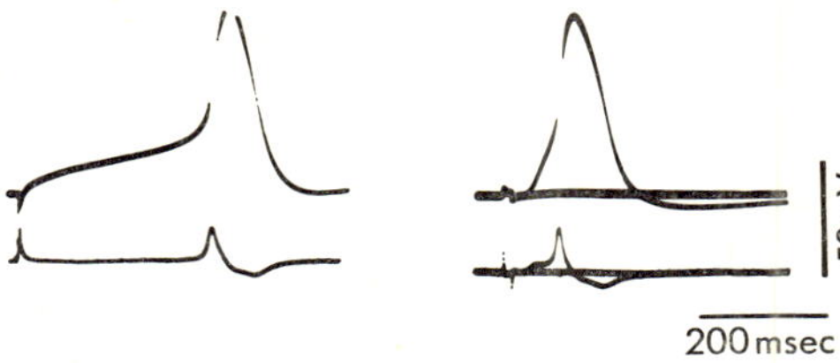

Fig. 7. Directly and antidromically evoked spike potentials in the neuron soma. Derivatives of the potential with respect to time were recorded under the potential records. A, The action potential was elicited by a direct current through the soma membrane. B, The action potential was elicited antidromically by stimulation of the sinus gland. Threshold membrane potentials were the same while the notch at the rising phase of the antidromic action potential was seen.

membrane at this critical depolarization. Since Mn ions abolish only the later part of the action potential (unpublished), the early part of the action potential must be generated by the activation of Na ions. The Ca system does exist in the soma membrane, but not in the axon as described below. From these facts, it can be deduced that the two activation systems coexist in a neurosecretory neuron soma.

In the normal medium, or in the living animal, it is quite natural that the Na is a primary and essential ion to induce the action potential, because (1) the conductance during the action potential in the normal medium is much larger than that in the Na-deficient or TTX medium (unpublished). (2) the threshold of the Na spike is much lower than that of the Ca spike (Figs. 4 B and 9 C). The Na system is activated in some way as a primary event, and the consequent depolarization of the membrane may trigger the Ca activation. Successive activation of the Na and Ca ion system may constitute the action potential of the neuron soma as a whole.

Some divalent cations other than the Ca ion may contribute to the action potentials, although they may rarely be found in the body fluid and the tissues under normal conditions. The neuron soma develops action potentials in the TTX medium in which Ca ions are replaced by Sr or Ba. In other words, these divalent cations play the same role as the Ca ion does. In Fig. 8, the $CaCl_2$ in the TTX medium was replaced by equimolar $SrCl_2$. Changes in the Sr concentration also modified the amplitude of this Sr spike potential. The Ca, Sr, or Ba spike potentials are completely and reversibly abolished by the addition of Co or Mn ions to the medium as seen in Fig. 8. The substitution ability of Sr and Ba and the abolition by Co or Mn seem to be common in all the Ca spike systems thus far

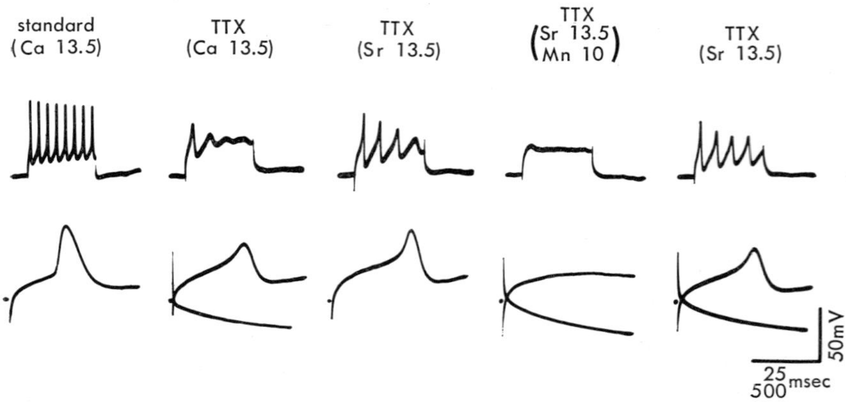

Fig. 8. Effects of divalent cations on the spike potentials in the TTX medium. Experiments were performed successively from left to right of the figure. $SrCl_2$ was substituted for $CaCl_2$ in the TTX medium. Action potentials appeared more vigorously after the substitution. These potentials were completely and reversively blocked by an addition of 10 mM of $MnCl_2$. $CoCl_2$ also blocked the Ca and Sr spikes. In the three faster records, the hyperpolarizing potential is superposed. Voltage calibration, 50 mV. Time calibration, 500 msec for the upper records and 25 msec for the lower records.

reported (Fatt & Ginsborg, 1958; Hagiwara & Nakajima, 1966; Geduldig & Junge, 1968; Brading, Bülbring & Tomita, 1969; Washio, 1972).

In recent years, evidence that the Ca activating system in excitable membranes either exists by itself or coexists with the Na activating system has been accumulated. In arthropoda, action potentials of muscle fiber are a typical example of the Ca spike potential (Fatt *et al.* 1958; Hagiwara, Chichibu & Naka, 1964; Hagiwara & Naka, 1964; Hagiwara *et al.* 1966; Washio, 1972), in which no cations other than Ca ions are necessary for the initiation of the spike potential. The membrane in the smooth muscle fiber of vertebrates also has a Ca activating system (Brading *et al.* 1969). Some neuron somata of Aplysia reveal double activating systems as in the X-organ of crayfish (Geduldig *et al.* 1968). Possible generalizations from these two kinds of membrane activating systems, the Na and Ca activating systems, are discussed below.

The axon membrane

It has been well established that the axon of the neurosecretory neuron conducts action potentials from the cell body to its terminal

TABLE 1. Axons of secretory neurons, in which the action potential has been recorded

Animals	Organs	References
Leech	Supraesophageal ganglion	Yagi & Bern, 1963*a*
Crab	Pericardial organ	Cooke, 1964
Crayfish	X-organ—sinus gland tract	Iwasaki & Satow, 1971
Fish	Caudal urophysis	Ishibashi, 1962; Bennett & Fox, 1962; Yagi & Bern, 1963*b*, 1965
	Hypothalamo-hypophysial tract	Potter & Loewenstein, 1955; Kandel, 1964; Bennett, Gimenez & Ravitz, 1968
Rat	Hypothalamo-hypophysial tract	Yagi, Azuma & Matsuda, 1966; Cross & Silver, 1966; Dyball & Koizumi, 1969; Ishida, 1970; Dyball & Dyer, 1971; Dreifuss & Kelly, 1972
Rabbit	Hypothalamo-hypophysial tract	Cross & Green, 1959; Cross, Novin & Sundsten, 1969; Novin, Sundsten & Cross, 1970
Cat	Hypothalamo-hypophysial tract	Ishikawa, Koizumi & Brooks, 1966; Brooks *et al.* 1966; Yamashita, Koizumi & Brooks, 1970; Barker, Crayton & Nicoll, 1971
Dog	Hypothalamo-hypophysial tract	Koizumi & Yamashita, 1972

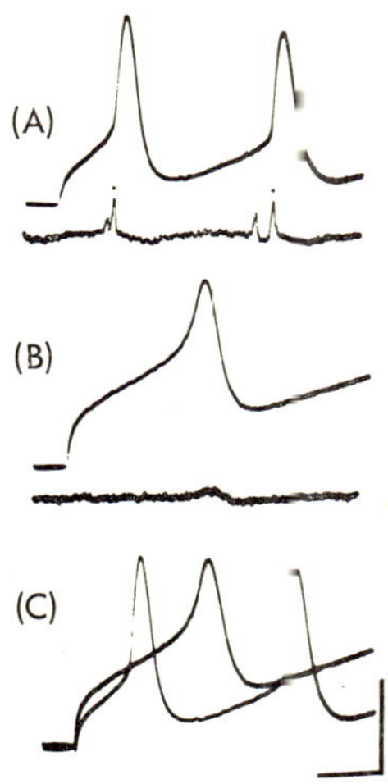

Fig. 9. Difference in the active components of action potentials between soma and axon. A, In the normal medium, the action potential of the soma (upper) elicited by a direct current injected in the neuron soma and the corresponding spike potential of the axon (lower) were seen as a one to one relationship (dotted above the spike potentials). Spontaneous spike potentials of other neurons are also seen (not dotted). B, After application of TTX, the spike potentials of the axon disappeared, while the spike potential of the neuron soma remained. C, Action potentials before and after the application of TTX are superposed. Difference in the threshold and the rate of rise was clearly seen. Calibration are 50 mV and 20 msec (from Iwasaki *et al.* 1971).

site where the neurohormone is believed to be stored and released into the circulating fluid as summarized in Table 1. The need for Na ion for the conduction of impulses in the neurosecretory axon was confirmed in rats (Yagi, Azuma & Matsuda, 1966) and in the X-organ of crayfish (Iwasaki *et al.* 1971). As mentioned above, the Na and Ca activation systems exist in the neuron soma of the X-organ. In this neuron, removal of Na ions or application of TTX, which blocks the Na activation system, results in the disappearance of the conductive spike potential of the axon, while the Ca spike potential still remains in the soma (Fig. 9). Therefore, it seems certain that the Na ion is the only ion playing a role in conduction of impulses of the excitable and conductive tissues. In the barnacle muscle fiber in which only the Ca spike can be elicited, the conduction of the Ca spike potential is not yet confirmed. A gradual local potential by distributed nerve innervation is the general response in this muscle fiber.

Nerve terminal membrane

Electrical activities of the nerve terminals of the neurosecretory neuron have been examined in the sinus gland of crabs (Cooke, 1971). After application of TTX, the terminal membrane exhibited regenerative action potentials, which depend on the amount of Ca ions outside the terminals. Cooke suggested that the neurosecretory axon terminal of crabs has the ability to develop regenerative depolarizations due to the inflow of Ca ions. The presence of Ca ions in the medium is the essential requirement for the release of the neurohormones; vasopressin and oxytocin from the neural lobe of the hypophysis of rats (Douglas & Poisner, 1964*a*; Ishida, 1968) and catecholamine from the adrenal medulla of cats (Douglas & Rubin, 1961, 1963). Ca uptake in association with the liberation of vasopressin was demonstrated in the neural lobe of rats (Douglas & Poisner, 1964*b*) and with the catecholamine liberation in the adrenal medulla in cats (Douglas & Poisner, 1962). During the application of Ach in the cultured cells of adrenal medulla, a depolarization depending on the Ca ions was evident (Douglas, Kanno & Sampson, 1967).

At the presynaptic nerve terminals of the giant synapse of squid and of the motor nerve of frogs during the application of TTX, regenerative spike potentials in the presence of TEA were recorded (Katz & Miledi, 1969*a*, *b*; Kusano, 1970). Their amplitude depended on the Ca ion concentration of the medium and also on the membrane potential of the nerve terminals. This fact suggests the regenerative inflow of Ca ions when the membrane is depolarized. Further study revealed that only this Ca dependent component is effective for liberation of the transmitter substance, and the Na component does not play a direct role.

Collective evidence concerning the liberation of neural substances through the neural membrane leads us to hypothesize the possible sequence of the release mechanism as follows. (1) Depolarization in the membrane of the nerve terminals either by arrival of the nerve action potential or by some other causes. Regenerative inflow of Na ions may occur at the terminal membrane by this depolarization. However, it may not be essential for the substance release. (2) Regenerative inflow of Ca ions when the depolarization attains the Ca threshold. (3) Release of transmitter or neurohormone by the unknown reaction of Ca ions with the transmitter or with the vesicles or granules. (4) Outflow of K ions which forces the Na and Ca activation to subside. In consequence, a fall in the release of the neural substance takes place (for further analysis, see Akiyama in this monograph).

Automatic and rhythmic activities of the neurosecretory neuron

It is well known that the cardiac muscle fibers of vertebrates have automatic and rhythmic activities that maintain the heartbeat. Their cyclic performance is supported by endogenous activities of the muscle cell. In the central nervous system of vertebrates and invertebrates, it is difficult to judge the possibility that the repetitive activities in a neuron are due to the endogenous automaticity of the neuron itself. Numerous interneurons may synapse on a neuron with

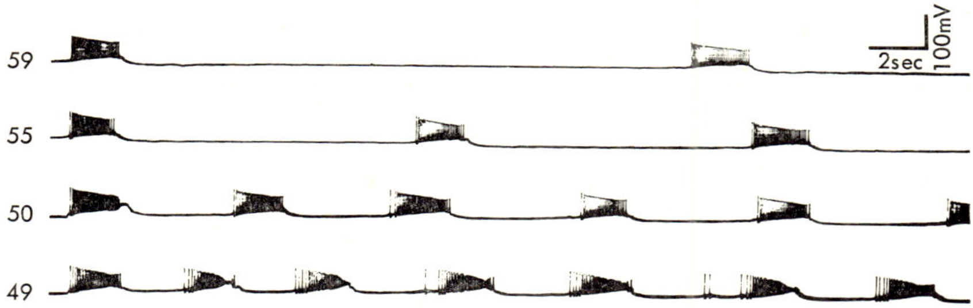

Fig. 10. Spontaneous burst discharges in the secretory neuron. The burst intervals are modified by the membrane potential level of the neuron soma. It was changed by the application of a current to the neuron soma. The membrane was depolarized to a new level (figures at left in mV) at 0.6 sec after the beginning of each record where the first grouped burst discharge took place (from Iwasaki *et al.* 1969).

excitatory or inhibitory nature and may generate repetitive firing of the postsynaptic neuron or regulate its firing rate. This is also true in secretory neurons. Many axons in the hypothalamo-hypophysial tract reported so far (see Table 1) fire spontaneously. This spontaneous firing may or may not be due to the inherent activities of the neuron.

In some neurons of invertebrates, however, the neuron itself seems to possess endogenous activities by which it fires regularly without any input signals. As seen in Fig. 10, intermittent burst discharges of the neuron soma of the X-organ were often observed. One burst discharge was composed of several to several tens of spike potentials. This cannot be due to injury by the insertion of the microelectrodes because in the sinus gland the same pattern of grouped discharges was observed before the insertion of the electrode. The characteristic nature of the burst discharge in the soma membrane of the X-organ can be described as follows.

(1) Burst intervals-membrane potential relationship. The interval of the grouped discharge increases when the membrane is hyperpolarized and decreases when the membrane is depolarized, and the duration and the number of spikes in one burst remain almost constant. It can be seen that the onset interval of the burst is a

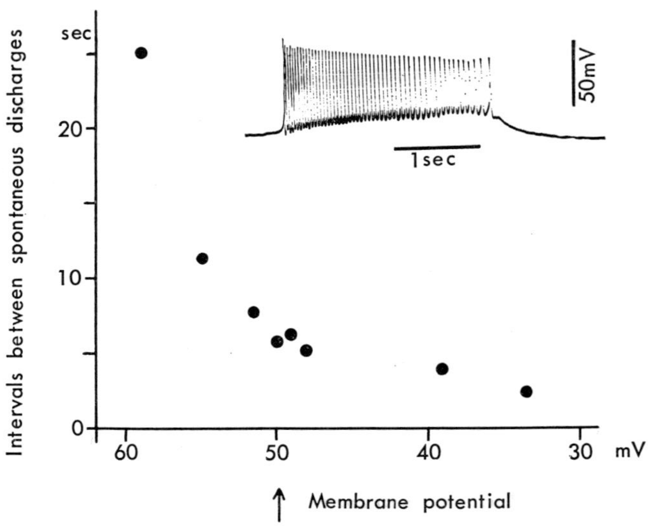

Fig. 11. Burst intervals and the membrane potential level of the neuron soma. The burst intervals were measured at the first spikes of every burst discharge. One of the burst discharges is enlarged in the inset (from Iwasaki *et al.* 1969).

continuous function of the membrane potential level, which is shifted by the current applied through the membrane (Fig. 11).

(2) Reset of the onset of the burst discharge. In Fig. 10, at 0.6 sec after the beginning of the records, the membrane potential is depolarized to a new potential level. The potential levels are shown in the left side of each record. A depolarizing change in the membrane potential results in the initiation of burst discharge and the following bursts appear with a shorter intervals. This suggests that the regular bursting cycle is reset by a change in the membrane potential of the neuron.

(3) Initiation of the burst discharge by a short depolarizing pulse. A short depolarizing pulse that was given during the pause after the burst (interburst period) triggers the initiation of a burst discharge, the pattern of which is not distinguishable from a spontaneous one. The next burst appears with a regular interval after this burst.

(4) Abolition of the burst discharge by a hyperpolarizing short pulse. As shown in Figs. 12 and 13, a hyperpolarizing short pulse (50 msec or less) applied during the burst interrupts further firing if it is applied at the last half of the burst. If it is applied at an earlier

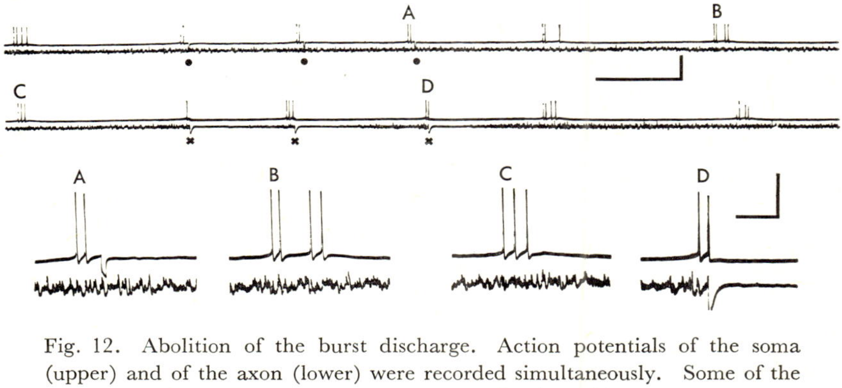

Fig. 12. Abolition of the burst discharge. Action potentials of the soma (upper) and of the axon (lower) were recorded simultaneously. Some of the potentials are enlarged below. The number of spikes within one spontaneous burst was three or four in this particular neuron (B, C). A hyperpolarizing pulse with 50 msec duration was applied after the second spike of the burst (dots under the record). The reset of the burst interval was obvious. In the same neuron, the optic peduncle was stimulated with a short pulse (crosses). Every burst was abolished by the IPSP. The burst interval after the abolition was shorter than that in the spontaneous occurrence. Calibration, 100 mV for the soma and 290 μV for the axon spikes. Calibration for the enlarged records, 50 mV for the soma and 145 μV for the axon. Time calibration, 5 sec and 500 msec respectively.

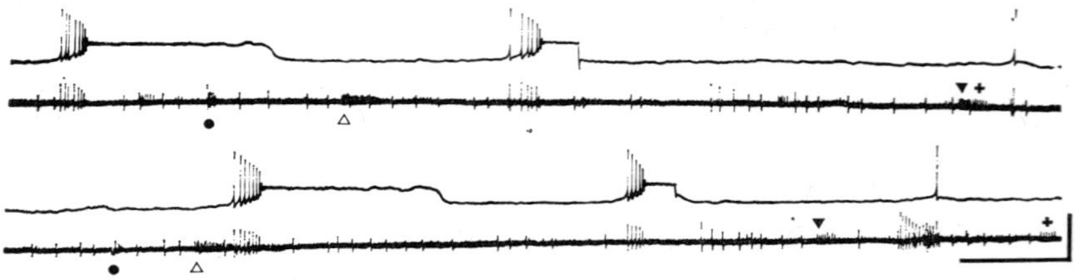

Fig. 13. Burst discharge with plateau (continuous record). Some neuron soma showed the plateau after several spikes (first and thirds burst potentials). It was abolished by the hyperpolarizing short pulse as seen in the figure (the second and fourth burst in the each row). Axon spikes corresponding to the soma spikes are seen in the lower records. It can be seen that several axons are bursting independently, besides the one that corresponds to the soma spike. The burst intervals are various. From the configuration of the burst discharge, four of the bursting axon potentials are marked by circles, triangles, crosses, and dots. Calibration, 50 mV for the soma and 120 μV for the axon potentials, time calibrationis 5 sec.

part of the burst, a stronger pulse is necessary. In general, the next burst appears with a shorter interval than the previous one. From the second burst after the abolition, the intervals revert to the original value.

(5) Reset of the burst by the IPSP (Fig. 12). This neuron is innervated by the inhibitory neuron from the brain through the optic peduncle. Stimulation of the optic peduncle brings about inhibitory postsynaptic potentials (IPSP) on the neuron soma. If it is applied during the burst discharge, the neuron ceases to fire immediately, and it begins to fire with the regular bursting cycle after some time. The time of reappearance of the burst depends on the period of the peduncle stimulation. This fact clearly shows the reset of the burst by the extrinsic IPSP.

(6) Extracellular record of the burst discharge. At the sinus gland, intermittent grouped discharges by several axons are recognized (Fig. 13). Sometimes they discharge synchronously (Fig. 14) and sometimes asynchronously. In the example shown in Fig. 13, several neurons are firing independently.

(7) Effect of picrotoxin on the burst intervals. IPSPs elicited by stimulation of the optic peduncle are eliminated by the application of picrotoxin, which has been believed to block the action of the inhibitory transmitter, γ-aminobutylic acid (GABA). When picrotoxin (10^{-5} g/ml) was applied to a neuron that had been firing repetitively with irregular intervals, it began to fire regularly. Sometimes,

several neurons which had fired independently with the proper intervals began firing synchronously after the application of picrotoxin. Modulation of the burst discharge by inhibitory innervation was evident.

(8) Conductance change during the pause. After the burst, the conductance measured by the short hyperpolarizing pulses is the highest. It decreases gradually with time and is the lowest just before the beginning of the burst. The I-V relation near the resting membrane potential was linear when the bursting was diminished by the application of TTX. Therefore, this gradual and cyclic change in the conductance is not attributed to the difference in the membrane potential after the spike potentials. It must result from the endogenous generator mechanism of the neuron. One may recall the cyclic conductance change similar to this in the cardiac muscle fibre (Weidmann, 1951). In the cardiac muscle membrane, this cyclic phenomenon of the conductance may be due to the change in K conductance.

Certain evidence stated above leads us to conclude that the repetitive burst discharge in some neurosecretory cells of the X-organ is caused by the endogenous automatic activity in its membrane. The reverberating circuit may exist in the central nervous system of vertebrates (Tsukahara, in this monograph). Abolition experiments, however, ruled out the possibility that the periodic bursting discharge, if any, in this organ are originated by the reverberating circuit.

Automaticity of the neuron was reported in the visceral ganglion of Aplysia (Frazier, Kandel, Kupfermann, Waziri & Coggeshall, 1967; Strumwasser, 1967; Alving, 1968), in the cardiac ganglion of Squilla (Hagiwara & Bullock, 1957; Bullock & Terzuolo, 1957; Watanabe, Obara & Akiyama, 1967), and in the ganglion cells of crayfish (Preston & Kennedy, 1962; Iwasaki *et al.* 1969). Further analysis of the mechanism of automaticity in the Aplysia neuron has been done (Alving, 1969; Carpenter & Gunn, 1970). It is interesting to notice that the autogenic neurons of Aplysia contain large granules, the diameter of which ranges from 1500–2500 A. These neurons are identified as L2, L3, L4, L6 and R15 (Frazier *et al.* 1967).

It must be stressed that the endogenous burst discharge does not always appear in all the neurosecretory cells of the X-organ of crayfish and does not always appear all the year around. The burst discharge in some neurons shows a different nature from that mentioned above: (1) Burst intervals do not depend on the membrane potential level.

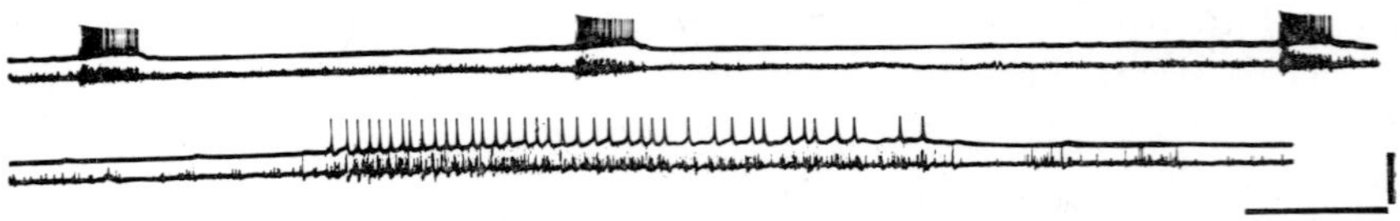

Fig. 14. Synchronous firing in the axons. Simultaneous record of the soma and axon spikes. During the burst discharge of the automatic neuron, a considerable number of axons are discharging synchronously. This is more evident in the expanded record (lower record). Calibration, 100 mV for the soma and 1.5 mV for the axon spikes. Time calibration, 10 sec for the upper and 1 sec for the lower record.

Changes in the membrane potential bring about no appreciable change in the burst intervals, except the onset of the first burst after the change in the membrane potential. (2) The slow depolarizing waves appear when the hyperpolarizing current is applied. The amplitude of the slow depolarization, however, does not increase by further hyperpolarization; it decreases. (3) A hyperpolarizing short pulse does not modify the occurrence of the bursting potential. (4) Usually, when the burst is observed in this type of cell, axon spikes recorded extracellulary from the sinus gland show synchronous bursting discharges of several axons, indicating the synchronous firing of many neurons (Fig. 14). (5) Picrotoxin has a tendency to synchronize the independent bursting discharge. (6) Reset of the burst onset in a neuron that has endogenous automaticity brings about the immediate reset of the synchronized burst discharges in the axons of the sinus gland. If the previous endogenous burst neurons are called *pacemaker* neurons, this type of neuron can be called a *follower* neuron.

The burst type of discharge is often observed from the early spring to the early summer, from April to July, in our experimental conditions. During these months, a considerable number of preparations exhibit the burst discharge, but in other seasons it can rarely be seen. The X-organ sinus gland system of crustacea is considered to be a system responsible for the production and liberation of the molt-inhibiting hormone (Gabe, 1966). Since the firing of the secretory neuron induces a release of the hormone into the circulating system from its terminal (Fridberg, Iwasaki, Yagi, Bern, Wilson & Nishioka, 1966; Ishida, 1970; Ishikawa, Koizumi & Brooks, 1966; Dyball, 1969, 1970; Dyball & Dyer, 1971), seasonal change in the activation of the neurosecretory neuron might be responsible for the molting behavior of crayfish. Further study on this point is necessary.

Innervation of the neurosecretory neuron

It is natural to expect that the neurosecretory neuron receives neural control in addition to humoral control, if one considers the neurosecretory neuron to be taking over a part of the endocrine control system in the living body. In neurons that have endogenous automaticity, mentioned above, initiation of the automaticity and the level of the activation (frequency of the burst) must be regulated by extrinsic factors. In the hypothalamic neuron soma, which sends the axon to the posterior pituitary lobe, a change in the firing frequency was evident after the injection of hypertonic NaCl or glucose, Ach, Carbachol, and $CaCl_2$ and by the mechanical stimulation of the nipple or uterus (Brooks, Ishikawa, Koizumi & Lu, 1966; Dyball & Koizumi, 1969; Cross & Green, 1959). These procedures must act on the hypothalamic neurosecretory neuron through neural innervation. In other words, neuronal control is performed through the synaptic contact on the neurosecretory neuron. Inhibitory innervation by catecholamine in the supraoptic nucleus of cats might be possible, since the iontophoretic application of dopamine, noradrenaline, or serotonin inhibits the spontaneous firing of this neuron (Baker, Crayton & Nicoll, 1971). Possible excitatory control on this neuron by Ach is not definite yet, because the reported action of Ach is not clear, sometimes exciting and sometimes inhibiting the neuron activity (Barker *et al.* 1971; Moss, Urban & Cross, 1971; Dreifuss & Kelly, 1972). EPSPs followed by IPSPs were recorded by the stimulation of several parts of the brain in cats and dogs (Koizumi & Yamashita, 1972). In the caudal neurosecretory neuron, stimulation of the spinal cord induces the excitatory synaptic potential which triggers the spike potentials in this neuron soma (Bennett *et al.* 1962; Ishibashi, 1962).

The secretory neuron in the X-organ of crayfish also receives inhibitory innervation through the optic peduncle. The inhibitory neuron may originate from the cerebral ganglion. The nature of this inhibitory transmission has been examined by electrophysiological techniques (Iwasaki *et al.* 1971). Stimulation of the optic peduncle evoked either a small depolarizing junction potential when the microelectrode was filled with KCl or a small hyperpolarizing junction potential when it was filled with K-citrate. This suggests that the permeability change in Cl ions is involved in the mechanism of electrogenesis. The inhibitory nature of the junction potential is convincing from the following evidence. (1) The equilibrium potential of the junction potential is very close to the resting membrane potential. (2) Stimulation of the optic peduncle eliminates the spontaneous burst discharge. One impulse wipes out not only the burst discharge of a neuron soma but also eliminates all the synchronized and unsynchronized discharges of the axon in the sinus gland (Fig. 12D). (3) Picrotoxin invalidates stimulation of the optic peduncle. (4) GABA mimics the effect of stimulation of the optic peduncle. However, one must be careful to regard GABA as an inhibitory transmitter in this neuron, because there is a certain difference in the equilibrium potentials between IPSPs and GABA potentials. Excitatory innervation to the X-organ has not been detected yet. It is not unreasonable to suppose that only inhibitory innervation controls the activities of the postsynaptic neuron. A decrease in inhibitory input may well result in an increase in the firing frequency if the neuron possesses automaticity. However, this may not always hold true in all neurosecretory neurons.

CONCLUSIONS

Recent work on the electrical nature of the neurosecretory neuron in various animals has provided a general picture of the neuron. It may be said that there are two main functions in the neuron, including the secretory neuron. One of the functions is reception of information from other neurons through synapses and transportation of the information to its destination. The reception of humoral information may also occur and may provide long-term regulation of the state of the neuron. However, decisive work on this point has not yet appeared. The other function of the neuron is the secretion of a specific substance from the terminals where neural information is translated into humoral information.

The secretory neuron receives inhibitory or excitatory information synaptically and transports it to the neuron terminals by means of impulses. At the terminals, the impulses may be changed to Ca inflow into the nerve terminals, which leads to the release of the specific substance through the terminal membrane. In this respect, no difference could be found between neurosecretory and ordinary neurons. The only difference, if any, is in the site of the action. One may be called a transmitter, which diffuses out within a short distance and acts on the restricted area of the postsynaptic membrane, and the others may be called neurohormone, which travel for long distances through the bloodstream to rather broad regions of the body. The action of the latter is usually slow and longlasting.

The regenerative activation of Ca ions (Ca spike) in the neuron soma mentioned above in the X-organ of crayfish might not be a general occurrence in all the neuron somata, although no prudent examination on this point has been done for other neurons. However, recent work on the membrane of the giant axon reveals two components of Ca inflow during action potentials. The late component depends on the membrane potential (Baker, Hodgkin & Ridgeway, 1971), and its behavior is similar to that of Ca activation in the nerve terminals (Katz & Miledi, 1969*a*, *b*). This evidence strongly suggests the general nature of the regenerative Ca inflow in all neurons. It is probably true that at the membrane where the release mechanism exists the Ca inflow can be used for the substance liberations while at the axon membrane it may not be utilized.

Automaticity of the secretory neuron is an interesting feature in conjunction with the secretory nature of the neuron. The neurons identified with large granules of Aplysia (L2, L3, L4, L6 and R15) possess automaticity (Frazier *et al.* 1967). It can be speculated that this kind of automaticity is one of the characteristics of the secretory neuron, although much more evidence is needed to confirm it.

REFERENCES

Alving, B. O. (1968). Spontaneous activity in isolated somata of Aplysia pacemaker neurons. *J. gen. Physiol.* **51**, 20–45.

Alving, B. O. (1969). Differences between pacemaker and nonpacemaker neurons of Aplysia on voltage clamping. *J. gen. Physiol.* **54**, 512–531.

Baker, P. F., Hodgkin, A. L. & Ridgeway, E. B. (1971). Depolarization and calcium entry in squid giant axons. *J. Physiol.* **218**, 709–755.

Barker, J. I., Crayton, J. W. & Nicoll, R. A. (1971). Noradrenaline and acetylcholine responses of supraoptic neurosecretory cells. *J. Physiol.* **218**, 19–32.

Bennett, M. V. L. & Fox, S. (1962). Electrophysiology of caudal neurosecretory cells in the skate and fluke. *Gen. Comp. Endocrinol.* **2**, 77–95.

Bennett, M. V. L., Gimenez, M. & Ravitz, M. J. (1968). Synaptically evoked impulse activity of morphologically identified neurosecretory cells. *Anat. Rec.* **160**, 313–314.

Bliss, D. E., Durand, J. B. & Welsh, J. H. (1954). Neurosecretory system in decapoda crustacea. *Z. Zellforsch. mikrosk. Anat.* **39**, 520–536.

Brading, A., Bülbring, E. & Tomita, T. (1969). The effect of sodium and calcium on the action potential of the smooth muscle of the guinea-pig taenia coli. *J. Physiol.* **200**, 637–654.

Brooks, C. McC., Ishikawa, T., Koizumi, K. & Lu, H-H. (1966). Activity of neurones in the paraventricular nucleus of the hypothalamus and its control. *J. Physiol.* **182**, 217–231.

Bullock, T. H. & Horridge, G. A. (1965). Structure and function in the nervous systems of invertebrates. San Fransisco, London: Freeman and Co.

Bullock, T. H. & Terzuolo, G. A. (1957). Diverse forms of activity in the somata of spontaneous and integrating ganglion cells. *J. Physiol.* **138**, 341–364.

Bunt, A. H. & Ashby, E. A. (1967). Ultrastructure of the sinus gland of the crayfish, *Procambarus clarkii*. *Gen. Comp. Endocrinol.* **2**, 334–342.

Carpenter, D. & Gunn, R. (1970). The dependence of pacemaker discharge of *Aplysia* neurons upon Na^+ and $Ca.^{2+}$ *J. Cell Physiol.* **75**, 121–127.

Cooke, I. M. (1964). Electrical activity and release of neurosecretory material in crab pericardial organs. *Comp. Biochem. Physiol.* **13**, 353–366.

Cooke, I. M. (1971). Calcium dependent depolarizing responses recorded from crab neurosecretory terminals. Proceedings of the Int. U. Physiol. Sci. IX, 119, XXV Int. Congress, Munich, 1971.

Cross, B. A. & Green, J. D. (1959). Activity of single neurones in the hypothalamus: Effect of osmotic and other stimuli. *J. Physiol.* **148**, 554–569.

Cross, B. A., Novin, D. & Sundsten, J. W. (1969). Antidromic activation of neurones in the paraventricular nucleus by stimulation in the neural lobe of the pituitary. *J. Physiol.* **203**, 68–70.

Cross, B. A. & Silver, I. A. (1966). Electrophysiological studies on the hypothalamus. *Br. med. Bull.* **22**, 254–260.

Douglas, W. W., Kanno, T. & Sampson, S. R. (1967). Influence of the ionic environment on the membrane potential of adrenal chromaffin cells and on the depolarizing effect of acetylcholine. *J. Physiol.* **191**, 107–121.

Douglas, W. W. & Poisner, A. M. (1962). On the mode of action of acetylcholine in evoking adrenal medullary secretion: Increased uptake of calcium during the secretory response. *J. Physiol.* **162**, 385–392.

Douglas, W. W. & Poisner, A. M. (1964*a*). Stimulus-secretion coupling in a neurosecretory organ: The role of calcium in the release of vasopressin from the neurohypophysis. *J. Physiol.* **172**, 1–8.

Douglas, W. W. & Poisner, A. M. (1964*b*). Calcium movement in the neurohypophysis of the rat and its relation to the release of vasopressin. *J. Physiol.* **172**, 19–30.

Douglas, W. W. & Rubin, R. P. (1961) The role of calcium in the secretory response of the adrenal medulla to acetylcholine. *J. Physiol.* **159**, 40–57.

Douglas, W. W. & Rubin, R. P. (1963). The mechanism of catecholamine release from the adrenal medulla and the role of calcium in stimulus-secretion coupling. *J. Physiol.* **167**, 288–310.

Dyball, R. E. J. (1969). The time relation between the secretory and electrical response of supraoptic neurones to osmotic stimuli. *J. Physiol.* **203**, 67–68.

Dyball, R. E. J. & Koizumi, K. (1969). Electrical activity in the supraoptic and paraventricular nuclei associated with neurohypophysial hormone release. *J. Physiol.* **201**, 711–722.

Dyball, R. E. J. (1970). Electrical discharge patterns in hypothalamic neurosecretory neurones associated with hormone release. In *Aspects of Neuroendocrinology*, ed. W. Bargmann & B. Scharrer, pp. 153–158. Berlin: Springer-Verlag.

Dyball, R. E. J. & Dyer, R. G. (1971). Plasma oxytocin concentration and paraventricular neurone activity in rats with diencephalic island and intact brain. *J. Physiol.* **216**, 227–235.

Dreifuss, J. J. & Kelly, J. S. (1972). The activity of identified supraoptic neurones and their response to acetylcholine applied by iontophresis. *J. Physiol.* **220**, 105–115.

Fatt, P. & Ginsborg, B. I. (1958). The ionic requirements for the production of action potentials in crustacean muscle fibres. *J. Physiol.* **142**, 516–543.

Frazier, W. T., Kandel, E. R., Kupfermann, I., Waziri, R. & Coggeshall, R. W. (1967). Morphological and function properties of identified neurons in the abdominal ganglion of *Aplysia californica*. *J. Neurophysiol.* **30**, 1288–1351.

Fridberg, G., Iwasaki, S., Yagi, K., Bern, H. A., Wilson, D. M. & Nishioka, R. S. (1966). Relation of impulse conduction to electrically induced release of neurosecretory material from the urophysis of the teleost fish *Tilapia mossambica*. *J. exp. Zool.* **161**, 137–150.

Gabe, M. (1966). Neurosecretion. Oxford: Pergamon Press.

Geduldig, D. & Junge, D. (1968). Sodium and calcium components of action potentials in the *Aplysia* giant neurone. *J. Physiol.* **199**, 347–365.

Hagiwara, S. & Bullock, T. H. (1957). Intracellular potentials in pacemaker and integrative neurons of the lobster cardiac ganglion. *J. cell. comp. Physiol.* **50**, 25–47.

Hagiwara, S., Chichibu, C. & Naka, K. (1964). The effects of various ions on resting and spike potentials of barnacle muscle fibres. *J. gen. Physiol.* **48**, 168–179.

Hagiwara, S. & Naka, K. (1964). The initiation of spike potential in barnacle muscle fibres under low intracellular Ca^{2+}. *J. gen. Physiol.* **48**, 141–162.

Hagiwara, S. & Nakajima, S. (1966). Differences in Na and Ca spikes as examined by application of tetrodotoxin, procaine and manganese ions. *J. gen. Physiol.* **49**, 793–806.

Ishibashi, T. (1962). Electrical activity of the caudal neurosecretory cells in eel *Anguilla japonica* with special reference to synaptic transmission. *Gen. Comp. Endocrinol.* **2**, 415–424.

Ishida, A. (1968). Stimulus-secretion coupling on the oxytocin release from the isolated posterior pituitary lobe. *Jap. J. Physiol.* **18**, 471–480.

Ishida, A. (1970). The oxytocin release and the compound action potential evoked by electrical stimulation on the isolated neurohypophysis of the rat. *Jap. J. Physiol.* **20**, 84–96.

Ishikawa, T., Koizumi, K. & Brooks, C. McC. (1966). Electrical activity recorded from the pituitary stalk of the cat. *Am. J. Physiol.* **210**, 427–431.

Iwasaki, S. & Satow, Y. (1969). Spontaneous grouped discharge of secretory neuron soma in X-organ of crayfish, *Procambarus clarki*. *J. Physiol. Soc. Japan* **31**, 629–630.

Iwasaki, S. & Satow, Y. (1970). Spike initiation of neurosecretory neuron soma in Na deficient or TTX medium. *J. Physiol. Soc. Japan* **32**, 37–38.

Iwasaki, S. & Satow, Y. (1971). Sodium-and calcium-dependent spike potentials in the secretory neuron soma of the X-organ of the crayfish. *J. gen. Physiol.* **57**, 216–238.

Kandel, E. R. (1964). Electrical properties of hypothalamic neuroendocrine cells. *J. gen. Physiol.* **47**, 691–717.

Katz, B. & Miledi, R. (1969*a*). Tetrodotoxin-resistant electric activity in presynaptic terminals. *J. Physiol.* **203**, 459–487.

Katz, B. & Miledi, R. (1969*b*). Spontaneous and evoked activity of motor nerve endings in calcium Ringer. *J. Physiol.* **203**, 689–706.

Koizumi, K. & Yamashita, H. (1972). Studies of antidromically identified neurosecretory cells of the hypothalamus by intracellular and extracellular recordings. *J. Physiol.* **221**, 683–705.

Kusano, K. (1970). Influence of ionic environment on the relationship between pre and postsynaptic potentials. *J. Neurobiol.* **1**, 437–457.

Moss, R., Urban, I. & Cross, B. A. (1971). Effect of acetylcholine and noradrenaline on neurons in the paraventricular nucleus of the rabbit hypothalamus. Proc. Int. Union Physiol., Sci. IX, 406.

Novin, D., Sundsten, J. W. & Cross, B. A. (1970). Some properties of antidromically activated units in the paraventricular nucleus of the hypothalamus. *Expl Neurol.* **26**, 330–341.

Potter, D. D. & Loewenstein, W. R. (1955). Electrical activity of neurosecretory cells. *Am. J. Physiol.* **183**, 652.

Preston, J. D. & Kennedy, D. (1962). Spontaneous activity in crustacean neurons. *J. gen. Physiol.* **45**, 821–836.

Strumwasser, F. (1967). Types of information stored in single neurons. In Invertebrate Nervous System, ed. A. G. Wiersma, pp. 271–319. Chicago: Univ. Chicago Press.

Washio, H. (1972). The ionic requirements for the initiation of action potentials in insect muscle fibers. *J. gen. Physiol.* **59**, 121–134.

Watanabe, A., Obara, S. & Akiyama, T. (1967). Pacemaker potentials for the periodic burst discharge in the heart ganglion of a stomatopod, *Squilla oratoria*. *J. gen. Physiol.* **50**, 839–862.

Weidmann, S. (1951). Effect of current flow on the membrane potential of cardiac muscle. *J. Physiol.* **115**, 227–236.

Yagi, K., Azuma, T. & Matsuda, K. (1966). Neurosecretory cell: Capable of conducting impulse in rats. *Science, N.Y.* **154**, 778–779.

Yagi, K. & Bern, H, A. (1963*a*). Action potentials of neurosecretory neurons in the leech *Theromyzon rude*. *Gen. Comp. Endocrinol.* **3**, 490–495.

Yagi, K. & Bern, H. A. (1963*b*). Electrophysiologic indications of the osmoregulatory role of the teleost urophysis. *Science, N.Y.* **142**, 491–493.

Yagi, K. & Bern, H. A. (1965). Electrophysiologic analysis of the response of the caudal neurosecretory system of *Tilapia mossambica* to osmotic manipulations. *Gen. Comp. Endocrinol.* **5**, 509–526.

Yamashita, T., Koizumi, K. & Brooks, C. McC. (1970). Electrophysiological studies of neurosecretory cells in the cat hypothalamus. *Brain Res.* **20**, 462–466.

DIVALENT CATIONS AND RELEASE OF HUMORAL SUBSTANCE FROM AXON TERMINALS

TOYOHIRO AKIYAMA

The relation between the amount of transmitter (GABA) released and $[Ca]_0$ and $[Mg]_0$ was determined electrophysiologically at the inhibitory synapse of *Squilla* heart ganglion. The increment in the membrane conductance during tetanic nerve stimulation was measured as an index of the amount of transmitter released, and it was subsequently converted into [GABA] with a dose-conductance curve of GABA. The reciprocal of concentrations of transmitter (1/[T]) estimated in this way was found to lie on a straight line when plotted against $[Mg]_0/[Ca]_0$ and/or $1/[Ca]_0$. That is, the following relations were found experimentally; $1/[T]=a\cdot([Mg]_0/[Ca]_0)+b$, and/or $1/[T]=a'\cdot(1/[Ca]_0)+b'$, where a, b, a' and b' are constants. These equations were shown to be identical to equations derived theoretically, assuming that Ca competes with Mg for a critical site in preterminals. The association constants of reactions of Ca and Mg with the site are $9.2/\alpha\times10^2$ and $4.0/\alpha\times10^2$ (M^{-1}) for one cell, respectively (α is an unknown constant; $0<\alpha\leqq1$). The possibility of a negatively charged site of phospholipids of the membrane as a candidate for the critical site is discussed.

INTRODUCTION

The release of chemical transmitter from axon terminals by nerve impulse does not occur in the absence of Ca at all synapses studied (Hubbard, 1970; Kuno, 1971); the secretion of hormones from a gland is also depressed when a bathing medium lacks Ca.

Sr and Ba added to a Ca-free saline can restore the release of transmitter from nerve terminals and also the secretion of hormones from a gland. However, Mg, the closely related alkaline earth to those mentioned above, antagonizes the action of Ca in the synaptic transmission (del Castillo & Engbaek, 1954; del Castillo & Katz, 1954*a*; Hutter & Kostial, 1954; Boyd & Martin, 1956; Jenkinson, 1957; Takeuchi & Takeuchi, 1962; Hubbard, Jones & Landau, 1968; Watanabe, Obara & Akiyama, 1968), as well as in the secretion of hormones (Douglas & Rubin, 1963 and 1964; Douglas & Poisner, 1964). These facts suggest that a similar mechanism may be involved in the secretion of substances from different types of cells and that a study of release of transmitter from axon terminals may contribute to the understanding of the process of hormone secretion.

The role of Ca in the release of transmitter has been extensively explored on frog neuromuscular junction, mammalian motor nerve terminals, and squid giant synapse (Eccles, 1964; Katz, 1969; Hubbard, 1970; Kuno, 1971; Katz & Miledi, 1967, 1969*a* and *b*, 1970 and 1971; Kusano, 1968 and 1970; Kusano, Livengood & Werman, 1967; Miledi, 1967; Miledi & Slater, 1966), but the nature of the antagonism between Ca and Mg has been studied less (del Castillo *et al.* 1954*a*; Jenkinson, 1957; Hubbard *et al.* 1968).

The work described below is a continuation of that by Watanabe *et al.* (1968), and it concerns mainly the antagonism of magnesium and calcium in the release of transmitter from inhibitory nerves (α-nerves; Alexandrowicz, 1934) in the heart ganglion of *Squilla oratoria* de Haan.

METHOD OF ESTIMATION OF THE AMOUNT OF TRANSMITTER RELEASED

In the present study, the amount of transmitter released was estimated electrophysiologically as follows: First, an increment in the membrane conductance (ΔG) during the tetanic nerve stimulation (70–100/sec) was obtained by subtracting the membrane conductance during an interburst period from the conductance at the peak level of summated IPSPs (the former conductance will be called the resting conductance hereafter and the latter the peak conductance; for the spontaneity of the cell, see Watanabe, Obara & Akiyama, 1967). Then ΔG was converted to a concentration of GABA with reference to a dose-conductance curve of GABA.

The increment (ΔG) is a direct consequence of the action of the transmitter on the subsynaptic membrane (Fatt & Katz, 1951; del Castillo & Katz, 1954*b*), and it may be a better index of the amount of transmitter released (Jenkinson, 1957) than the size of the postsynaptic potential (PSP); PSP is brought about secondarily as the result of a conductance increase in the subsynaptic membrane, and a linear relation cannot be expected between the size of PSP and ΔG (Martin, 1955). Although ΔG reflects the output of transmitter more faithfully than the size of PSP, ΔG is not necessarily in direct proportion to the released amount of transmitter. The second step, therefore, had to be employed to discover the relation between them.

GABA is established as the inhibitory transmitter substance in several synapses in crustacea (Iversen, Kravitz & Otsuka, 1966; Otsuka, Iversen, Hall & Kravitz, 1966; Otsuka, Kravitz & Potter, 1967; Takeuchi & Takeuchi, 1965, 1966*a* and *b*, 1967*a* and *b*), to which *Squilla* belongs. Furthermore, the effect of GABA on the *Squilla* heart ganglion cell is quite similar to the effect of tetanic stimulation on inhibitory nerves (Watanabe *et al.* 1968). These facts suggest that there may be a relation between ΔG and the released amount of transmitter such as that between the increment in the membrane conductance brought about with GABA and the amount of GABA applied. The dose-conductance curve of GABA (Takeuchi *et al.* 1967*a*) was therefore determined, and it was used as a calibration curve in converting ΔG to [GABA], the concentration of transmitter released. Such a procedure may not be valid at the cholinergic synapse, because the desensitization develops rather fast at the synapse (Katz & Thesleff, 1957). The GABA synapse, however, shows little or no desensitization (Takeuchi *et al.* 1967*b*), so the increment in conductance obtained from prolonged soaking (10–15 min) with GABA can be directly compared with ΔG obtained by tetanic stimulation to α-nerves (duration 1 to 1.5 sec).

EXPERIMENTS WITH GABA

Experimental conditions

As shown in Fig. 1a, the sensitivity of the cell to a certain amount of GABA is not affected by changing the ratio of [Ca] and [Mg] in the saline for *Squilla* when their sum is kept constant (each saline used is designated by the ratio of the concentration of Ca and Mg: 15/30 saline, for example, means a saline with 15 mM Ca and 30 mM Mg. The expressions Ca/Mg and Mg/Ca will be used instead of [Ca]/[Mg] and [Mg]/[Ca] in the text. For the composition of the original saline for *Squilla*, see Watanabe, Obara & Akiyama, 1968 and 1969).

A_1 and A_2 of Fig. 1 present I-V relations without (circles) and with

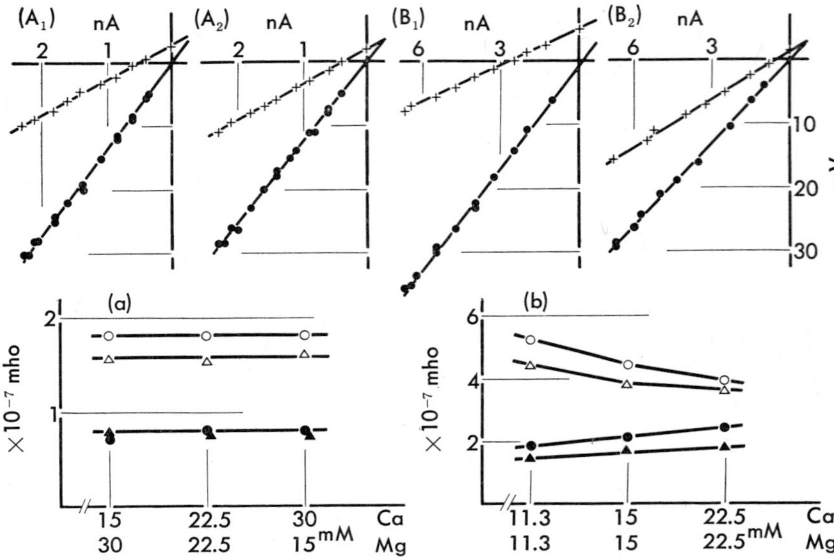

Fig. 1. Sensitivity of cell to GABA. A_1, A_2, and a, Sensitivity in salines with different ratios (Ca/Mg) but with the constant sum of 45 mM (Gc 5, Feb. 15, 1971). A_1 and A_2, I-V relations without (filled circles) and with (crosses) 6×10^{-5}M GABA in 15/30 (A_1) and 30/15 (A_2) salines. a, Membrane conductances obtained from such relations as shown above are plotted against [Ca] and [Mg] in salines (filled symbols, conductances without GABA; hollow symbols, with 6×10^{-5}M GABA). B_1, B_2, and b, Sensitivity in salines with a fixed ratio (Ca/Mg) but with different sums (Gc 5, Mar. 18, 1971). B_1 and B_2, I-V relations without (filled circles) and with (crosses) 6×10^{-5}M GABA in 11.3/11.3 (B_1) and 22.5/22.5 (B_2) salines. b, Membrane conductances without (filled symbols) and with (hollow symbols) 6×10^{-5}M GABA are plotted against [Ca] and [Mg] in salines. The units of the ordinate in B_2 are the same as those in A_1, A_2, and B_1. Ordinates in a and b, membrane conductance; abscissae in a and b, [Ca] and [Mg] in saline.

(crosses) 6×10^{-5}M GABA in 15/30 and 30/15 saline, respectively. Conductances obtained from such relations are plotted as filled (without GABA) and hollow (with GABA) circles in Fig. 1a. The experiment was repeated on the same cell and data of the second run are shown as filled (without GABA) and hollow (with GABA) triangles in the same graph. It is apparent from Fig. 1a that the sensitivity of the cell is not affected by changing the ratio as long as the sum is kept constant. Watanabe *et al.* (1968) have also reported that the conductance increase with 10^{-4}M GABA in saline with 30 mM Ca and 0 mM Mg is the same as that in saline with 0 mM Ca and 30 mM Mg.

The exact cause of the small difference between the increments in conductance of the first run (average 0.1×10^{-6} mho) and those of the second run (0.08×10^{-6} mho) is not known. This type of discrepancy in results was frequently observed (see also Fig. 1b) when the same series of solutions was applied again to the same cell in the same sequence (the sequence in the experiment of Fig. 1a was 15/30, 22.5/22.5, 30/15, 15/30, 22.5/22.5, and 30/15, without and with 6×10^{-5} M GABA alternately).

The sensitivity of the subsynaptic membrane to a transmitter is known to be affected by the concentration of divalent cations in a medium (del Castillo *et al.* 1954; Takeuchi, 1963; Katz *et al.* 1969*b*). The same is also true at this synapse. B_1 and B_2 of Fig. 1 show I–V relations without (circles) and with (crosses) 6×10^{-5} M GABA in 11.3/11.3 and 22.5/22.5 saline, respectively. In 11.3/11.3 saline the frequency of spontaneous firing was increased considerably and current pulses had to be injected at the bottom of afterhyperpolarization. Namely, the zero point of the potential axis is not the same in B_1 and B_2. This is the reason for the apparent difference in the reversal potential (crossing level of two lines) in B_1 and B_2. Conductances obtained are plotted with filled (without GABA) and hollow (with GABA) circles in Fig. 1b. Data of the second run on the same cell are shown with filled (without GABA) and hollow (with GABA) triangles.

Increments in conductance with GABA were: 0.34 (in 11.3/11.3 saline), 0.24 (in 15/15), and 0.15 (in 22.5/22.5) $\times 10^{-6}$ mho in the first run, and 0.29, 0.21, and 0.19×10^{-6} mho in the second run. Taking values in 22.5/22.5 saline as 100%, increments are expressed as follows: first, 226, 156, and 100%; second, 153, 110, and 100%. Though the increments in each saline are not identical in the first and second runs, a qualitative conclusion can be drawn from the results that the sensitivity of the cell to GABA is increased in a saline with less divalent cation concentration. The data also indicate that the sensitivity of the cell was less influenced when the divalent cation concentration was decreased from 45 to 30 mM than when it was decreased from 30 to 22.6 mM (half decrease, 7.4 mM, compared to the former decrease of 15 mM).

The results in Fig. 1 indicate that changing [Ca] and [Mg] in such a manner as to keep the sum always constant can simplify the conditions of experiments on the release of transmitter; the method excludes the contribution of change in the sensitivity of the cell from the cause of change in ΔG obtained by nerve stimulation.

Dose-conductance curve of GABA: calibration curve for converting ΔG to [GABA]

Figure 2 shows one example of the dose-conductance curve. A_1, A_2, and A_3 of Fig. 2 present I–V relations in 15/15 saline without (shown with filled circles in all graphs) and with 2.5 (A_1), 5 (A_2), and 10×10^{-5} M (A_3) GABA (shown with crosses). The increments in

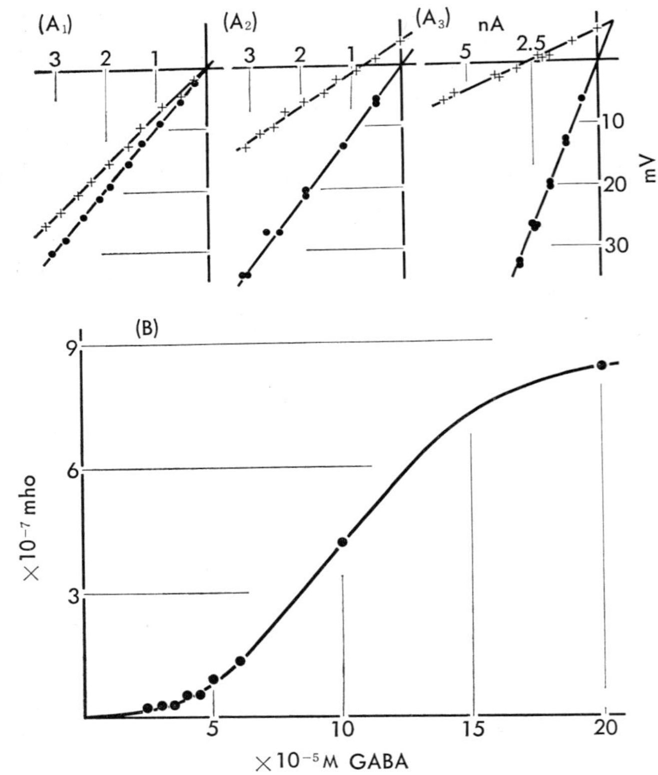

Fig. 2. Dose-conductance curve of cell in 15/15 saline. A_1, A_2, and A_3, I–V relations without (filled circles) and with (crosses) GABA; concentrations of GABA were 2.5×10^{-5}M (A_1), 5×10^{-5} (A_2), and 10^{-4} (A_3). B, Increments in membrane conductance due to GABA obtained from such relations as shown in A_1, A_2, and A_3 are plotted against concentrations of GABA. Units of the ordinate in A_3 are the same as those in A_1 and A_2. Ordinate in B, increment in membrane conductance; abscissa in B, concentration of GABA applied. Gc 5, Feb. 9, 1971.

conductance with GABA obtained from such relations are plotted against [GABA] in Fig. 2B, and a smooth curve with a typical sigmoid shape is drawn through the data.

The curve shows that if tetanic stimulation to α-nerves produces ΔG of 0.3×10^{-6} mho in this cell, that stimulation is equivalent to the application of GABA of 8.4×10^{-5} M; the concentration of released transmitter in synaptic clefts during tetanic stimulation may be taken as 8.4×10^{-5} M, and so forth for other values of ΔG.

Dose-conductance curves of two other cells are presented in A_2 and

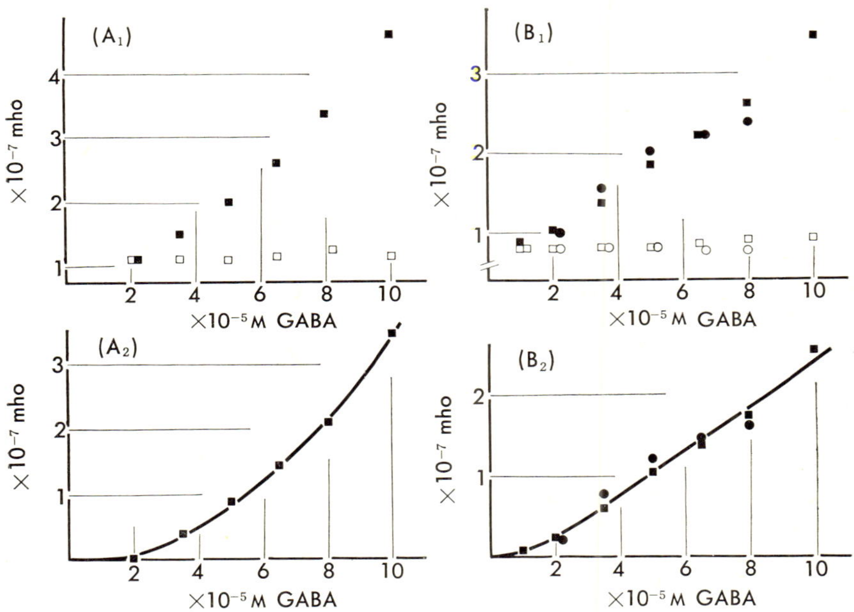

Fig. 3. A_1 and A_2: Dose-conductance curves of cell in 15/15 saline (Gc 5, June 25, 1971). A_1, Membrane conductances without (hollow squares) and with (filled squares) GABA. A_2, Increments in membrane conductance (differences between filled and hollow squares in A_1) plotted against concentrations of GABA. B_1 and B_2, Dose-conductance curve of the other cell (Gc 5, June 23, 1971) in salines with sums of 30 and 45 mM, data for 30 mM saline are shown with circles, and those for 45 mM saline are indicated with squares. B_1, Membrane conductances without (hollow symbols) and with (filled symbols) GABA, the cell was first bathed in 45 mM saline without and with GABA alternately, and next in 30 mM saline in the same way. B_2, Increments in membrane conductance (differences between filled and hollow symbols in B_1) plotted against concentrations of GABA. Ordinates, membrane conductance (A_1 and B_1) or incremen tin membrane conductance (A_2 and B_2); abscissae, concentration of GABA applied.

B_2 of Fig. 3. The range of [GABA] examined was less than 10^{-4} M in both cases. These curves resemble the lower part of the curve in Fig. 2B in shape, suggesting that their shape may also be sigmoid for a wider range of GABA. Data in A_2 of Fig. 3 were obtained from a cell in 15/15 saline without and with GABA. Data in B_2 of Fig. 3 were obtained from a cell in 15/15 and also in 22.5/22.5 saline without and with GABA. Data in B_2 show that increments in conductance due to GABA were roughly identical in both salines (filled squares, 15/15 saline; filled circles, 22.5/22.5 saline), indicating that the sensitivity of the cell was little affected by the difference of 15 mM in divalent cation concentration. This agrees with results of the second run in Fig. 1b; increments in conductance in 15/15 and 22.5/22.5 saline were fairly close, 0.19 and 0.21×10^{-6} mho.

The most exact calibration curve is apparently the dose-conductance curve obtained from a cell on which ΔG is also measured. Such an experiment has not been successful yet due to the difficulty of keeping electrodes in position. The curve in Fig. 2B was, therefore, employed as the calibration curve in converting ΔG to [GABA], the concentration of transmitter in synaptic clefts.

EFFECT OF CHANGING [Ca] AND [Mg] ON THE RELEASE OF TRANSMITTER

Determination of ΔG

When Ca/Mg in the saline was increased, maintaining a constant sum of [Ca] and [Mg], tetanic stimulation of α-nerves produced a larger increment in the membrane conductance. Specimen records and examples of I–V relations of such an experiment are shown in Fig. 4. The cell was bathed in salines with the sum of 30 mM. Each record taken by triple exposure shows potential changes (lower traces) due to a certain amount of current injected (upper traces) without, with 10/sec and with 100/sec stimulation. The ratio (Ca/Mg) in salines was 6/24 (A), 10/20 (B), 15/15 (C), 20/10 (D), 27/3 (E), and 6/24 (F). The order in the experiment was from E to F, A, B, C, and D. The recording in 6/24 saline was made twice (F and A) with an interval of 15 min to check the efficacy of washing for 5 min (between E and F and between F and A). I–V relations in both cases (a and f of Fig. 4) gave the same resting and peak conductances (Fig. 5A), indicating that 27/3 saline in E was completely replaced with 6/24 saline during the first flush.

Unitary IPSPs were barely seen in 6/24 saline (A and F), even

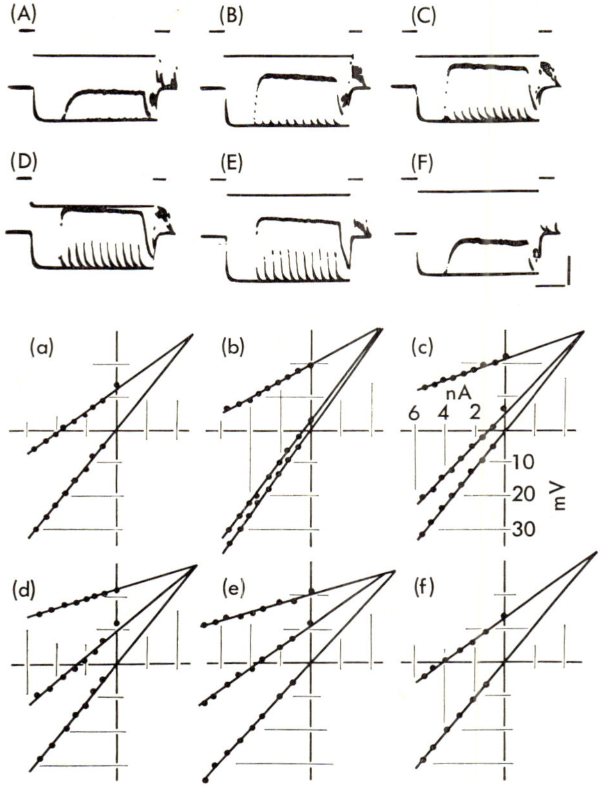

Fig. 4. Specimen records and examples of I–V relation of cell in salines with sums of 30 mM. Conductances obtained are shown in Fig. 5 A. A to F, A certain amount of current was injected into the cell without, with 10/sec, and with 100/sec stimulation of α-nerves, and traces were superimposed three times in each record. Upper traces, current (amplification in E and F was different from that in A to D). Lower traces, potential; summated IPSPs and trains of unitary IPSPs are shown on basic hyperpolarization (in A and F, unitary IPSPs are barely seen), rising phases of unitary IPSPs are not reproduced, rebound firings are seen to have followed summated IPSPs. Horizontal bar below F indicates 0.4 sec. Vertical bar, 20 mV for all potential records and 5 nA (A to D) or 12.5 nA (E and F) for current records. a to f, I–V relations corresponding to above records. Upper relations, at peaks of summated IPSPs; middle ones, at peaks of unitary IPSPs (not determined in a and f); lower ones, without nerve stimulation. Units of abscissa and ordinate in c are the same in all graphs. A and a, in saline with 6/24 (Ca/Mg); B and b, 10/20; C and c, 15/15; D and d, 20/10; E and e, 27/3; F and f, 6/24. Results were obtained in the following sequence: from E (e) to F (f), A (a), B (b), C (c) and D (d). Gc 5, Jan. 26, 1971.

though the membrane potential was hyperpolarized about 30 mV. Their size increased considerably as Ca/Mg was increased (B to E). In all cases they showed marked facilitation. The peak level was usually attained at the third or fourth IPSP and remained almost constant for the following three or four IPSPs. After that, the peak level was decreased a little (depression). Further analysis on unitary IPSPs was not made, because the main object of this work was to establish the relation at the steady state between released amount of transmitter and divalent cation concentration.

A summated IPSP grew to a fairly large size on a basic hyperpolarization even in 6/24 saline (A and F). In salines with high Ca/Mg, a summated IPSP rose steeply to the peak, due to larger size of unitary IPSPs and also due to the facilitation, and declined slowly probably due to depression of unitary IPSPs (for desensitization, see page 113).

Potential differences were measured in such records from the standard level (the level of membrane potential during an interburst period; the level just prior to the current injection) to the bottom of hyperpolarization, to the peak of summated IPSP, and/or to the average peak of unitary IPSPs in some cases. Potential differences obtained in this way are plotted against the current in a to f (the saline used in a corresponded to that in A, and similarly for the rest). In 6/24 saline, unitary IPSPs were so small that I–V relations at their average peak were not plotted. Therefore, a and f lack the middle line that is seen in other graphs. Upper lines in all graphs indicate I–V relations at peaks of summated IPSPs, while lower lines represent relations without nerve stimulation. The potential level at the crossing of I–V relations in each graph shows the level of reversal potential measured from the standard potential. The reversal potential of this cell, therefore, lies at about 30 mV depolarized level from the standard membrane potential. This rather high reversal potential might be due to the leakage of KCl from or around the tips of intracellular electrodes. The reversal potential in f appears to differ markedly from those in others. The reason for this is the same as that given for B_1 of Fig. 1.

The set of conductances obtained from such relations as shown in Fig. 4 are plotted in Fig. 5A against [Ca] (the cell was subjected to the following changes in Ca/Mg: from 15/15 to 20/10, 25/5, 27/3, 6/24, 8/22, 10/20, 12/18, 15/15, and 20/10). The results in the first four salines are shown with hollow circles (resting conductances), filled triangles (conductances at the average peak of unitary IPSPs),

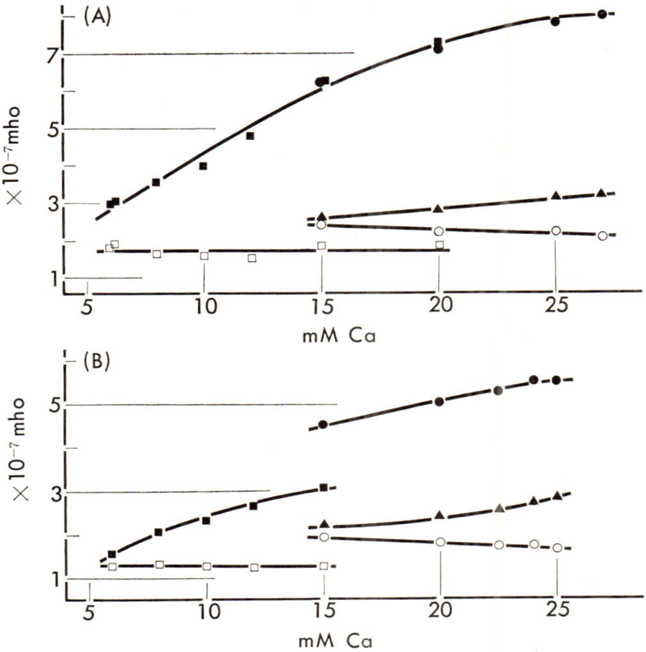

Fig. 5. Relation between peak and resting conductance and [Ca]. Filled squares and circles in A and B, peak conductances; hollow squares and circles, resting conductances; filled triangles, conductances at average peak of unitary IPSPs. A, Conductances were determined in salines with 15/15 (Ca/Mg), 20/10, 25/5, 27/3, 6/24 (twice), 8/22, 10/20, 12/18, 15/15, and 20/10 in this order; triangles and hollow and filled circles indicate results of the early part of the experiment in 15/15, 20/10, 25/5 and 27/3 salines; hollow and filled squares, results in the remainder. Some results are shown in Fig. 4. B, Hollow and filled squares show the results of the other cell (Gc 5, Jan. 27, 1971) in salines with 6/24 (Ca/Mg), 8/22, 10/20, 12/18, and 15/15; triangles and hollow and filled circles represent data of another cell (Gc 5, Jan. 28, 1971) in salines with 15/15, 20/10, 22.5/7.5, 24/6, and 25/5. Conductances at an average peak of unitary IPSPs were also measured in experiments represented with squares in A and B, but they are not shown to simplify the graphs. Straight lines and smooth curves were drawn through data arbitrarily in each graph. Abscissae, [Ca] in saline; ordinates, membrane conductance.

and filled circles (peak conductances). Data in the following six salines are presented with hollow squares (resting conductances) and with filled squares (peak conductances). To clarify the figure, the conductances at the average peak of unitary IPSPs are omitted for the latter series, and smooth lines are drawn through each group of symbols.

Two other examples in salines with the sum of 30 mM are also shown in Fig. 5B. All data presented in this figure were obtained from three fifth ganglion cells (for the nomenclature of ganglion cell, see Alexandrowicz, 1934, and also Watanabe, Obara, Akiyama & Yumoto, 1967) with 100/sec stimulation to both of a pair of in-

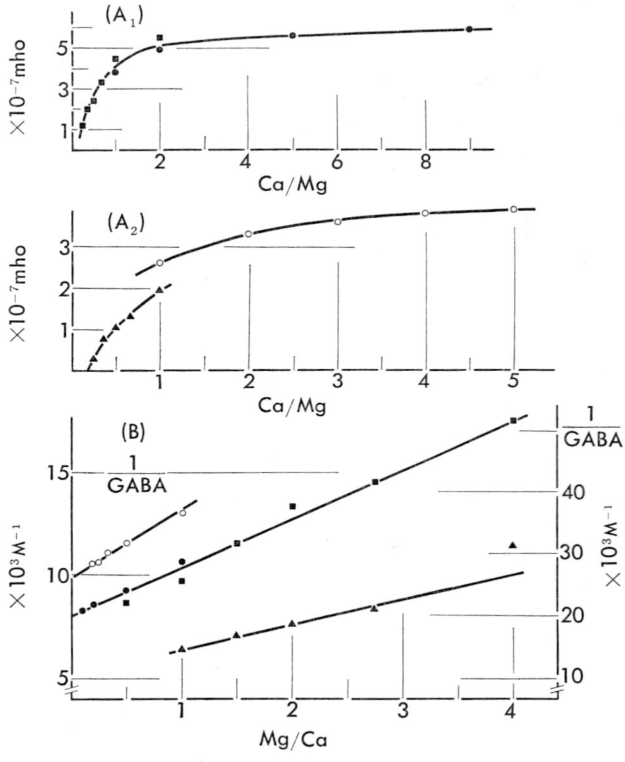

Fig. 6. A_1 and A_2, ΔGs obtained from results in Fig. 5 are plotted against Ca/Mg. A_1, Circles and squares represent ΔGs obtained by subtracting hollow circles and squares from filled circles and squares in Fig. 5A, respectively. A_2, Triangles and hollow circles indicate ΔGs obtained by subtracting hollow squares and circles from filled squares and circles in Fig. 5B, respectively. Abscissae, ratio (Ca/Mg) in saline; ordinates, increment in membrane conductance due to tetanic stimulation to α-nerves. B, Linear relation between Mg/Ca and reciprocal of [GABA] which was estimated with the dose-conductance curve as described in the text. Hollow circles were obtained from hollow circles in A_2, filled circles and squares came from corresponding symbols in A_1 and triangles, from triangles in A_2. Left ordinate, for relations shown with hollow circles and with filled circles and squares. Right ordinate, for relation shown with triangles. Both ordinates, 1/[GABA]; abscissae, [Mg]/[Ca] in saline.

hibitory nerves (in Fig. 5A and B shown with circles and triangles) or to one of a pair (in Fig. 5B shown with squares).

The results suggest that if Ca is changed within a narrow range under the condition of a constant sum of [Ca] and [Mg], the conductance increase due to tetanic stimulation may be taken as linearly related to [Ca] (see the almost linear increment in peak conductance at fairly steady resting conductance from 6 to 20 mM of Ca in Fig. 5A). Peak conductances against 25 and 27 mM Ca, however, deviate markedly from a straight line which can be drawn through data obtained with Ca below 20 mM, showing that ΔG is not linearly related to [Ca] at higher concentrations. In other similar experiments in salines with the sum of 45 mM, peak conductances also attained a ceiling at high Ca. The saturation of peak conductance seemed to originate presynaptically because an application of GABA of more than 2×10^{-4} M increased the membrane conductance to a value that exceeded the saturation value of peak conductance.

To relate ΔG not only to [Ca] but also to [Mg], differences between peak and resting conductances were next plotted against Ca/Mg in A_1 and A_2 of Fig. 6 (A_1 from data of Fig. 5A; A_2 from data of Fig. 5B). Both plots show that ΔG increased steeply as Ca/Mg was increased to unity (15/15) from 0.25 (6/24), but the increment in ΔG was less marked when the ratio was further increased.

TABLE 1. Converting ΔG to [GABA] with a dose-conductance curve

Ca (mM)	Mg (mM)	Ca/Mg	Mg/Ca	Resting G* ($\times 10^{-6}$ mho)	Peak G* ($\times 10^{-6}$ mho)	ΔG** ($\times 10^{-6}$ mho)	GABA† ($\times 10^{-5}$ M/liter)	1/GABA ($\times 10^{3}$ liter/M)
15	15	1	1	0.24	0.62	0.38	9.4	10.6
20	10	2	0.5	0.22	0.71	0.49	10.9	9.2
25	5	5	0.2	0.22	0.78	0.56	11.8	8.5
27	3	9	0.11	0.21	0.80	0.59	12.2	8.2
6	24	0.25	4	0.19	0.31	0.12	5.7	17.5
6	24	0.25	4	0.18	0.30	0.12	5.7	17.5
8	22	0.36	2.75	0.16	0.36	0.20	6.9	14.5
10	20	0.5	2	0.16	0.40	0.24	7.5	13.3
12	18	0.67	1.5	0.15	0.48	0.33	8.7	11.5
15	15	1	1	0.18	0.63	0.45	10.3	9.7
20	10	2	0.5	0.18	0.73	0.55	11.7	8.6

* These membrane conductances are also shown in Fig. 5A. ** These membrane conductances are also shown in A_1 of Fig. 6. † Concentrations of GABA that produce the same increments in membrane conductance as those shown in the seventh column. The calibration curve is that in Fig. 2B.

Conversion of ΔG to [GABA]

The increments in conductance shown in A_1 of Fig. 6 were next converted to concentrations of GABA with a dose-conductance curve as described before. It was assumed that a dose-conductance curve of this cell was identical to that in Fig. 2B. The ordinate of Fig. 2 B was, therefore, regarded as representing ΔG of A_1 of Fig. 6, and the corresponding [GABA] was obtained for each value of ΔGs. Results are summarized in Table 1. The fifth, sixth, and seventh columns recapitulate the data shown in Figs. 5A and $6A_1$. The eighth column presents concentrations of GABA that produced the same increments in conductance in Fig. 2B as those ΔGs shown in the seventh column. Reciprocals of those concentrations of GABA are shown in the ninth column, and data are analyzed in the following, section.

Relation between divalent cation concentration and amount of transmitter released

The antagonism between Ca and Mg has been shown to be competitive in nature (Jenkinson, 1957; Hubbard *et al.* 1968). Keeping this in mind, the double reciprocal plot of [GABA] and Ca/Mg was attempted. The plot gave a fairly linear relation as seen in Fig. 6B. Filled circles and squares in the figure represent reciprocals of [GABA] plotted against Mg/Ca. The straight line drawn through filled circles and squares has a slope of 2.4×10^3 (M^{-1}), and it intersects the ordinate (left one) at 8.0×10^3 (M^{-1}).

Increments in conductance shown in A_2 were similarly converted to concentrations of GABA with the dose-conductance curve of Fig. 2B. Data were also plotted in Fig. 6B double-reciprocally: hollow circles, from hollow circles of A_2 (the left ordinate); filled triangles, from filled triangles of A_2 (the right ordinate). Straight lines drawn through those symbols intersect ordinates at $9.8 \times 10^3\ \text{M}^{-1}$ (hollow circles) and 10.0×10^3 (filled triangles). Their slopes are $3.3 \times 10^3\ \text{M}^{-1}$ (circles) and 4.0×10^3 (triangles).

These results clearly demonstrate that there exists the following relation between the divalent cation concentration and the steady state concentration of transmitter released into synaptic clefts during tetanic nerve stimulation;

$$\frac{1}{[\text{T}]} = a\frac{[\text{Mg}]}{[\text{Ca}]} + b \qquad (1)$$

where *a* and *b* are constants, and [T] is the concentration of transmitter in synaptic clefts.

Validity of the empirical equation

Equation (1) indicates that the steady concentration of transmitter in synaptic clefts is determined by the ratio of [Ca] and [Mg] in salines. The experiment shown in Fig. 7 was performed to check this point further by measuring peak and resting conductances in two series of salines with different sums (30 and 45 mM) of [Ca] and [Mg] ([Na] was kept constant at 425.5 mM; the tonicity was adjusted with TrisCl).

Peak and resting conductances obtained are plotted against [Ca] in Fig. 7D, in which hollow symbols indicate results obtained before the dislodgement of the current electrode while filled symbols represent data from the same cell after the repenetration. Lines are drawn for clarity through peak conductances in salines with the sum of 30 and 45 mM obtained after the repenetration.

Examples of I–V relations are shown in the graphs above (A, B, and C, in salines with the sum of 30 mM; a, b, and c, in salines with 45 mM). Graphs marked with corresponding capital and small letters show relations in salines with the same ratio (Ca/Mg) (A and a, 0.364=8/22=12/33; B and b, 0.667=12/18=18/27; C and c, 2.0=20/10=30/15). These graphs roughly indicate that when the ratio is same, conductance increase is nearly equal in salines with the sums of 30 and 45 mM.

Plotting increments in conductance against ratios shows more clearly that ΔGs in salines with the same ratio are almost identical, regardless of the difference in the sum (Fig. 8A: hollow and filled circles and squares were obtained from data shown with corresponding symbols in Fig. 7D).

These ΔGs were simply converted to concentrations of GABA with the curve of Fig. 2B. The sensitivity of the cell in salines with the sum of 30 mM is not so different from that in salines with the sum of 45 mM (see the dose-conductance curve in B_2 of Fig. 3, also the data of the second run in Fig. 1b). This procedure of conversion, therefore, may not introduce serious errors into estimated amounts of released transmitter.

The reciprocals of GABA determined in this way are plotted against Mg/Ca in Fig. 8B (hollow and filled circles and squares were obtained from data of corresponding symbols in A). It is apparent from the figure that the linear relation as expressed in Eq. (1) is also valid for this cell.

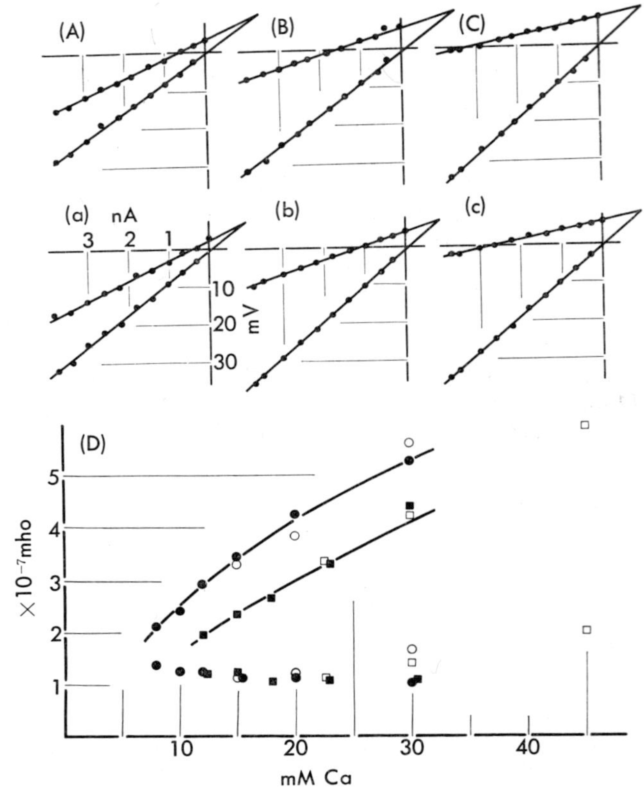

Fig. 7. Upper six graphs, I–V relations of one cell in salines with sums of 30 (A to C) and 45 mM (a to c). Ratio (Ca/Mg) was identical in A and a (8/22=12/33=0.364), B and b (12/18=18/27=0.667), and C and c (12/10=30/15=2.0). Units of ordinate and abscissa in a are the same for all graphs. D, Results of the entire experiment plotted against [Ca] in saline. When two identical symbols are given at the same [Ca], the upper one indicates peak conductance, and the lower one shows resting conductance. Hollow squares and circles represent results of the first run in salines with sums of 45 and 30 mM, respectively, before the dislodgement of the current electrode. Salines were applied as follows: from 45/0 to 30/0, 30/15, 20/10, 22.5/22.5 and 15/15. Filled squares and circles indicate results obtained from the same cell after the repenetration in salines with sums of 45 and 30 mM, respectively. The sequence of saline change was from 8/22 to 12/33, 10/20, 15/30, 12/18, 15/15, 22.5/22.5, 20/10, 30/15 and 30/0. Two curves were drawn through data (peak conductances) in salines with 30 mM (upper curve) and with 45 mM (lower one) obtained after repenetration. I–V relations shown above were also those obtained after repenetration. Gc 5, June 16, 1971.

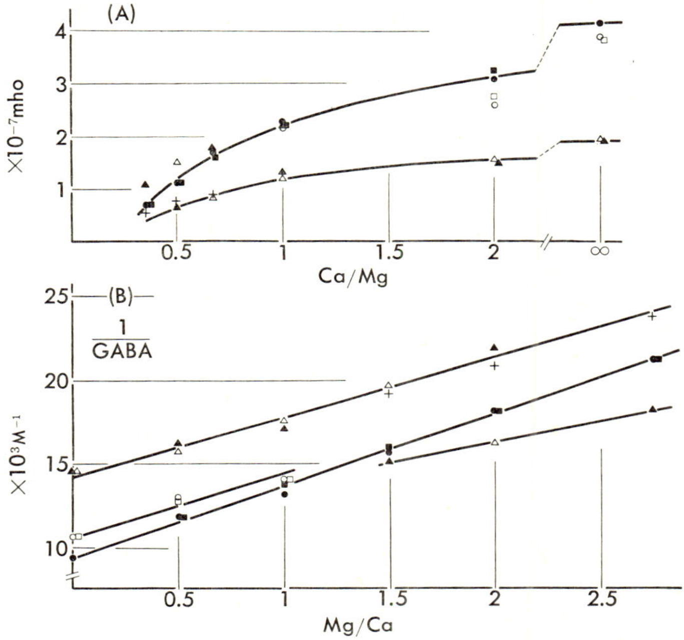

Fig. 8. A, ΔG plotted against ratio (Ca/Mg). Hollow and filled squares and circles were obtained from results shown with corresponding symbols in Fig. 7D. Triangles and crosses are data of the other cell (Gc 5, June 17, 1971). For further details see text. Ordinate, increment in membrane conductance due to tetanic stimulation to α-nerves; abscissa, ratio (Ca/Mg) in saline; 30/0 and 45/0 are indicated with ∞, and interruption is posed between ratio 2.0 and ∞; corresponding parts of curves are connected with dotted lines. B, Linear relation between [Mg]/[Ca] and reciprocal of [GABA]. For details see text.

Triangles and crosses in Fig. 8A represent data from the other cell (both cells in this figure were fifth ganglion cells). In this case hollow and filled triangles show experimentally determined ΔGs in salines with the sums of 30 and 45 mM, respectively. Crosses were obtained by halving values shown with a filled triangle at Ca/Mg= 0.364, with a hollow one at Ca/Mg=0.5, and with a filled one at Ca/Mg=0.667. Triangles, other than those halved, and crosses seem to lie on or near a smooth curve of similar shape to that drawn through circles and squares in this figure. Those ΔGs on or near the curve may be regarded as reflecting the activity of only one nerve of a pair (the other nerve might receive some injury during the dissection, and it could work only during the early stage of the experiment; see also low values of ΔG in the case shown with squares in Fig. 5B).

These results confirm that Eq. (1) describes satisfactorily the relation between divalent cation concentration in the saline and the steady concentration of transmitter in synaptic clefts during tetanic nerve stimulation, at least for the concentration range between 30 and 45 mM.

The straight lines in Fig. 8B intersect the ordinate at 14.3×10^3 M^{-1} (line drawn through triangles and crosses), 10.5×10^3 (hollow circles and squares), 9.4×10^3 (filled circles and squares), and about 11×10^3 (filled and hollow triangles). Their slopes are 3.5×10^3, 3.8×10^3, 4.3×10^3, and about 3×10^3 M^{-1}, respectively.

Another form of empirical equation: plotting 1/[GABA] against 1/[Ca]

In the present experiments the sum of [Ca] and [Mg] was always kept constant. Therefore [Mg] in Eq. (1) can be expressed as follows;

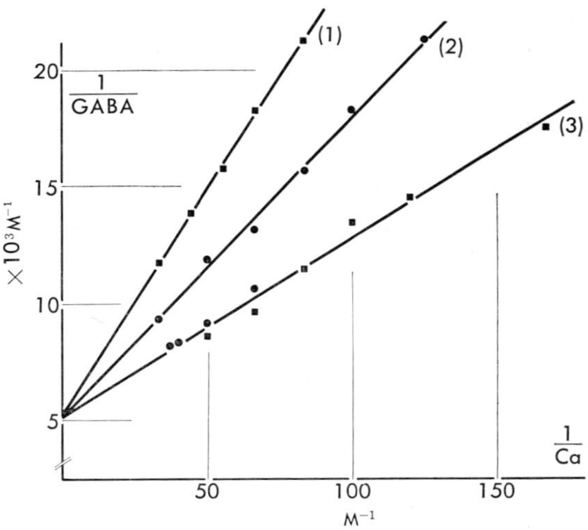

Fig. 9. Relation between 1/[GABA] and 1/[Ca]. Squares on line (1) and circles on or near line (2) came from filled squares (data in salines with the sum of 45 mM) and from filled circles (data of the same cell in salines with the sum of 30 mM) in Fig. 8B, respectively. Circles and squares on or near line (3) were obtained from the same symbols in Fig. 6B (data of the other cell in salines with the sum of 30 mM). Ordinate, reciprocal of [GABA]; abscissa, reciprocal of [Ca] in the saline.

$$[\mathrm{Mg}] = S - [\mathrm{Ca}]\,, \tag{2}$$

where S is the sum of [Ca] and [Mg], and it is constant. Inserting (2) into (1) gives

$$\frac{1}{[\mathrm{T}]} = \frac{aS}{[\mathrm{Ca}]} + (b-a)\,. \tag{3}$$

Equation (3) indicates that the relation between 1/[T] (namely 1/[GABA]) and 1/[Ca] should be linear. Figure 9 shows that this is the case. In the figure, data in Figs. 6B, and 8B, are replotted against 1/[Ca] [squares on line (1) and circles on or near line (2) are from the corresponding symbols in Fig. 8B; circles and squares on or near line (3) are from the corresponding symbols in Fig. 6B]. Three straight lines intersect the ordinate at $5.2 \times 10^3\,\mathrm{M}^{-1}$, and their slopes are 190 (line 1), 130 (line 2), and 75 (line 3). Line (1), obtained in salines with the sum of 45 mM, has a steeper slope than line (2) of the same cell in salines with the sum of 30 mM. This satisfies the requirement for the slope of the relation between 1/[T] and 1/[Ca], which is a function of S. Furthermore lines (1) and (2) intersect the ordinate at the same level as expected from the equation.

Nature of the antagonism between Ca and Mg

The antagonism between Ca and Mg was first proposed to be competitive in nature by del Castillo *et al.* (1954*a*), and their hypothesis has been proved valid at the frog neuromuscular junction (Jenkinson, 1957). The essence of the hypothesis, competition of Ca and Mg on the same critical site, molecule, or substance, is also included in a more elaborate explanation for the antagonism at the mammalian motor nerve terminals (Hubbard *et al.* 1968).

In the following description it will be shown that the relations found experimentally between the amount of released transmitter (GABA) and [Ca] and [Mg] [Eqs. (1) and (3)] are in good agreement with the original hypothesis put forward by del Castillo *et al.* (1954*a*) for the release of a different transmitter, Ach.

The hypothesis may be stated as follows. (1) Ca in a bathing medium flows into preterminals as well as prefibers during an action potential (Hodgkin & Keynes, 1957; Baker, Hodgkin & Ridgway, 1971; Katz *et al.* 1967, 1969*a*, 1970), and Ca in a unit volume of preterminals ($[\mathrm{Ca}]_i$) during tetanic nerve stimulation is

$$[\mathrm{Ca}]_i = \alpha[\mathrm{Ca}]_0 \tag{a}$$

where $[\mathrm{Ca}]_0$ is the concentration of Ca in the medium, and α is a constant (the range of α is $0<\alpha\leqq 1$). (2) $[\mathrm{Mg}]_i$ can be expressed in the same way,

$$[\mathrm{Mg}]_i=\alpha[\mathrm{Mg}]_0 \tag{b}$$

(3) Ca and Mg flowing in make complexes (CaX and MgX) with an unknown site X in the following manner:

$$\mathrm{Ca}+\mathrm{X} \rightleftharpoons \mathrm{CaX} \tag{c}$$

$$\mathrm{Mg}+\mathrm{X} \rightleftharpoons \mathrm{MgX} \tag{d}$$

(4) The total amount of X (bound or unbound, expressed in terms of concentration $[\mathrm{X}_0]$) in the unit volume of the preterminal is

$$[\mathrm{X}_0]=[\mathrm{X}]+[\mathrm{CaX}]+[\mathrm{MgX}] \tag{e}$$

where [X], [CaX], and [MgX] represent concentrations of free X, X bound by Ca, and X bound by Mg. (5) The concentration of released transmitter ([T]) in the unit volume of synaptic clefts during tetanic nerve stimulation is in direct proportion to the concentration of CaX formed;

$$[\mathrm{T}]=\beta\,[\mathrm{CaX}]\,, \tag{f}$$

where β is a constant.

Equations (c), (d), and (e) give the following relations under steady state conditions;

$$K_1[\mathrm{Ca}]_i([\mathrm{X}_0]-[\mathrm{CaX}]-[\mathrm{MgX}])=[\mathrm{CaX}]$$

and

$$K_2[\mathrm{Mg}]_i([\mathrm{X}_0]-[\mathrm{CaX}]-[\mathrm{MgX}])=[\mathrm{MgX}]\,,$$

where K_1 and K_2 are the association constants for Eqs. (c) and (d), respectively. Hence

$$[\mathrm{CaX}]=\frac{K_1[\mathrm{X}_0][\mathrm{Ca}]_i}{K_1[\mathrm{Ca}]_i+K_2[\mathrm{Mg}]_i+1}\,,$$

replacing $[\mathrm{Ca}]_i$, $[\mathrm{Mg}]_i$, and [CaX] with Eqs. (a), (b), and (f), respectively, and rearranging yield

$$\frac{1}{[\mathrm{T}]}=\frac{K_2}{\beta K_1[\mathrm{X}_0]}\cdot\frac{[\mathrm{Mg}]_0}{[\mathrm{Ca}]_0}+\frac{1}{\beta[\mathrm{X}_0]}+\frac{1}{\alpha\beta K_1[\mathrm{X}_0]}\cdot\frac{1}{[\mathrm{Ca}]_0}\,. \tag{g}$$

The last term of (g) can be rewritten as follows;

$$\frac{1}{\alpha\beta K_1[\mathrm{X}_0]}\cdot\frac{1}{[\mathrm{Ca}]_0}=\frac{1}{\alpha\beta K_1 S[\mathrm{X}_0]}\cdot\frac{[\mathrm{Mg}]_0+[\mathrm{Ca}]_0}{[\mathrm{Ca}]_0}\,,$$

where S represents the sum of $[\mathrm{Ca}]_0$ and $[\mathrm{Mg}]_0$ (see page 128, Eq. (2) in which [Mg] and [Ca] mean $[\mathrm{Mg}]_0$ and $[\mathrm{Ca}]_0$). Replacing the last term of (g) with the above expression and rearranging give

$$\frac{1}{[\mathrm{T}]}=\frac{1+\alpha K_2 S}{\alpha\beta K_1 S[\mathrm{X}_0]}\cdot\frac{[\mathrm{Mg}]_0}{[\mathrm{Ca}]_0}+\frac{1+\alpha K_1 S}{\alpha\beta K_1 S[\mathrm{X}_0]}, \tag{h}$$

or

$$\frac{1}{[\mathrm{T}]}=a\frac{[\mathrm{Mg}]_0}{[\mathrm{Ca}]_0}+b. \tag{h$'$}$$

Expressing $[\mathrm{Mg}]_0$ in (h) with Eq. (2) produces

$$\frac{1}{[\mathrm{T}]}=\frac{1+\alpha K_2 S}{\alpha\beta K_1[\mathrm{X}_0]}\cdot\frac{1}{[\mathrm{Ca}]_0}+\frac{K_1-K_2}{\beta K_1[\mathrm{X}_0]}, \tag{i}$$

or

$$\frac{1}{[\mathrm{T}]}=a'\frac{1}{[\mathrm{Ca}]_0}+b'. \tag{i$'$}$$

Namely, the hypothesis predicts the relation between the concentration of liberated transmitter in synaptic clefts and $[\mathrm{Ca}]_0$ and $[\mathrm{Mg}]_0$ with Eqs. (h) and (i) or (h′) and (i′).

It is apparent now that the theoretically derived equations are quite identical in form to the equations experimentally found. This clearly demonstrates that Ca and Mg exert their effects on the release of transmitter (GABA as well as Ach) by competing for a critical site (or substance) as described in the hypothesis.

Simple calculations using values of slopes and intersections of lines (1) and (2) in Fig. 9 (see page 129) give $\alpha K_1=9.2\times10^2\ \mathrm{M}^{-1}$, $\alpha K_2=4.0\times10^2\ \mathrm{M}^{-1}$ and $\beta[\mathrm{X}_0]=1.1\times10^{-4}\,\mathrm{M}$. Inserting these values into Eq. (h) yields

$$\frac{1}{[\mathrm{T}]}=4.3\times10^3\frac{[\mathrm{Mg}]_0}{[\mathrm{Ca}]_0}+9.4\times10^3 \qquad (S=30\times10^{-3}\mathrm{M})$$

and

$$\frac{1}{[\mathrm{T}]}=4.2\times10^3\frac{[\mathrm{Mg}]_0}{[\mathrm{Ca}]_0}+9.3\times10^3 \qquad (S=45\times10^{-3}\,\mathrm{M}).$$

Slopes and intersections of the above relations are almost identical to the experimentally found values on page 128.

The values of αK_1, αK_2, and $\beta[\mathrm{X}_0]$ for the cell shown with line (3) in Fig. 9 can not be determined rigorously because we have only two relations (slope and intersection) for three unknown quantities. Combining these two relations, however, we can obtain the following relations:

$$K_1 = \frac{10^2}{2.31 \times 10^4 \cdot \beta[X_0] - 3}$$

and

$$\frac{K_2}{K_1} = 1 - 5.2 \times 10^3 \times \beta[X_0]\,,$$

where K_1 and K_2/K_1 are both positive quantities. The range of $\beta[X_0]$, therefore, is

$$\frac{10}{5.2} \times 10^{-4} > X_0 > \frac{3}{2.31} \times 10^{-4}\,.$$

If we assume $\beta[X_0] = 3.1/2.31 \times 10^{-4} \doteqdot 1.34 \times 10^{-4}$ M, αK_1 and αK_2 become 10×10^2 and 3×10^2 M^{-1}, respectively. Inserting these values into (h) produces

$$\frac{1}{[T]} = 2.5 \times 10^3 \frac{[Mg]_0}{[Ca]_0} + 7.8 \times 10^3\,,$$

which is close to the equation determined experimentally with data in Fig. 6 B (filled circles and squares);

$$\frac{1}{[T]} = 2.4 \times 10^3 \frac{[Mg]_0}{[Ca]_0} + 8.0 \times 10^3\,.$$

Values of αK_1, αK_2 and $\beta[X_0]$ of this cell are close to those obtained before (page 131). The same type of calculation for the other cells gives αK_1, αK_2, and $\beta[X_0]$ of the same order of magnitude (the ratio, K_1/K_2, seems to be about 3, and $\beta[X_0]$ appears to be roughly 10^{-4} M). Jenkinson (1957) has reported the dissociation constant of the Mg-site complex of the order of 4×10^{-3} M. This corresponds to the association constant of 2.5×10^2 M^{-1}, and it is in good agreement with αK_2 in this work.

As for the value of α, only rough estimation based on relevant reports is possible at present. Those reports are: The location of the critical site for transmitter release may be on or within the membrane (Miledi *et al.* 1966); an association constant of the Ca-plasma membrane of rat liver cells is 4.0×10^3 M^{-1} (Shlatz & Marinetti, 1972). If the association constant (K_1) of the Ca-site X, which is probably located on (within) the membrane of preterminals, is assumed to be equal to the value described above, α becomes 0.23 for the cell shown in Fig. 9 [lines (1) and (2)]. The α gives $[Ca]_i$ of about 10 mM for the saline with 45 mM Ca and 0 mM Mg (relation a). This rather high $[Ca]_i$ may not be absurd if $[Ca]_i$ reflects the concentration of [Ca] within the membrane.

Shlatz *et al.* (1972) have also reported the number of binding site of the plasma membrane for Ca of 22,000 pmole/mg of membrane protein. If 1 kg of membrane protein is assumed to occupy 1 liter, and if $\beta[X_0]$ is taken as 10^{-4} M, then β becomes about 5×10^{-3}.

Phospholipids, the main components of the membrane other than proteins, are known to have negatively charged sites and to bind Ca and Mg in a reversible electrostatic manner. The apparent association constants of Ca (Mg)-phospholipid complexes are of the order of $10^3\,M^{-1}$ (Abramson, Colacicco, Curci & Rapport, 1968; Barton, 1968), and the separation factors (p. 153 of Helfferich, 1962) of phospholipids are 2.4 (phosphatidylinositol; Abramson *et al.* 1968) and 4.2 (phosphatidylserine; Blaustein, 1967). The factor is expressed in the form

$$\frac{[Ca]_1}{[Ca]_0}\cdot\frac{[Mg]_0}{[Mg]_1},$$

where $[Ca]_1$ and $[Mg]_1$ indicate amounts of Ca and Mg bound by phospholipid, and $[Ca]_0$ and $[Mg]_0$ represent concentrations in a bulk solution. Namely, the factor may be taken as representing the ratio of the association constant of the Ca complex to that of the Mg complex. The values reported are close to the K_1/K_2 of about 3 of the present work. The negatively charged site of phospholipids of the membrane, therefore, might be responsible for the site X. Blioch, Glagoleva, Lieberman & Nenashev (1968) have also proposed that negative charges of phospholipids of the membrane play a critical role in release of transmitter.

ACKNOWLEDGEMENTS

I wish to thank Professor A. Watanabe for helpful criticism and suggestions during the course of this work. I also wish to thank Dr. E. Ozawa for reading the manuscript and making valuable suggestions.

REFERENCES

Abramson, M. B., Colacicco, G., Curci, R. & Rapport, M. M. (1968). Ionic properties of acidic lipids. Phosphatidylinositol. *Biochemistry, N.Y.* **7**, 1692–1698.

Alexandrowicz, J. S. (1934). The innervation of the heart of crustacea. II. Stomatopoda. *Q. J. microsc. Sci.* **76**, 511–548.

Baker, P. F., Hodgkin, A. L. & Ridgway, E. B. (1971). Depolarization and calcium entry in squid giant axons. *J. Physiol.* **218**, 709–755.

Barton, P. G. (1968). The influence of surface charge density of phosphatides on the binding of some cations. *J. biol. Chem.* **243**, 3884–3890.

Blaustein, M. P. (1967). Phospholipids as ion exchangers: Implications for a possible role in biological membrane excitability and anesthesia. *Biochim. biophys. Acta* **135**, 653–668.

Blioch, Z. L., Glagoleva, I. M., Lieberman, E. A. & Nenashev, V. A. (1968). A study of the mechanism of quantal transmitter release at a chemical synapse. *J. Physiol.* **199**, 11–35.

Boyd, I. A. & Martin, A. R. (1956). The end-plate potential in mammalian muscle. *J. Physiol.* **132**, 74–91.

del Castillo, J. & Engbaek, L. (1954). The nature of the neuromuscular block produced by magnesium. *J. Physiol.* **124**, 370–384.

del Castillo, J. & Katz, B. (1954*a*). The effect of magnesium on the activity of motor nerve endings. *J. Physiol.* **124**, 553–559.

del Castillo, J. & Katz, B. (1954*b*). The membrane change produced by the neuromuscular transmitter. *J. Physiol.* **125**, 546–565.

Douglas, W. W. & Poisner, A. M. (1964). Stimulus-secretion coupling in a neurosecretory organ: The role of calcium in the release of vasopressin from the neurohypophysis. *J. Physiol.* **172**, 1–18.

Douglas, W. W. & Rubin, R. P. (1963). The mechanism of catecholamine release from the adrenal medulla and the role of calcium in stimulus-secretion coupling. *J. Physiol.* **167**, 288–310.

Douglas, W. W. & Rubin, R. P. (1964). The effects of alkaline earths and other divalent cations on adrenal medullary secretion. *J. Physiol.* **175**, 231–241.

Eccles, J. C. (1964). *The Physiology of Synapses.* Berlin: Springer-Verlag.

Fatt, P. & Katz, B. (1951). An analysis of the endplate potential recorded with an intracellular electrode. *J. Physiol.* **115**, 320–370.

Helfferich, F. (1962). *Ion Exchange.* New York: McGraw-Hill.

Hodgkin, A. L. & Keynes, R. D. (1957). Movement of labelled calcium in squid giant axons. *J. Physiol.* **138**, 253–281.

Hubbard, J. I. (1970). Mechanism of transmitter release. In *Progress in Biophysics and Molecular Biology*, vol. 21, ed. Butler, J. A. V. & Noble, D., pp. 35–124. Oxford: Pergamon.

Hubbard, J. I., Jones, S. F. & Landau, E. M. (1968). On the mechanism by which calcium and magnesium affect the spontaneous release of transmitter from mammalian motor nerve terminals. *J. Physiol.* **194**, 355–380.

Hutter, O. F. & Kostial, K. (1954). Effect of magnesium and calcium ions on the release of acetylcholine. *J. Physiol.* **124**, 234–241.

Iversen, L. L., Kravitz, E. A. & Otsuka, M. (1966). Release of gamma-aminobutyric acid (GABA) from lobster inhibitory neurons. *J. Physiol.* **188**, 21.

Jenkinson, D. H. (1957). The nature of the antagonism between calcium and magnesium ions at the neuromuscular junction. *J. Physiol.* **138**, 434–444.

Katz, B. (1969). *The Release of Neural Transmitter Substances.* Springfield: Charles C. Thomas.

Katz, B. & Miledi, R. (1967). A study of synaptic transmission in the absence of nerve impulses. *J. Physiol.* **192**, 407–436.

Katz, B. & Miledi, R. (1969*a*). Tetrodotoxin-resistant electric activity in presynaptic terminals. *J. Physiol.* **203**, 459–487.

Katz, B. & Miledi, R. (1969*b*). Spontaneous and evoked activity of motor nerve endings in calcium Ringer. *J. Physiol.* **203**, 689–706.

Katz, B. & Miledi, R. (1970). Further study of the role of calcium in synaptic transmission. *J. Physiol.* **207**, 789–801.

Katz, B. & Miledi, R. (1971). The effect of prolonged depolarization on synaptic transfer in the stellate ganglion of the squid. *J. Physiol.* **216**, 503–512.

Katz, B. & Thesleff, S. (1957). A study of the 'desensitization' produced by acetylcholine at the motor end-plate. *J. Physiol.* **138**, 63–80.

Kuno, M. (1971). Quantum aspects of central and ganglionic synaptic transmission in vertebrates. *Physiol. Rev.* **51**, 647–678.

Kusano, K. (1968). Further study of the relationship between pre- and postsynaptic potentials in the squid giant synapse. *J. gen. Physiol.* **52**, 326–345.

Kusano, K. (1970). Influence of ionic environment on the relationship between pre- and postsynaptic potentials. *J. Neurobiol.* **1**, 435–457.

Kusano, K., Livengood, D. R. & Werman, R. (1967). Correlation of transmitter release with membrane properties of the presynaptic fiber of the squid giant synapse. *J. gen. Physiol.* **50**, 2579–2601.

Martin, A. R. (1955). A further study of the statistical composition of the end-plate potential. *J. Physiol.* **130**, 114–122.

Miledi, R. (1967). Spontaneous synaptic potentials and quantal release of transmitter in the stellate ganglion of the squid. *J. Physiol.* **192**, 379–406.

Miledi, R. & Slater, C. R. (1966). The action of calcium on neuronal synapses in the squid. *J. Physiol.* **184**, 473–498.

Otsuka, M., Iversen, L. L., Hall, Z. W. & Kravitz, E. A. (1966). Release of gamma-aminobutyric acid from inhibitory nerves of lobster. *Proc. natn. Acad. Sci. U.S.A.* **56**, 1110–1115.

Otsuka, M., Kravitz, E. A. & Potter, D. D. (1967). Physiological and chemical architecture of a lobster ganglion with particular reference to gamma-aminobutyrate and glutamate. *J. Neurophysiol.* **30**, 725–752.

Shlatz, L. & Marinetti, G. V. (1972). Hormone-calcium interactions with the plasma membrane of rat liver cells. *Science, N.Y.* **176** 175–177.

Takeuchi, A. & Takeuchi, N. (1962). Electrical changes in pre- and postsynaptic axons of the giant synapse of Loligo. *J. gen. Physiol.* **45**, 1181–1193.

Takeuchi, A. & Takeuchi, N. (1965). Localized action of gamma-aminobutyric acid on the crayfish muscle. *J. Physiol.* **177**, 225–238.

Takeuchi, A. & Takeuchi, N. (1966*a*). A study of the inhibitory action of γ-aminobutyric acid on neuromuscular transmission in the crayfish. *J. Physiol.* **183**, 418–432.

Takeuchi, A. & Takeuchi, N. (1966*b*). On the permeability of presynaptic terminal of the crayfish neuromuscular junction during synaptic inhibition and the action of γ-aminobutyric acid. *J. Physiol.* **183**, 433–449.

Takeuchi, A. & Takeuchi, N. (1967*a*). Anion permeability of the inhibitory post-synaptic membrane of the crayfish neuromuscular junction. *J. Physiol.* **191**, 575–590.

Takeuchi, A. & Takeuchi, N. (1967*b*). Electrophysiological studies of the action of GABA on the synaptic membrane. *Fedn Proc.* **26**, 1633–1638.

Takeuchi, N. (1963). Effects of calcium on the conductance change of the end-plate membrane during the action of transmitter. *J. Physiol.* **167**, 141–155.

Watanabe, A., Obara, S. & Akiyama, T. (1967). Pacemaker potentials for the periodic burst discharge in the heart ganglion of a stomatopod, *Squilla oratoria*. *J. gen. Physiol.* **50**, 839–862.

Watanabe, A., Obara, S. & Akiyama, T. (1968). Inhibitory synapses on pacemaker neurons in the heart ganglion of a stomatopod, *Squilla oratoria*. *J. gen. Physiol.* **52**, 908–924.

Watanabe, A., Obara, S. & Akiyama, T. (1969). Acceleratory synapses on pacemaker neurons in the heart ganglion of a stomatopod, *Squilla oratoria*. *J. gen. Physiol.* **54**, 212–231.

Watanabe, A., Obara, S., Akiyama, T. & Yumoto, K. (1967). Electrical properties of the pacemaker neurons in the heart ganglion of a stomatopod, *Squilla oratoria*. *J. gen. Physiol.* **50**, 813–838.

SECRETORY MECHANISM IN ADRENAL MEDULLARY CELLS

Subcellular Catecholamine Pools in Relation to Secretion, Storage, and Synthesis of the Amine in Adrenal, Pheochromocytoma, and Neuroblastoma.

RITSUJI YAMADA

The subcellular distribution of catecholamine in rat adrenal, 17 pheochromocytomas, and 2 neuroblastomas was studied by the ultracentrifuge method. The subcellular catecholamine pool was divided into 4 fractions: coarse (the heaviest granules), mitochondrial (heavy granules), microsomal (light granules), and supernatant (cytoplasmic sap). Catecholamine granules of adrenal and pheochromocytomas were studied by the ultracentrifuge method of continuous density gradient of sucrose. The investigation was pursued for the purpose of studying the relation between the subcellular pools and secretion or synthesis of catecholamines in rat adrenal after administration of insulin, reserpine, L-DOPA, MAO inhibitor, and α-MT.

The subcellular pools of catecholamines in adrenal medulla were not affected by a change of secretion but were altered by a change in synthesis of the amine. The role of each subcellular catecholamine pool in adrenal is discussed in relation to the mechanism of secretion and synthesis of the amine. The catecholamine granules showed widespread buoyant density. The lighter portion of the granules was lost in an early stage of secretion, and the heavier portion increased by acceleration of synthesis of catecholamines. In pheochromocytomas and neuroblastoma, the relation between the subcellular pools and secretion of catecholamines was also studied. The subcellular distribution of catecholamines in these tumors was similar to that in adrenal after acceleration of synthesis of catecholamines.

INTRODUCTION

Oliver & Schaffer (1895) found that extracts of the adrenal medulla possess pressor activity. Abel & Crawford (1897) proved that the agent responsible for the pressor activity of crude adrenal extract was adrenaline (A). Adrenaline was at first thought to be the neurohumor of the sympathetic nervous system. However, certain differences were detected between the action of the neurohumor and A. This led to the discovery of noradrenaline (NA) and its identification as the active neurohumor (von Euler, 1946) and as an adrenal medullary hormone (Holtz & Schümann, 1948; James, 1948). Noradrenaline is the immediate precursor of A in biosynthesis of catecholamines. Noradrenaline, therefore, has a dual role, not only as a hormone (and neurohumor), but also as an intermediate in A formation. Various stimulation, such as administration of pyrogen and insulin, resulted in the secretion of varying amounts of either A or NA or both. However, the increase of A secretion was far more than it of NA at the time of significant acceleration of catecholamines secretion from the adrenal medulla, and the proportion of A to NA found in adrenal venous blood became similar to that of the contents of both catecholamines in the adrenal medulla (Yamada, 1959).

Catecholamines of ox adrenal medulla were largely stored within intracellular particles, which could be sedimented in the mitochondrial fraction obtained by differential centrifugation of homogenates of ox adrenal medulla (Blaschko & Welch, 1953). Later, amine storage granules, *i.e.*, heavy granules, were separated from mitochondria. However, not all medullary hormones were found in heavy granules. Some part of the catecholamine pool exists in cytoplasmic sap. Blaschko *et al.* (1953) suggested that stimulation releases amines from the pool in cytoplasmic sap, and the granules serve as reserved stores. Hillarp (1960*a*) argued that a functional pool was " free " catecholamines in the cytoplasmic sap of the chromaffin cells.

Yamada (1966, 1967, 1968*a*, *b*) found that administration of insulin and reserpine did not increase the so-called functional pool of catecholamines in the cytoplasmic sap in the adrenal medulla when catecholamines were lost from the pool in heavy granules, and that the subcellular distribution of catecholamines remained almost unchanged during acceleration of secretion of catecholamines from the adrenal medulla. These results conflict with the conception, provided by many authors, that catecholamines are secreted from heavy granules to cytoplasm and hence, presumably by diffusion, across cytoplasm to the cell exterior.

My results suggest that the process of medullary secretion involves a complex chain of events during which the secretory heavy granules become fragile and are first attached to the cell membrane and then their contents directly extrude through it, probably by exocytosis or some closely related process (Yamada, 1966, 1967, 1968*a*). At almost the same time, some other laboratories independently reported evidence that catecholamines were secreted from the adrenal medulla by means of exocytosis or a similar process. In the first studies, the substances that appeared in perfusates of isolated adrenal gland following chemical or nervous stimulation were determined, and the proportion of amounts of each constituent was compared with that in the catecholamine storage granules (Douglas & Poisner, 1966; Kirshner, Sage, Smith & Kirshner, 1966; Blaschko, Comline, Schneider, Silver & Smith, 1967).

Intracellular localization of several enzymes involved in the biosynthesis of catecholamines was investigated in the adrenal medulla. Tyrosine hydroxylase, DOPA decarboxylase, and phenylethanolamine-N-methyltransferase were found in the cytoplasmic sap, and dopamine-ß-hydroxylase was contained in the heavy granules (catecholamine granules). In the biosynthetic pathway of catecholamine, the two first steps (tyrosine→DOPA, DOPA→dopamine) occur in the cytoplasmic sap, and after being taken up into granules, dopamine is converted to NA. Finally, a certain amount of NA, in becoming extragranular, becomes N-methylated and thereafter stored again in the granules (Laduron & Belpaire, 1968).

During the past decade, a number of laboratories have reported *in vitro* studies dealing with isolated catecholamine granules. There are, however, only a few *in vivo* studies on the subcellular pools of catecholamines. In this paper the subcellular distribution of catecholamines in the adrenal medulla in rats after injection of insulin, reserpine, DOPA, monamine oxidase inhibitor (MAO-I), and alpha-methyl-para-tyrosine(α-MT) is discussed and compared with that in pheochromocytomas and neuroblastomas.

SUBCELLULAR POOLS IN RAT ADRENAL GLAND

The homogenate of the rat adrenal in 0.25 M (isotonic) sucrose was separated by differential centrifugation into a coarse fraction (sediment obtained by centrifuging the sample for 10 min at $800 \times g$: N), heavy granules (sediment obtained by centrifuging the supernatant of previous centrifugation for 15 min at $5{,}000 \times g$: mitochondrial fraction: Mit), light granules (sediment obtained by centrifuging the

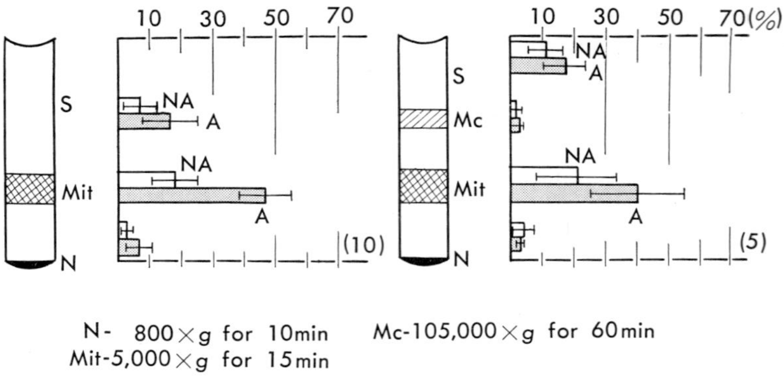

Fig. 1. Subcellular distribution of catecholamines in normal rat adrenal. S, Supernatant; Mc, microsomal fraction; Mit, mitochondrial fraction; N, coarse fraction. S in left figure includes S and Mc in right figure. The number of animals in each group is shown in parentheses.

supernatant of previous centrifugation for 60 min at $105{,}000\times g$: microsomal fraction: Mc), and cytoplasmic sap (the supernatant fluid of high-speed centrifugation: S). In some experiments, subcellular fractions were separated into three fractions, the coarse fraction (N), heavy granules (Mit), and supernatant (S).

Biochemical analyses of the fractions of rat adrenal (Fig. 1) demonstrated clearly that a large part of catecholamine is present in heavy granules (Mit). This result is consistent with those reported by other studies (Blaschko *et al.* 1953; Hillarp, 1960*a*; Eade, 1958; Bertler, Hillarp & Rosengren, 1960; Hillarp, 1960*b*; Euler & Lischajko, 1961; Winkler, 1969*a*; Viveros, Arqueros, Connett & Kirshner, 1969*a*; Viveros, Arqueros & Kirshner, 1971*a*; Burack, Avery, Draskoczy & Weiner, 1962). There are 5.0 ± 0.8 (S.E.)$\cdot10^{12}$ granules/g weight of fowl adrenal gland and individual heavy granules contain about 8 million molecules of catecholamines (Burack *et al.* 1962). A single cell of cat adrenal medulla contains an average of 13,000 catecholamine granules (Kirshner & Viveros, 1970). Catecholamine-containing granules contain a specific protein and 1 mole of ATP for every 4 moles of amine (Falck, Hillarp & Högberg, 1956). The proteins are named chromogranin by Blaschko *et al.* (1967).

The catecholamine pool in cytoplasmic sap is present in the extragranular space, *i.e.*, free form, and one-third of the catecholamines was found in this pool. Although the content in the pool is smaller than in the heavy granules, the pool contains a large quantity of catecholamines as compared with the amount in light granules microsomal fraction) and the coarse fraction (Fig. 1). It seems that a small part of the catecholamine granules (the heaviest part) is included in the coarse fraction. There are nuclei and other cell fragments in the coarse fraction.

The pool of catecholamines in light granules (microsomal fraction) is only a small part of the catecholamine content in the adrenal medulla, but most parts of catecholamine are included in the light granules in other organs: the vas deferens, submaxillary gland, pineal gland and heart (Potter & Axelrod, 1963). Even in other organs, subcellular distribution of catecholamines may be divided into the particulate fraction and the soluble fraction. The particulate-bound NA account for 46–56% of the sum of the particulate soluble fractions in uteri and is similar to that found in the heart. There is no significant change in the percentage at different phases of the estrous cycle (Gutman & Weil-Malherbe, 1967). This indicates the possibility that alteration of the physiological condition does not influence the subcellular distribution of catecholamines.

The partial isolation and characterization of catecholamine granules (heavy granules) was done with the method of continuous density gradient of sucrose. By this method catecholamine granules were separated from mitochondria (Potter *et al.* 1963; Hagen & Barrnet, 1960). Futher detailed investigation was done by some modified methods, *i.e.*, changes in density gradient and duration of centrifugation. Two narrow bands became visible in the mid or upper part of the tube after centrifugation. The rings were composed

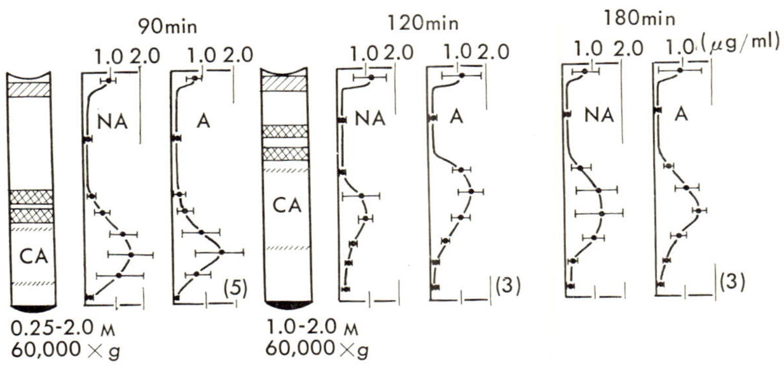

Fig. 2. Subcellular distribution of catecholamines in normal rat adrenal by centrifugation in continuous density gradient of sucrose. Two visual narrow bands in mid part of tube indicate mitochondria from cortex and medulla of adrenal. Band of catecholamine granules is invisible.

of mitochondria from the adrenal cortex and medulla. The bands containing catecholamine granules were below the visual bands of mitochondria in the centrifuge tube, but their spread was invisible (Fig. 2).

The specific gravity of catecholamine-containing granules in the adrenal medulla generally was heavier than mitochondria and varied widely from the specific gravity of 2.0 M sucrose to that of mitochondria. This diversity of the granules indicates the possibility that some physiological alteration always occurs among them: In other words, some granules are newly formed and have potent ability to synthesize catecholamines, while other granules mature and are ready for secretion.

INFLUENCE OF SECRETION UPON SUBCELLULAR POOLS OF CATECHOLAMINE IN RAT ADRENAL MEDULLA

Bertler *et al.* (1960) studied the content and intracellular distribution of A and NA in the adrenal medulla of rabbits at varying periods of time after injection of insulin. They estimated catecholamines in the large granule fraction, containing amine granules and mitochondria (sediment obtained by centrifugation at $3{,}800 \times g$ for 30 min) and cytoplasmic sap, containing the soluble cytoplasm and microsomes (supernatant obtained by previous centrifugation). This large granule fraction corresponds to Mit, and this cytoplasmic sap corresponds to S in the left side in Fig. 1. According to Bertler *et al.* (1960), the adrenal medulla of normal, untreated rabbits contained no, or very small amounts of, NA. Noradrenaline was very rapidly formed in the rabbit adrenal medulla when insulin-induced hypoglycemia stimulated to secretion. The newly synthesized NA was slightly recovered in the cytoplasmic sap but was mainly bound to particles in the large granule fraction. Large amounts of bound NA again accumulated during the recovery phase in spite of new-synthesis and restorage of A (Bertler *et al.* 1960). A slight induced increase of A in the supernatant indicated that subcellular pools of catecholamine in rabbit adrenal medulla were changed by administration of insulin (Bertler *et al.* 1960).

In many other animals, such as rats, NA is always present in the adrenal medulla. The turnover time of adrenal NA is 7 days (Brodie, Costa, Dlabac, Neff & Smookler, 1966) and the significant recovery of the catecholamine content was found 48 hr after insulin

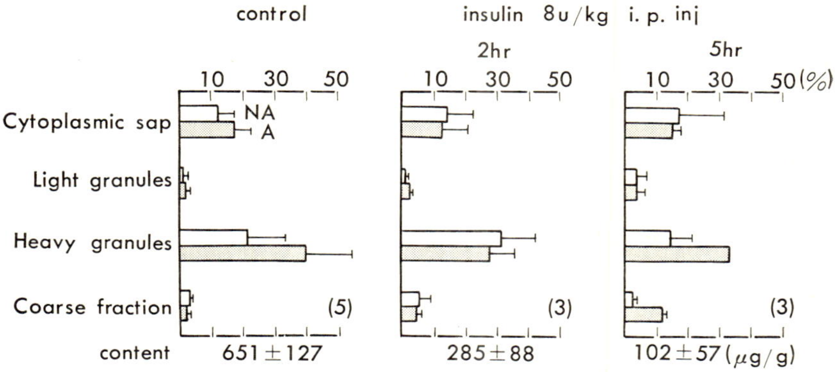

Fig. 3. Influence of insulin on the subcellular pools of catecholamines in rat adrenal. Cytoplasmic sap, supernatant (S); light granules, microsomal fraction (Mc); heavy granules, mitochondrial fraction (Mit); coarse fraction, nuclear fraction (N).

injection in rats (Yamada, 1968; Viveros, Arqueros, Connett & Kirshner, 1969*b*). It seems that there is a significant difference in the synthesis of catecholamine between rabbits and many other animals. Rabbits may be not suitable for investigation of the general relation between secretion and synthesis of catecholamines, since many other animals always contain NA in the adrenal and have not so rapid a turnover time of NA in the adrenal as rabbits.

The subcellular distribution of catecholamine in rat adrenal was investigated after administration of insulin. Although a relative increase of the catecholamine pool in the cytoplasmic sap might be expected from the previous experiment by Bertler *et al.* (1960), no remarkable change in the subcellular distribution in the rat adrenal occurred, in spite of the definite decrease in absolute content of catecholamine in the adrenal by insulin injection (Fig. 3). As stated above, the results suggest that catecholamines in heavy granules are not secreted through the cytoplasmic sap into the extracellular spaces, and resynthesis of catecholamine presumedly does not begin to become so active at the time of secretion.

The whole homogenate of adrenal from insulin-treated rats was carefully placed on the surface of the sucrose gradient (0.25–2.0 M and 1.0–2.0 M), and the tubes were centrifuged by the Spinco preparative ultracentrifuge at an average force of $60{,}000 \times g$ for varying periods of time. The distribution of the granules in the centrifuge tube shifted to the bottom.

This suggests that the lighter portion of catecholamine granules is

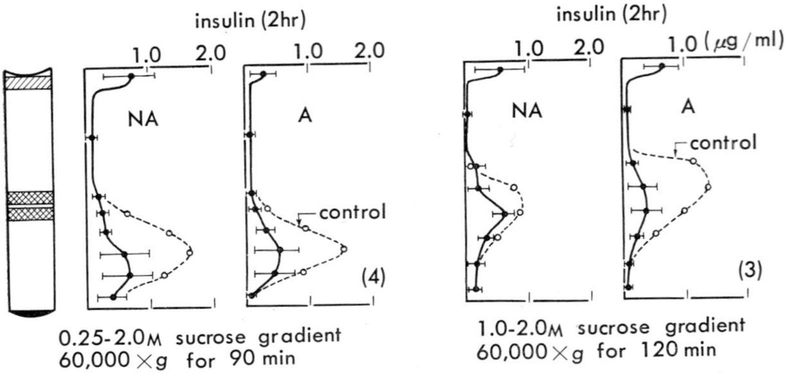

Fig. 4. Influence of insulin on the buoyant density of the catecholamine granules in adrenal. The distribution of the catecholamine granules shifted to the bottom of the tube, since the lighter part of granules was lost and the heavier part remained.

lost at the early stage of secretion but the heavier portion remains: In other words, the lighter portion of catecholamines in adrenal is a rapidly releasable pool and the heavier portion is a slowly releasable pool (Fig. 4).

The investigation of the rat adrenal by the same method as above disclosed that the catecholamine content did not begin to decrease at an early stage, that is, 60 min after injection of insulin. The distribution of adrenaline granules in the centrifuge tube slightly shifted to the lighter portion of the gradient.

It seems that the granules transform the property and are ready for secretion (Yamada, 1971) and that there are two pools in the adrenal medullary granules. The alteration in property of the granules increases the content of catecholamine in the rapidly releasing pool and decrease its content in the slowly releasing pool (Slotkin, Ferris & Kirshner, 1971).

It is a well known phenomenon that reserpine induces the depletion of the catecholamine content in many tissues. A single injection of reserpine decreases urinary excretion of NA at an early stage but does not affecte the excretion of A. However, urinary A excretion shows transient, remarkable increase at a later stage, and NA excretion slightly increases at that time (Yamada, 1959). Hyperglycemia induced by reserpine becomes manifest several hours after administration of the drug. The depletion of tissue content of catecholamine was found before a transient increase in urinary excretion of catecholamines. The rate of disappearance of catecholamine contents might correspond to the turnover rate of catecholamine in each tissue: For example, the rate of depletion of the content in the brain, heart, and stomach was far faster than in the adrenal medulla (Yamada, 1968*a*).

The uptake of dopamine and NA into chromaffine granules is blocked by reserpine. As a consequence of this inhibition of dopamine uptake, the conversion of dopamine to NA, which occurs within the granules, is also inhibited (Bertler, Hillarp & Rosengren, 1961; Kirshner, 1962). The persistent depletion of NA by small doses of reserpine is induced and maintained by the physical presence of minute amounts of very highly bound reserpine. Calculations reveal that about 20 molecules of reserpine persist for each affected catecholamine-containing granule (Alpers & Shore, 1969).

Although the reserpine treatment caused a marked decrease in the catecholamine content of the adrenal medulla, about the same intracellular distribution of the amine was found after injection of reserpine in rabbits (Hillarp, 1960*b*) and rats (Yamada, 1966, 1968*b*). The inhibition of the accumulation of catecholamines in cytoplasmic sap as a consequence of the blockade of uptake in granules by reserpine indicated that free catecholamines in cytoplasmic sap preferentially underwent oxidative deamination by monoamine oxidase (MAO).

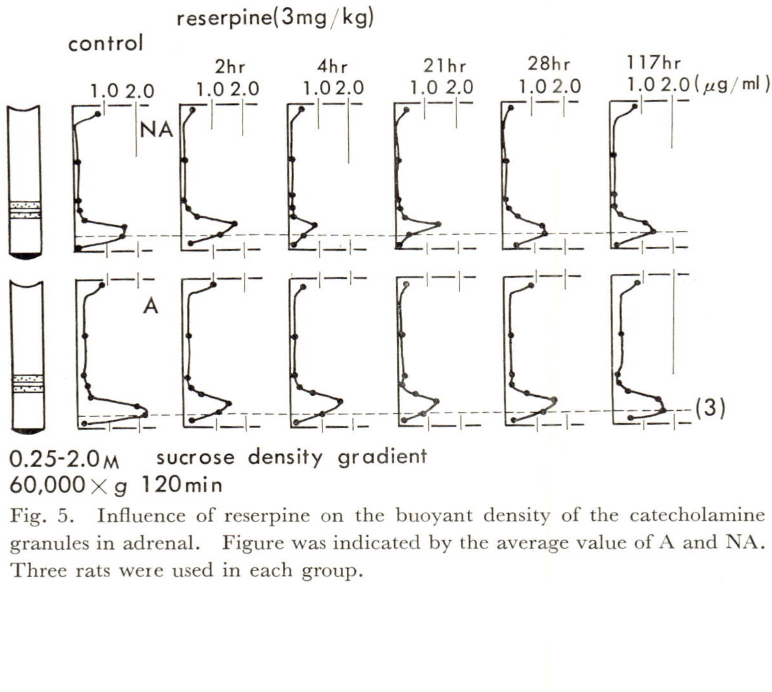

Fig. 5. Influence of reserpine on the buoyant density of the catecholamine granules in adrenal. Figure was indicated by the average value of A and NA. Three rats were used in each group.

The whole homogenate of adrenal from the reserpine-treated rat was centrifuged with the method of continuous density gradient of sucrose. In this experiment the author (Yamada, 1968*a*) showed that the catecholamine granules from the reserpine-treated rat equilibrated at a lower density than did granules from the control rats. In that paper the change of property of the granules is studied in rats killed at varying periods of time after injection of reserpine. This treatment induced a clear decrease in the buoyant density of the granules for 28 hr and it was completely recovered 117 hr after administration of reserpine (Fig. 5). Similar results were found in the study of reserpine-treated rabbits (Viveros, Arqueros & Kirshner, 1969, 1971*b*).

Neither insulin nor reserpine altered the subcellular pools of catecholamines in the adrenal medulla but they did change the buoyant density of the catecholamine granules. Further studies were made in an *in vitro* experiment to reveal the meaning of the change of the buoyant density. When the adrenal glands were kept frozen, or for too long in cold storage or in a warm bath, catecholamine granules were destroyed and lost their contents and the liberated catecholamines were found in the cytoplasmic sap. However, the amount of catecholamines in the coarse fraction (the heaviest granules) was little changed; that is, the heaviest granules were resistant to storage in bad conditions (Yamada & Oda, 1970).

After partial destruction by the incubation of adrenal homogenate

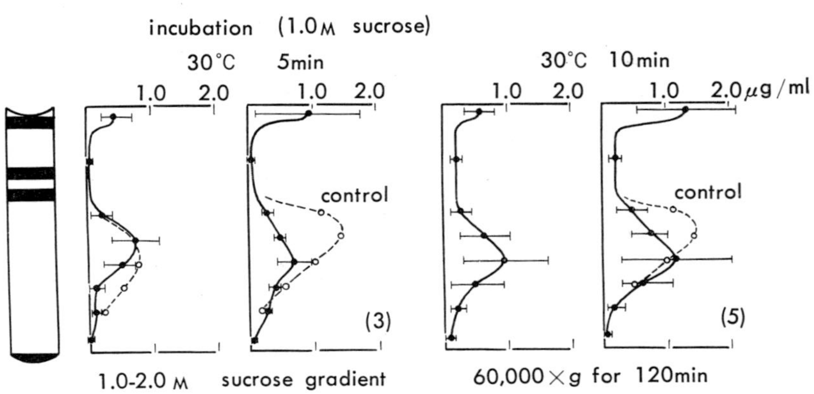

Fig. 6. Influence of partial destruction on the buoyant density of the catecholamine granules. After partial destruction by the incubation of adrenal homogenates in a bath of 1.0 M sucrose kept at 30°C for 5 or 10 min, the homogenates were centrifugated.

in a bath of 1.0 M sucrose kept at 30°C for 5 or 10 min, the homogenate was investigated by the method of continuous density gradient of sucrose (Yamada, 1971). The results are shown in Fig. 6. The lighter part of the catecholamine granules was lost, but the heavier part was not destroyed. The alteration of buoyant density of the catecholamine granules was similar to that after administration of insulin (Fig. 4).

The results obtained from experiments with insulin *in vivo* and partial destruction *in vitro* suggest that the lighter part of the catecholamine granules may correspond to the rapid releasing pool, and the heavier part may correspond to the slowly releasing pool. Slotkin *et al.* (1971) found two pools of catecholamine in isolated bovine adrenal medullary granules, and one was named the rapid releasing pool and the other the slowly releasing pool.

INFLUENCE OF BIOSYNTHESIS UPON SUBCELLULAR POOLS OF CATECHOLAMINE IN RAT ADRENAL MEDULLA

The adrenal medullary cell that synthesizes catecholamines receives L-tyrosine from the blood and converts it to L-DOPA by tyosine hydroxylase (Nagatsu, Levitt & Udenfriend, 1964). Tyrosine hydroxylase is the key enzyme and this hydroxylation is the rate-limiting step in catecholamine formation. Alpha-methyl-para-tyrosine (α-MT) is a potent inhibitor of tyrosine hydroxylase and lowers endogenous levels of catecholamines by a blockade of synthesis at the tyrosine hydroxylase step (Spector, Sjoerdsma & Udenfriend, 1965).

The subcellular distribution of catecholamines in adrenal from α-MT-treated rats was studied by different types of centrifugation. As mentioned above, the subcellular pools were divided into 4 fractions: a coarse fraction, heavy granules, light granules, and cytoplasmic sap.

The alteration of catecholamine content in each fraction was investigated at various times after treatment with α-MT. Although the content of catecholamine in rat adrenal slightly decreased, the subcellular distribution of catecholamine was significantly affected by α-MT. The NA pool in heavy and light granules increased, and the A pool in heavy granules and cytoplasmic sap decreased. The NA pool in cytoplasmic sap showed a tendency to decrease. The most remarkable change in the subcellular distribution was a lowering of total catecholamine (A + NA) in cytoplasmic sap; the percentage reached one-third of the normal level after injection of α-MT. The

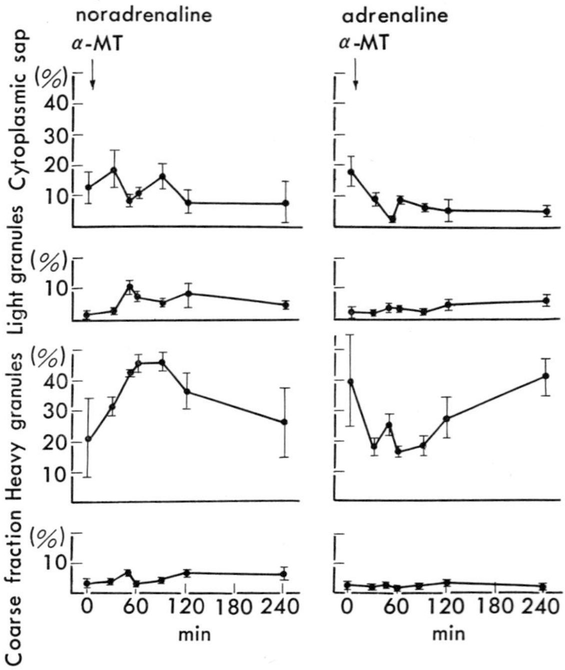

Fig. 7. Alteration of subcellular distribution of catecholamines in rat adrenal after intraperitoneal injection of α-methyl-*p*-tyrosine (200 mg/kg). The number of rats is 5 at control and 60 min, 6 at 120 min, and 3 at other periods.

duration of the significant alteration was 120 min, and the tendency to be normalized appeared 240 min, after injection of α-MT (Fig. 7).

Fuller & Hunt (1967) suggested that the A content of the adrenals normally suppresses the activity of phenyl-ethanolamine-N-methyl-transferase, but that increased production of NA will overcome the inhibitory effect and therefore stimulate renewed synthesis of A.

It seems that liberation of NA from the granules into cytoplasmic sap is suppressed by decrease of synthesis of NA after injection of α-MT, and therefore renewed synthesis of A is not stimulated. Uptake of A in heavy granules also decreases while A is secreted from adrenal by exocytosis. As a result of the reduction of A, the relative increase of NA in the heavy granules is found in the subcellular distribution of catecholamines in the adrenal medulla. In brief, the decrease of the catecholamine pool in cytoplasmic sap is the most important and significant effect of suppression of catecholamine synthesis.

L-DOPA is not influenced by tyrosine hydroxylase. The DOPA

decarboxylase is present in the adrenal medulla but DOPA itself was never seen in experiments with radioactive precursors. This is explained by the high rate of conversion of DOPA to dopamine (Holtz, 1959). Administration of L-DOPA causes a significant increase in the contents of catecholamine in tissues.

Some alterations were found in the subcellular distribution of catecholamine in adrenal after L-DOPA injection. Thirty min after an L-DOPA injection, the relative amount of NA increased in the cytoplasmic sap. The pool of NA in the heavy granules began to be restored and increase again 45 min or more after injection of L-DOPA. Since the absolute quantity of NA excessively increased and exceeded that of A, especially at the early stages, the pool of A in heavy granules and cytoplasmic sap indicated the relative diminution. Catecholamines in the coarse fraction (the heaviest part of the granules) moderately increased after administration of L-DOPA (Fig. 8).

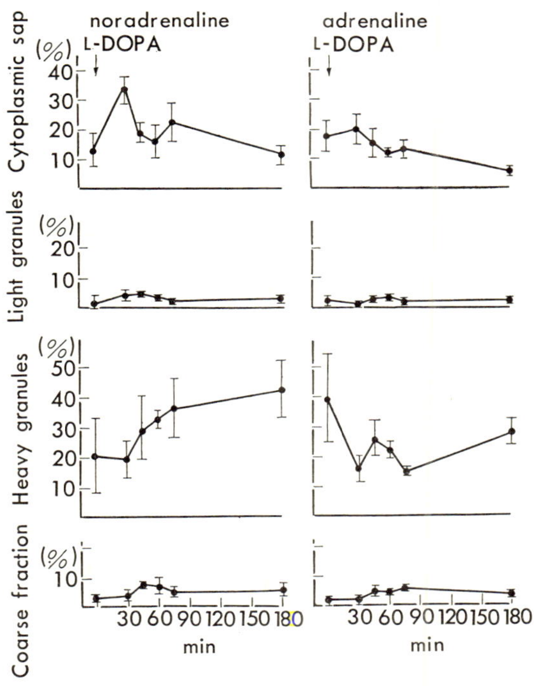

Fig. 8. Alteration of subcellular distribution of catecholamines in rat adrenal after intraperitoneal injection of L-DOPA (100 mg/kg). A number of rats is 4 at 45 min and 3 at other period.

The properties of the catecholamine granules were investigated by the use of the sucrose continuous density gradient in adrenal from L-DOPA-treated rats. Most of the granules showed a tendency to shift to near the bottom of the centrifuge tube. The increase of buoyant density of the catecholamine granules probably corresponds to an increase of catecholamines in the coarse fraction (Fig. 9).

Catecholamines are destroyed by catechol-*o*-methyl transferase (COMT) in cytoplasmic sap after liberation, but stored catecholamines are gradually metabolized by monoamine oxidase (MAO) in mitochondria. COMT may act to inactivate free catecholamine,

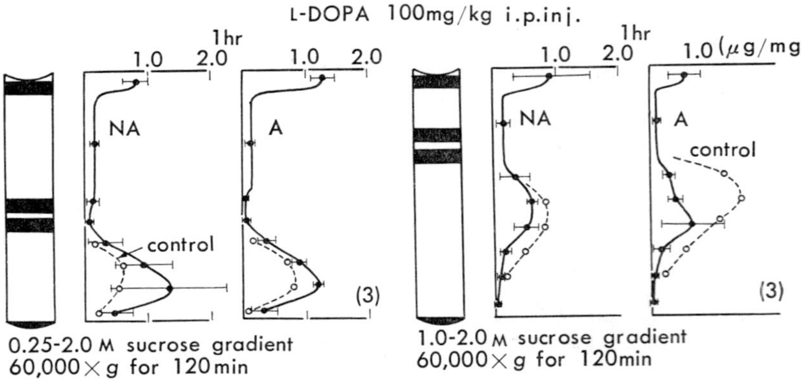

Fig. 9. Influence of L-DOPA on the buoyant density of catecholamine granules in rat adrenal.

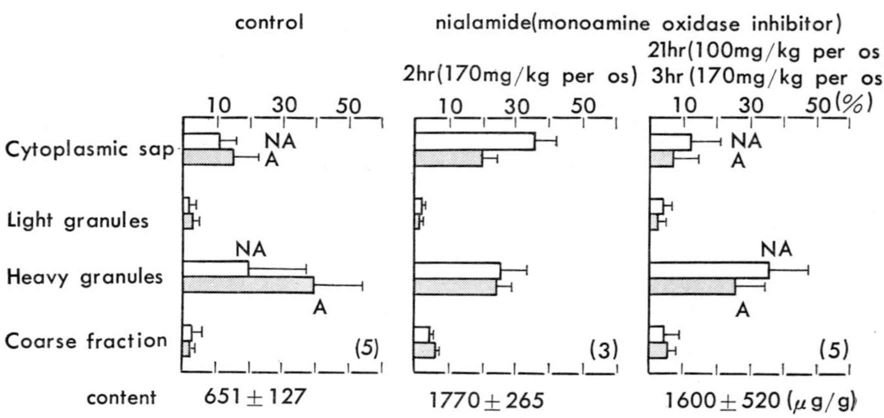

Fig. 10. Influence of MAO inhibitor on subcellular distribution of catecholamines in rat adrenal. Nialamide was used as MAO inhibitor.

and MAO may regulate the amount of stored catecholamines. A MAO inhibitor caused the catecholamine content in adrenal to increase. The subcellular distribution of catecholamines in adrenal showed, in the early stages, an increase in the cytoplasmic sap, a decrease in the heavy granules, and a slight increase in the coarse fraction. This phenomenon resembled the conditions found after L-DOPA injection (Fig. 10).

In rabbit adrenal, too, very large amounts of dopamine and NA rapidly (within 30 min) appeared after injection of L-DOPA. The newly formed NA was partly recovered in the fraction of crude cytoplasmic sap, in which the microsomal fraction (light granules) and cytoplasmic sap were included as stated above. The newly formed dopamine was rapidly incorporated in the heavy granules (Bertler *et al.* 1960). Since, as shown in Fig. 8, NA is synthesized from dopamine in the granules and liberated into the cytoplasmic sap very rapidly, the newly formed dopamine must be rapidly taken up in the granules after injection of L-DOPA.

Noradrenaline is a much poorer substrate for MAO than dopamine (Weiner, 1960), and dopamine is much more rapidly deaminated by this enzyme than NA. Twenty-five percent of the infused dopamine was synthesized into catecholamine (Goodal & Alton, 1968). A large proportion of dopamine is catabolized by MAO, and MAO inhibitors prevent the rapid destruction of dopamine. The dopamine accumulated in cytoplasmic sap by MAO inhibitors is naturally taken up in the catecholamine granules and then transformed into NA. The newly formed NA is again extruded into cytoplasmic sap. Both L-DOPA and MAO inhibitors increased the catecholamine content in tissues and caused the accumulation of a NA pool in the cytoplasmic sap.

The catecholamine pool in the coarse fraction showed a tendency to increase after administration of each agent. In the density gradient experiment, the catecholamine granules shifted toward the bottom of the centrifuge tube. The granules with heavier specific gravity increased after biosynthesis of catecholamine was enhanced by these agents. As shown by the experiment with insulin (Fig. 4) and partial destruction (Fig. 6), the lighter portion of the granules was easily released, and the heavier portion of the granules and the coarse fraction were difficult to release or catabolize.

These results indicate that the subcellular catecholamine pool in the adrenal is not affected by acceleration of catecholamine secretion but by the change of biosynthesis of catecholamines. The pool of catecholamines in the cytoplasmic sap in adrenal is connected with catecholamine synthesis but not with the secretion mechanism. The lighter portion of the catecholamine pool in the granules may be a rapidly releasable particle bound form, and the heavier portion may be a slowly releasable particle bound form, which presumedly has a more profound relation to biosynthesis of catecholamines than the lighter portion.

COMPARISON OF SUBCELLULAR POOLS OF CATECHOLAMINES IN PHEOCHROMOCYTOMAS AND NEUROBLASTOMAS TO POOLS IN THE ADRENAL MEDULLA

There have been a few studies of catecholamine secretion and storage in human pheochromocytoma published. However, most of them studied a limited number of tumors. The content and secretion of catecholamines in pheochromocytoma show great variety, and the macroscopic and microscopic appearance of pheochromocytoma shows many different cellular patterns. Indeed, I have seen a great variation in the consistency, from a soft sponge-like ball to a solid rubber-like ball, of about 50 pheochromocytomas collected for 17 years. A large number of pheochromocytomas must be investigated to understand the release and storage of catecholamines in these tumors. It seems necessary to compare the subcellular pools of catecholamines in the adrenal medulla to the pools in pheochromocytomas. The purpose of the present study is to reveal the relation between secretion or biosynthesis of catecholamines and the subcellular pools, since various alterations of that relation may be naturally present in pheochromocytoma.

Relation between urinary excretion of catecholamines and the storage of the amines in pheochromocytoma

It is questionable whether the cells of pheochromocytoma are characterized by a high rate of secretion of catecholamines compared with normal adrenal medullary cells. The content of A and NA in tumors varies widely and, in some of them, greatly exceeds that in the adrenal medulla. The amount of urinary excretion of catecholamines never corresponded to the content, which was expressed as micrograms per gram weight of tissue (Fig. 11).

Although pheochromocytomas have a high content of catecholamines per gram wet weight, and the weight of tumors are much

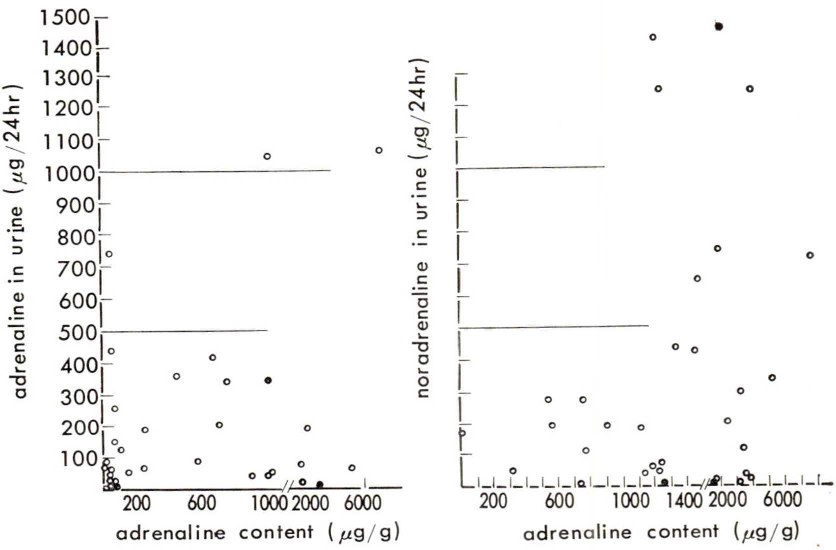

Fig. 11. Relation between urinary excretion of catecholamines and the contents of tumors (32 cases of pheochromocytoma). ●, Malignant; ○, beneign.

heavier than it of adrenal medulla, the amount of NA excreted in urine is very small compared with the value calculated from the secretion rate of the adrenal medulla. It seems, therefore, that most pheochromocytomas are characterized by a slow rate of secretion of catecholamines and by a large amount of storage, which may be induced by uncontrolled excessive synthesis of catecholamines (Roth, Stjärne, Levine & Girman, 1968) or by any abnormal storage mechanism for the hormone.

Subcellular distribution of catecholamines in pheochromocytoma and the properties of catecholamine granules

Most catecholamines in pheochromocytomas were found in the heavy granules, and the amines formed a complex with ATP. This evidence is similar to that regulary found in the adrenal medulla, although some differences between the tumors and the adrenal were found by some authors (Stjärne, Euler & Lishajko, 1964; Winkler, Ziegler & Strieder, 1967; Winkler & Smith, 1968; Blaschko, Jerrome, Robb-Smith, Smith & Winkler, 1968). In these reports, however, attention was directed only to the pool in heavy granules, and other pools were practically left out of consideration.

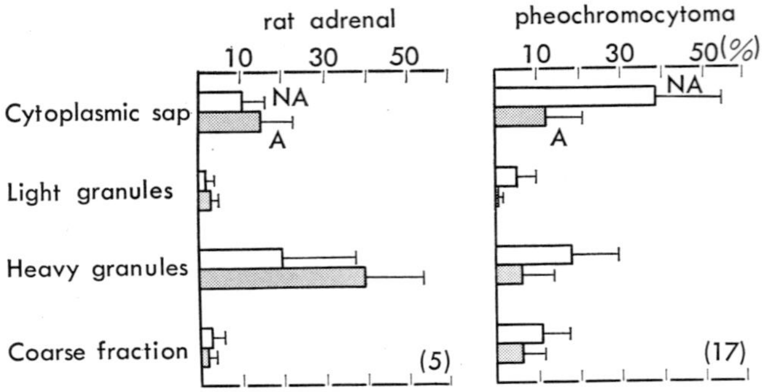

Fig. 12. Subcellular distribution of catecholamines in 17 pheochromocytomas.

The subcellular pools of 17 pheochromocytomas were compared with those of the adrenal medulla. The amounts of catecholamines (mainly NA) were greater in the cytoplasmic sap than in the heavy granules in pheochromocytomas. The catecholamine pools in both light granules and the coarse fraction consisted of a greater share of the subcellular distribution in the tumors than in adrenal (Fig. 12). Of course, a large part of catecholamines was found in the heavy granules in some pheochromocytomas, which were kept in good condition, but a small amount of catecholamines in cytoplasmic sap was present.

An increase of catecholamines in cytoplasmic sap of tumors might have occurred partly because of breakdown of the cell before or after the operation. The subcellular distribution of catecholamines in Fig. 12 resembles that of the frozen adrenals which were partially destroyed. Storage of tissues in cold for a long period of time did not induce remarkable changes in the catecholamine pool in the coarse fraction but did induce a marked increase in the pool within cytoplasmic sap. Catecholamine pools in the coarse fraction of tumors, too, were in a slowly releasable particle bound form.

An alternative explanation for the catecholamine increase in cytoplasmic sap of the tumor is the excessive synthesis of catecholamines. As stated above, the catecholamine pool in cytoplasmic sap accumulated in rats and rabbits by overproduction of the amine after administration of L-DOPA and MAO inhibitor. The increase of catecholamines in the coarse fraction (the heaviest granules) was found in both pheochromocytoma and L-DOPA-treated rats. The alteration of the pool in cytoplasmic sap may be brought about as a result of the above two mechanisms, partial destruction and excessive

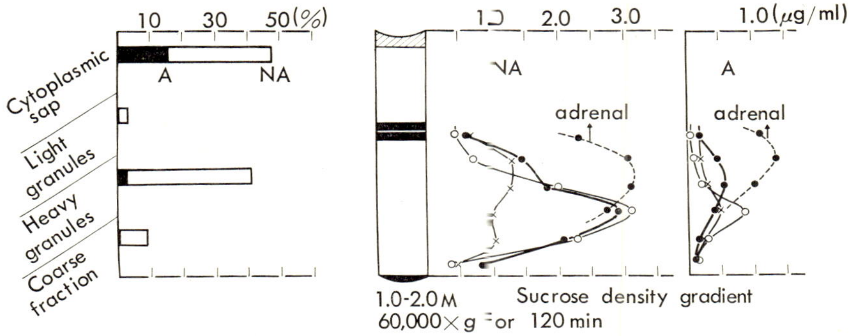

Fig. 13. Subcellular distribution of catecholamines in pheochromocytomas and the buoyant density of the catecholamine granules. Samples obtained from different parts of the same tumor did not always show the same pattern of the distribution or density.

synthesis. Therefore, it became of interest to see whether the buoyant desity of the catecholamine granules in pheochromocytoma is similar to that of the adrenal granules obtained from rats treated with L-DOPA.

The homogenate of pheochromocytoma was divided into two portions. One was used for the investigation with differential centrifugations, and the other was used for ultracentrifugation with the continuous density gradient of sucrose. The subcellular distribution of catecholamines in the tumor is shown in the left half of Fig. 13, and the buoyant density of the granules is shown in the right half of Fig. 13. The subcellular distribution was similar to that in adrenal obtained from rats treated with L-DOPA or MAO inhibitor (Fig. 13). The subcellular pool of catecholamines and the properties of catecholamine granules in pheochromocytoma are similar to those in rat adrenal, that is, secretion of catecholamines does not increase, but synthesis of the amines is accerelated.

Subcellular distribution of catecholamines in neuroblastomas

Some of the patients with neuroblastoma excrete abnormally large quantities of catecholamine metabolites in their urine. In contrast with pheochromocytoma, patients with neuroblastoma do not exhibit significant clinical signs of catecholamine oversecretion even in patients whose daily excretion of catecholamine metabolite exceeds the range usually associated with pheochromocytoma

(Voorhess & Gardner, 1962; Voorhess, 1968). Catecholamine metabolites have no biological activity, and so these secretions do not induce any clinical signs.

In my experience with 14 cases of neuroblastoma, all daily excretion of A and NA remained within normal, but some cases showed high excretion of metaadrenaline (including normetaadrenaline) without any clinical signs due to physiological effects of catecholamines.

In our previous paper, it was reported that these tumors contained catecholamine and cytoplasmic osmiophilic granules that resembled adrenal medullary and pheochromocytoma granules. The contents varied greatly from case to case (Yokoyama, Okada, Tokue, Takayasu & Yamada, 1971). The catecholamine found in neural tumors was present in a rather large concentration in the granule fraction (sediment after centrifugation 14,000 × *g* for 30 min) (Page & Jacoby, 1964).

The subcellular distribution of catecholamines was studied in 2 cases of neuroblastoma. One patient had a large amount of catecholamines in the tumor, and the urinary excretion of dopamine (DA), NA, and A was normal and that of metaadrenaline exceed the normal range. The other patient had a small amount of catechol-

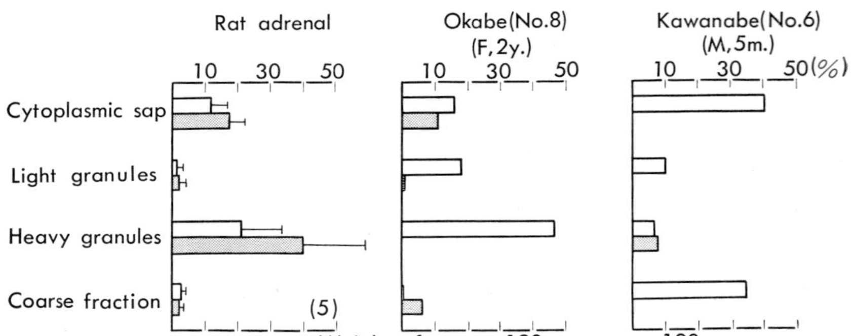

		Okabe (No.8)	Kawanabe (No.6)
Weight of tumor		190g	120g
Content µg/g	DA	0	0
	NA	256	12
	A	51	1.03
	MA	3	0
Excretion µg/day	DA	60	2.6
	NA	1.4	19.2
	A	1.1	3.3
	MA	330	15

Fig. 14. Subcellular distribution of catecholamines in neuroblastomas. Weight of tumor, content of catecholamines and urinary excretion are shown under the figure.

amines in the tumor. The urinary excretion of catecholamines and their derivates was normal. The subcellular pool of catecholamines in the tumor of the first case, presented in the middle in Fig. 14, was similar to that of adrenal except for a considerable accumulation of NA in light granules. The pool in the tumor of the second case, presented in the right side in Fig. 14, resembled that of some pheochromocytomas, that were partially destroyed. In this case the content of catecholamines decreased, and NA accumulated in the cytoplasmic sap. NA in the coarse fraction was firmly held (Fig. 14).

It may be concluded from the results from neuroblastoma that the lack of secretory ability of catecholamines has no relation to the variation of their subcellular pools.

THE ROLE OF SUBCELLULAR CATHECHOLAMINE POOLS IN ADRENAL MEDULLA

A number of previous biochemical studies on the secretory mechanism were made by *in vitro* experiments on the efflux of catecholamines of the adrenal into perfusate. However, biochemical investigations *in vivo* are very few. Most laboratories have investigated the nature of catecholamine granules and have been fascinated by the granules. Thus, other catecholamine pools have been left out of consideration. Therefore, it may be interesting to study the role of each subcellular pool of catecholamines in relation to the mechanisms of secretion, storage, and biosynthesis of the amine.

Role of the catecholamine pool in the coarse fraction (the heaviest granules)

The catecholamine content in the coarse fraction increases at the time of acceleration of biosynthesis of the amine in adrenal or excessive synthesis in tumors, and catecholamines in this fraction are rarely liberated. The buoyant density of the catecholamine granules becomes heavier on accerelation of biosynthesis of catecholamines. Thus, the coarse fraction is likely to include a type of catecholamine granules that has relation to biosynthesis of the amine.

Role of the catecholamine pool in the mitochondial fraction (heavy granules)

The largest amount of catecholamines in adrenal was contained in the mitochondrial fraction. The depletion of the whole catecholamine content in the adrenal was mainly due to decrease of catecholamine in this fraction. The remarkable alteration of its content occurred following a change in biosynthesis, for example, due to administration of L-DOPA or MAO inhibitor. The lighter portion of the heavy granules was rapidly lost by the accerelation of secretion of catecholamines. The heavier portion significantly increased by the excessive production of the amines. The lighter portion of the heavy granules, therefore, would have a relation preferably with secretion, and the heavier portion may have a relation with biosynthesis of catecholamines.

Role of the catecholamine pool in light granules

In other organs, *i.e.*, the heart, vas deferens, submaxillary gland, pineal gland and brain, most parts of catecholamines (mainly NA) are included in the light granules. But the adrenal medulla contains only small amounts of catecholamine in this fraction. The percentage of catecholamine in this fraction increased after administration of α-MT (inhibition of catecholamine biosynthesis) before the depletion of the whole catecholamine content in the adrenal medulla. However, the role of this pool remains uncertain.

Role of the catecholamine pool in cytoplasmic sap

The pool of catecholamines has been regarded as a functional pool even before the secretory mechanism of catecholamine from the adrenal medulla was disclosed. Since it became clear that the secretion was done by release of the soluble content of the storage granules directly to the exterior of the cell (not across the cytoplasm), the pool of catecholamines in cytoplasmic sap has been thought to be unimportant and even to be an artificial one made by vigorous homogenation. On the other hand, it has been found that the biosynthetic process of dopamine from tyrosine occurs in this cytoplasmic sap, and also in this space, NA is converted into A. Noradrenaline in cytoplasmic sap suppresses the activity of tyrosine hydroxylase and regulates the synthesis of catecholamines (Stjärne, Lishajko & Roth, 1967). Moreover, the rate of synthesis of A depends upon the concentration of NA in the cytoplasm, as mentioned above.

Indeed, accumulation of NA in cytoplasmic sap was found during an increase of catecholamine synthesis after administration of L-DOPA or MAO inhibitor, and the NA content in this pool was depleted by suppression of the synthesis after injection of α-MT. These results suggest that NA in this pool plays a very important role in the regulation of catecholamine biosynthesis.

THE ROLE OF THE SUBCELLULAR CATECHOLAMINE POOLS IN PHEOCHROMOCYTOMA AND NEUROBLASTOMA

It seems to come into question whether the alterations of subcellular catecholamine distribution in adrenal medulla induced by experiments are present in physiological conditions. Since excessive production of catecholamines was found in pheochromocytoma and neuroblastoma, the tumors may be suitable for comparison with the experiments with adrenal after administration of L-DOPA and MAO inhibitor.

A number of similar changes in the subcellular distribution of catecholamine pools adrenal were found in tumors and experiments. For example, it was confirmed by investigation of pheochromocytoma that the heavier part of the heavy granules (slowly releasable pool in granules) existed in physiological conditions, and a part of them (rapidly releasable pool in granules) was lost after partial destruction in tumors.

The results obtained by experiments with rats conformed with the findings in pheochromocytoma. This suggests that excessive catecholamine synthesis is found in pheochromocytoma and, as a result, the subcellular distribution of catecholamine in these tumors shows some difference from that of normal adrenal medulla.

REFERENCES

Alpers, H. S. & Shore, P. A. (1969). Specific binding of reserpine-association with norepinephrine depletion. *Biochem. Pharmac.* **18**, 1363–1372.

Bertler, A., Hillarp, N. Å. & Rosengren, E. (1960). Some observations on the synthesis and storage of catecholamines in the adrenaline cells of the suprarenal medulla. *Acta physiol. scand.* **50**, 124–131.

Bertler, A., Hillarp, N. Å. & Rosengren, E. (1961). Effect of reserpine on the storage of new-formed catecholamines in the adrenal medulla. *Acta physiol. scand.* **52**, 44–48.

Blaschko, H. & Welch, A. D. (1953). Localization of adrenaline in cytoplasmic particles of the bovine adrenal medulla. *Arch. exp. Path. Pharmak.* **219**, 17–22.

Blaschko, H., Comline, R. S., Schneider, F. H., Silver, M. & Smith, A. D. (1967). Secretion of chromaffine granule protein, chromograin, from the adrenal gland after splanchnic stimulation. *Nature, Lond.* **215** 58–59.

Blaschko, H., Jerrome, D. W., Robb-Smith, A. H. T., Smith, A. D. & Winkler, H. (1968). Biochemical and morphological studies on catecholamine storage in human pheochromocytoma. *Clin. Sci.* **34**, 453–465.

Brodie, B. B., Costa, E., Dlabac, A., Neff, N. H. & Smookler, H. H. (1966). Application of steady state kinetics to the estimation of synthesis rate and turnover time of tissue catecholamines. *J. Pharmac. exp. Ther.* **154**, 493–498.

Burack, W. R., Avery, E., Draskoczy, R. & Weiner, N. (1962). The number of catecholamine granules in adrenal glands. *Biochem. Pharmac.* **9**, 85–91.

Douglas, W. W. & Poisner, A. M. (1966). Evidence that the secreting adrenal medullary chromaffine cell releases catecholamines directly from ATP-rich granules. *J. Physiol.* **183**, 236–248.

Eade, N. R. (1958). The distribution of the catecholamines in homogenates of the bovine adrenal medulla. *J. Physiol.* **141**, 183–192.

Euler, U.S.v. (1946). The presence of a substance with sympathin E properties in spleen extracts. *Acta physiol. scand.* **11**, 168–186.

Euler, U. S. v. & Lischajko, F. (1961). Effect of reserpine on the release of catecholamines from isolated nerve and chromaffine cell granules. *Acta physiol. scand.* **52**, 137–145.

Falck, B., Hillarp, N. Å & Högberg, B. (1956). Content and intercellular distribution of adenosine-triphosphate in cow adrenal medulla. *Acta physiol. scand.* **36**, 360–376.

Fuller & Hunt (1967). *Life Sci.* **6**, 1107. Cited by Sandler, M. & Ruthven, C. R. J. (1969). *Prog. Med. Chem.* **6**, 200–265.

Goodall, McC. & Alton, H. (1968). Metabolism of 3-hydroxytryramine (dopamine) in human subjects. *Biochem. Pharmac.* **17**, 905–917.

Gutman, Y. & Weil-Malherbe, H. (1967). Subcellular distribution of norepinephrine in uteri of some species. *Nature, Lond.* **214**, 108–109.

Hagen, P. & Barrnet, R. J. (1960). The storage of amines in the chromaffine cell. A Ciba Foundation Symposium on Adrenergic Mechanisms., ed. Vane, J. R., Wolstenholme, G. E. W. & O'Connor, M., pp. 83–99. London: Chuchill.

Hillarp, N-Å. (1960*a*). Different pools of catecholamines stored in the adrenal medulla. *Acta physiol. scand.* **50**, 8–22.

Hillarp, N. Å. (1960*b*). Effect of reserpine on the adrenal medulla of sheep. *Acta physiol. scand.* **49**, 376–382.

Holtz, P. & Schümann, H. J. (1948). Arterenolein neues Hormon des Nebennierenmarks. *Naturwissenschaften* **5**, 159.

Holtz, P. (1959). Role of L-DOPA decarboxylase in the biosynthesis of catecholamines in nervous tissue and the adrenal medulla. Symposium on Catecholamines, Amer. Soc. Pharmacol. Exp. Ther., pp. 317–329. Baltimore: Williams and Wilkinson.

James, W. O. (1948). Demostration and separation of noradrenaline, adrenaline and methyladrenaline. *Nature, Lond.* **161**, 851–852.

Kirshner, N. (1962). Uptake of catecholamines by a particulate fraction of the adrenal medulla. *J. biol. Chem.* **237**, 2311–2317.

Kirshner, N., Sage, H. J., Smith, W. G. & Kirshner, A. G. (1966). Release of catecholamines and specific protein from adrenal glands. *Science, N.Y.* **154**, 529–531.

Kirshner, N. & Viveros, O. H. (1970). Quantal aspects of the secretion of catecholamines and dopamine-β-hydroxylase from the adrenal medulla. Bayer-Symposium. II, ed. Schümann H. J. & Kroneberg, G., pp. 78–88. Berlin, Heiderberg, New York: Springer-Verlag.

Laduron, P. & Belpaire, F. (1968). Intracellular distribution patterns of tyrosine hydroxylase, DOPA decarboxylase, dopamine-β-hydroxylase, phenylethanolamine-N-methyltransferase and monoamine oxidase in adrenal medulla. *Biochem. Pharmac.* **17**, 1127–1140.

Nagatsu, T., Levitt, M. & Udenfriend, S. (1964). Tyrosine hydroxylase-The initial step in norepinephrine biosynthesis. *J. biol. Chem.* **239**, 2910–2917.

Page, L. B. & Jacoby, G. A. (1964). Catecholamine metabolism and storage granules in peochromocytoma and neuroblastoma. *Medicine* (Balt) **43**, 379–386.

Potter, L. T. & Axelrod, J. (1963). Subcellular localization of catecholamines in tissues of the rat. *J. Pharmac. exp. Ther.* **142**, 291–298.

Roth, R. H., Stjärne, L., Levine, R. J. & Girman, N. J. (1968). Abnormal regulation of catecholamine synthesis in pheochromocytoma. *J. Lab. clin. Med.* **72**, 397–403.

Slotkin, T. A., Ferris, R. M. & Kirshner, N. (1971). Compartmental analysis of amine storage in bovine adrenal medullary granules. *Mol. Pharmacol.* **7**, 308–316.

Spector, S., Sjoerdsma, A. & Udenfriend, S. (1965). Blockade of endogenous norepinephrine synthesis by α-methyl-tyrosine, an inhibitor of tyrosine hydroxylase. *J. Pharmac. exp. Ther.* **147**, 86–95.

Stjärne, L., Euler, U. S. von & Lishajko, F. (1964). Catecholamine and nucleotides in pheochromocytoma. *Biochem. Pharmac.* **13**, 809–818.

Stjärne, L., Lishajko, F. & Roth, R. H. (1967). Regulation of noradrenaline biosynthesis in nerve tissue. *Nature, Lond.* **215**, 770–772.

Viveros, O. H., Arqueros, L., Connett, R. J. & Kirshner, N. (1969*a*). The fate of the storage vesicles following insulin and reserpine administration. *Mol. Pharmacol.* **5**, 69–82.

Viveros, O. H., Arqueros, L., Connett, R. J. & Kirshner, N. (1969*b*). The fate of the storage vesicles following insulin and reserpine administration. *Mol. Pharmacol.* **5**, 69–82.

Viveros, O. H., Arqueros, L. & Kirshner, N. (1969). Quantal secretion from adrenal medulla: All-or-none release of storage vesicle content. *Science, N.Y.* **165**, 911–913.

Viveros, O. H., Arqueros, L. & Kirshner, N. (1971*a*). Effect of insulin administration on the buoyant density, dopamine β-hydroxylase and catecholamine content of adrenal storage vesicles. *Mol. Pharmacol.* **7**, 444–454.

Viveros, O. H., Arqueros, L. & Kirshner, N. (1971*b*). Effect of reserpine on the dopamine β-hydroxylase and catecholamine content and on the buoyant density of adrenal storage vesicles. *Mol. Pharmacol.* **7**, 434–443.

Voorhess, M. L. (1968). The catecholamines in tumor and urine from patients with neuroblastoma, ganglio-neuroblastoma and pheochromocytoma. *J. Ped. Surg.* **3**, 147–148.

Voorhess, M. L. & Gardner, L. I. (1962). Studies of catecholamine excretion by children with neural tumors. *J. clin. Endocr. Metab.* **22**, 126–133.

Weiner, N. (1960). Substrate specificity of brain amine oxidase of several mammals. *Arch. Biochem.* **91**, 182–188.

Winkler, H. (1969). Isolierung und Charakterisierung von chromaffinen Noradrenalin-Granula aus Schweine-Nebennierenmark. *Naunyn-Schmiedebergs Arch. Pharmak. Exp. Path.* **263**, 340–357.

Winkler, H., Ziegler, E. & Strieder, N. (1967). Katecholaminspeichernde Granula eines Phäochromocytos. *Klin. Wochenschr.* **24**, 1238–1241.

Winkler, H. & Smith, A. D. (1968). Catecholamines in pheochromocytoma, normal storage but abnormal release? *Lancet* **13**, 793–795.

Yamada, R. (1959). Studies of adrenaline and noradrenaline in blood and urine. *J. Jap. Soc. Internal. Med.* **48**, 1097–1127 (in Japanese).

Yamada, R. (1966). Biological active amine and hypertension. *Metab. Disease* **3**, 724–730 (in Japanese).

Yamada, R. (1967). Hormone—(adrenomedullary hormone)—. *Metab. Disease* **4**, 608–614 (in Japanese).

Yamada, R. (1968*a*). Catecholamine metabolism in adrenal medulla. *Metab. Disease* **5**, 607–612 (in Japanese).

Yamada, R. (1968*b*). Autonomic nervous system and catecholamine. *The Autonomic Nervous System.* **5**, 14–17 (in Japanese).

Yamada, R. & Oda, T. (1970). Catecholamine granules (pheochromocytoma, neuroblastoma and adrenal). *Proc. Symp. Chem. Physiol. Path.* **10**, 22–26.

Yamada, R. (1971). The investigated method of catecholamine metabolism. *Sōgorinshō* **20**, 1148–1158 (in Japanese).

Yokoyama, M., Okada, K., Tokue, A., Takayasu, H. & Yamada, R. (1971). Ultrastructual and biochemical study of neuroblastoma and ganglioneuroblastoma. *Invest. Urol.* **9**, 156–164.

NEURAL CONTROL OF ENDOCRINE GLANDS

INNERVATION OF PANCREATIC ISLET CELLS

Nakazo Watari

Nervous elements in the pancreatic tissue, especially in the islets of Langerhans, are reviewed and summarized. Electron microscopically, the nerve endings are divided into four types: Type 1-a containing only synaptic vesicles and Type 1-b containing a variable number of large-cored vesicles along with synaptic vesicles. Types 1-a and -b nerve endings may be cholinergic (parasympathetic) and come in contact with A cells in the pancreatic islets of the rat, bat, and yellow-tail fish. Type 2-a is characterized by small-cored vesicles, as well as some large-cored ones, and by coming in contact with B cells of the monkey, bat, dog and snake. The fourth type of nerve ending is Type 2-b containing small vesicles with a crystalline core, which comes in contact with the wall of the blood vessels and B cells of the cat. Types 2-a and -b may be adrenergic (sympathetic). The nerve endings are usually in loose contact with islet cells, except some have the membrane thickening.

Physiologically, glucagon-secreting A cells are regulated by sympathetic and insulin-secreting B cells by parasympathetic nerves. However, the ultrastructural innervation does not completely agree with this theory, *i.e.*, some A cells come in contact with the cholinergic nerve endings, but others come in contact with the adrenergic ones, and vice versa. This may be explained by the concept that the transmitter substance is liberated freely into the intercellular space, and then flows to the effector cells. Ganglion-like cells, along with ganglion cells, were observed in some animals and may suggest the neurosecretory activities.

INTRODUCTION

Recently electron microscopical attention has been focused on the nervous elements in the pancreatic islets (Bencosme, 1959; Winborn,

1963; Stahl, 1963; Legg, 1967; Watari, 1968; Esterhuizen, Spriggs & Lever, 1968; Kobayashi & Fujita, 1969; Shorr & Bloom, 1970; Klein, 1971). Watari (1968) recently has observed the nerve endings in the pancreatic islets of a variety of animals, and classified them into four types. Types 1-a and -b, which are cholinergic (parasympathetic), and Types 2-a and -b, which are adrenergic (sympathetic).

Ganglion cells and nerve fibers occurring in the pancreatic tissue have also been studied by many investigators. Using a light microscope, some authors have pointed out an intimate contact between ganglion cells and islets of Langerhans and named these relationships *complexes sympathico-insulaires* or *complexes neuroinsulaires* (Simard, 1935, 1937, 1942). On the other hand, using an electron microscope, Shorr *et al.* (1970) observed the existence of ganglion cells in the islet tissue, and Watari (1972) reported ganglion-like cells in the carp pancreatic islets, postulating the possibility of neurosecretory activity in these organs.

This paper summarizes the islet cell innervation in the pancreas in some vertebrates, including humans, monkeys, dogs, cats, rabbits, guinea pigs, rats, mice, doves, snakes, carp, yellow-tail fishes and *Xiphophorus helleri* H. Some new data will also be added.

NERVE SUPPLY IN THE PANCREATIC ISLETS

Nerve fibers

The pancreatic islet is one of the endocrine organs not directly regulated by the anterior pituitary hormones. It is thought to be regulated mainly by the fluctuation of the blood sugar level, although this may be an indirect action. Recently, it has been well established that the pancreatic islet is innervated and controlled by the autonomic nervous system (Watari, 1968).

Light microscopically, the nerve fibers, which are stained with silver impregnation, are abundantly observed in the pancreas of certain animals. Most nerve fibers entering into the pancreatic tissue are unmyelinated, although a few myelinated fibers are observed. These nerve fibers terminate at the perivascular, periacinar and peri-insular portions forming the nerve plexuses (Honjin, 1956; Hagen, 1956). In some animals, the nerve fibers entering the islet tissue make a complex with islet cells (Fig. 12) (Kobayashi *et al.* 1969).

Electron microscopically, the nerve fibers (Fig. 3) entering into the islet tissue still have the Schwann cell sheath, and before entering between the islet cell space they loose the Schwann cell (Fig. 6). At the periphery of the islets, it is sometimes observed that the axons come directly in contact with the islet cell surface, whereas the axons facing the connective tissue have an incomplete covering of Schwann cell cytoplasm (Figs. 4 and 11). However, Winborn (1963) and Kobayashi *et al.* (1969) observed axons covered with a complete Schwann cell sheath entered in between the islet cells in the monkey and dog pancreas, respectively.

Ganglion cells

The intimate relationship between the ganglion cells and islet cells was first observed by Van Campenhout (1925, 1927) using a light microscope and then by many other authors, such as Ritter (1946, newborn cat), Creutzfeldt (1949, dog), Ferner (1952, dog), Coujard & Theret (1952, guinea pig), Hagen (1956, dog), Honjin (1956, mouse) and Coupland (1958, rat and cat). Simard (1935, 1937) also observed such relationships, and called them *complexes sympathico-insulaires* and *complexes neuro-insulaires*. A corresponding structure was described by Bargmann (1937, 1939) where he found nerve cells mingled with islet cells within the teleostian Brockmann body. Electron microscopically, Shorr *et al.* (1970) observed the ganglion cells in the pancreatic islets of both the normal and alloxan-treated rats, and Kern, Hofmann & Kern (1971) observed the ganglion cells in the pancreatic islets of nutria (*Myocaster coypus*).

The ganglion cells in the pancreatic tissue are characterized as accompanied by satellite cells, and contain a well-developed Golgi apparatus, an abundant rough-surfaced endoplasmic reticulum, mitochondria and peculiar pigment granules, along with some large-cored vesicles (Figs. 1 and 2). An abundance of delicate unmyelinated nerve endings come in close contact with the surface of the ganglion cells. These nerve endings usually contain synaptic vesicles and some large-cored vesicles (cholinergic type). Rarely nerve endings contain small-cored vesicles (adrenergic type).

Nerve endings observed in the pancreatic islets

The nerve endings occurring within the islet tissue are usually classified into four types (Watari, 1968). The first is Type 1-a, which contains synaptic vesicles (about 500 Å in diameter) (De Robertis & Bennett, 1955), along with mitochondria (Fig. 5). The second is Type 1-b, containing synaptic vesicles and large-cored vesicles (Fig. 2). The latter are 800 to 2,000 Å in diameter. These Types 1-a and -b nerve endings are cholinergic (parasympathetic) (Richardson, 1964), and come in contact with the A cells in the pancreatic islets of the bat (Fig. 2) (Watari, 1968), rat (Shorr *et al.* 1970) and yellow-tail fish (Watari, Tsukagoshi & Honma, 1970), and B cells of the carp (Fig. 20), yellow-tail fish (Watari *et al.* 1970), cat (Legg, 1967), monkey (Winborn, 1963) and rabbit (Sakagami, 1972).

The third is Type 2-a, containing small vesicles (500 Å in diameter) with an electron dense core (small-cored vesicles); it is usually observed in intimate contact with B cells of the bat, dog (Fig. 7) (Watari, 1968) and monkey (Winborn, 1963). The fourth type of nerve ending, Type 2-b, contains small vesicles with a bar-shaped core, and is usually observed in close contact with the blood vessels in the pancreatic tissue of bats (Fig. 8) (Watari, 1968), or in contact with B cells in the cat pancreatic islets (Legg, 1967). These nerve endings, Types 2-a and -b, are adrenergic (sympathetic) (Richardson, 1964). Some of these adrenergic nerve endings also contain a variable number of large-cored vesicles (Figs. 7 and 11) (Watari *et al.* 1970; Hökfelt, 1967).

Some islet cells simultaneously come in contact with both the adrenergic and cholinergic nerve endings (Kobayashi & Fujita, 1969). Other islet cells, except glucagon-secreting A cells and insulin-secreting B cells, also come in contact with the nerve endings, *i.e.*, unknown nerve endings to D cells of the guinea pig (Kobayashi *et al.* 1969), and Types 1-a or -b to W cells and 1-b to F cells in the bat pancreas (Watari, unpublished data, 1972).

In the exocrine pancreas, on the other hand, the majority of nerve endings are Types 1-a and -b which loosely come in contact with the acinar and ductular cells (Watari, 1968), and some Types 2-a and -b come in contact with the wall of the blood vessels (Lever & Esterhuizen, 1961; Watari, 1968) (Fig. 8).

The synaptic vesicles were first observed by De Robertis and Bennett (1955), and they found that the vesicles contain acetylcholine (De Robertis, 1967). The synaptic vesicles are thought to be derived from neurotubules (De Robertis, 1967) or the membrane of the mitochondria (Csillik & Joó, 1967). On the other hand, small-cored vesicles seem to contain catecholamine (Iversen, 1967). Wolf and his coworkers (1962*a*, *b*) indicated the presence of labeled norepinephrine in the terminal and preterminal parts of the sympathetic axons containing small-cored vesicles in the rat pineal. The rat had received an intravenous injection of tritiated norepinephrine, as demonstrated by electron microscopic radioautography.

Population of the nerve endings in the islet tissue

The number of nerve endings observed in the islet tissue is quite variable. These nerve endings are frequently encountered in bats (*Myotis lucifugus lucifugus*), dogs and carp (*Cyprinus carpio*) (Fig. 9). In the monkey, guinea pig, hamster, rat and cat, they are sometimes observed. On the other hand, in the human, rabbit, dove, snake (*Elaphe climacophora*), mouse (*Mus musculus*) and yellow-tail fish (*Seriola quinqueradiata*), they are seldom observed Winborn, 1963; Legg, 1967; (Watari, 1968; Watari *et al.* 1970).

The specialization of the apposed plasma membrane, such as desmosomal structures or membrane thickenings observed in either the neuromuscular junction of the skeletal muscle or the central nervous system (Uchizono, 1966), is very rarely observed in the lower vertebrates, such as bony fishes, including carp (*Cyprinus carpio*) (Fig. 20) (Watari, 1971*b*), yellow-tail fish (Watari *et al.* 1970) and *Xiphophorus helleri* H. (Klein, 1971).

Physiologically, the glucagon-secreting A cells are regulated by sympathetic nerves, and the insulin-secreting B cells are regulated by parasympathetic nerves (Kurosumi, Kobayashi & Watari, 1970). However, the ultrastructural innervation does not completely agree with this physiological opinion. For example, some nerve endings reaching A cells are cholinergic and some to B cells are adrenergic, and vice versa.

A large number of the nerve endings entering the pancreatic tissue loosely come in contact with the cell surface (Figs. 4–7, 11, 12 and 18), as those reported in the salivary glands (Scott & Pease, 1959), bronchial glands (Bensch, Gorden & Miller, 1965), liver (Yamada, 1965) and nasal gland (Yamamoto, 1966*a*), and in the endocrine organs, such as mammarian islets (Legg, 1967; Esterhuizen *et al.* 1968; Watari, 1968; Shorr *et al.* 1970), adrenal cortex (Unsicker, 1969), avian testis (Baumgarten & Holstein, 1968), and ultimobranchial body of amphibians (Robertson, 1967) and birds (Stoeckel & Porte, 1967), as well as the parafollicular cells of a mammal (Young & Harrison, 1969). This may be explained as reported by Yamamoto (1966*a*), that these glandular cells do not possess specific receptor sites on the cell surface for the nerve impulse, but the transmitter substance may be liberated freely into the intercellular space and reach the effector cells.

Large-cored vesicles in the nervous elements

The large-cored vesicles contained in the nerve endings of Types 1-b and 2 vary in number according to the animal species. In the higher vertebrates, there are relatively fewer large-cored vesicles (Figs. 6 and 7). However, in lower vertebrates the endings contain a large number of large-cored vesicles (Figs. 10–14). In the case of the carp, such nerve endings are abundantly observed. The size of the large-cored vesicles is from 800 to 2,000 Å in diameter. The vesicles are spherical or slightly elongated (Figs. 10–14) and contain a core of variable density. These nerve endings sometimes continue to the axons that contain neural tubules (Figs. 10, 12 and 13).

Similar cytoplasmic process containing many large-cored vesicles was observed in the mouse pancreas (Fig. 21). These large-cored vesicles may be elaborated in the perikaryon of the ganglion cells. The ganglion cells observed in bat pancreatic tissue contain some large-cored vesicles (Fig. 2). At first, the raw secretory substance may be elaborated in the rough-surfaced endoplasmic reticulum (Nissl's substance), and then it is transported to the Golgi apparatus like other protein-secreting cells (Watari, 1970). The secretory substance is condensed in the Golgi apparatus and packaged into the large-cored vesicles, which may flow down to the nerve endings via the axons. Therefore, some large-cored vesicles are observed in the course of the axons (Figs. 3 and 12).

Similar large-cored vesicles were observed in the nerve endings in certain places other than in islet tissues, such as neuromuscular junction (Uchizono, 1966; Yamamoto, 1966*b*) and central nervous tissue (Wood, 1966). Some investigators (Ishii, 1967) reported that the transmitter substance in the large-cored vesicles is noradrenalin which is depleted with the treatment of reserpine. However, other investigators (Bargmann, Lindner & Andres, 1967) thought that such nerve endings belong to the peptidergic.

Ganglion-like cells observed within the carp pancreatic tissue

In the connective tissue and at the periphery of the carp pancreatic islets, a number of ganglion-like cells are observed (Watari, 1972) (Fig. 15). The cells are sometimes irregular in shape and have an irregular nucleus. The cytoplasm contains an abundance of mitochondria, a well-developed rough-surfaced endoplasmic reticulum and an enlarged Golgi apparatus, along with a variable number of large-cored vesicles like the ganglion cells (Figs. 15–17 and 19). Relatively abundant nerve endings are in contact with the ganglion-like cells, and some cytoplasmic processes are projected from the cell body, some of which are axons containing the neurotubules and synaptic vesicles (Fig. 22). However, when compared with usual ganglion cells, the ganglion-like cells have some differences. For instance, they have no satellite cells, and the ribosomes are evenly arranged on the surface of the rough-surfaced endoplasmic reticulum.

These cells are morphologically changed in some cases, such as after administration of excess glucose and in back muscle atrophy of the carp (Sekoke disease, Yokote, 1970). The ganglion-like cells, 2 hr after the intraperitoneal injection of excess glucose (10 mg/g BW), showed that cisternae of the rough-surfaced endoplasmic reticulum became enlarged and separated into small vesicles, and the large-cored vesicles decreased in number, suggesting cell hyperfunction. A similar phenomenon is also observed in the case of carp with back muscle atrophy (Sekoke disease) (Fig. 16). The carp with back muscle atrophy show glucosuria like diabetic animals (Yokote, 1970). The B cells in such cases are usually degranulated. The function of A cells is also activated, such as the degranulation and the vesiculation of the rough-surfaced endoplasmic reticulum. This is in contrast with the case of diabetic animals, where the A cells usually have no morphological changes (Watari, 1971*b*).

After the injection of excess glucose, some ganglion-like cells show unusual shapes, *i.e.*, the Golgi apparatus is atrophied, some of the large-cored vesicles contain crystalline cores, and rough-surfaced endoplasmic reticulum is irregularly enlarged and contains an electron dense colloidal substance (Fig. 17). This may result from the disturbance of the transport mechanism of the secretory substance from the rough-surfaced endoplasmic reticulum to the Golgi apparatus as reported by Kurosumi (1967). This may be one of the very special intracisternal granules (Palade, 1956).

Recently, Kanerva (1972) observed similar cells in the rat paracervical ganglion in postnatal development. He called it a small granule-containing cell, which contains the large-cored vesicles varying between 800 and 2,500 Å, and without the satellite cells.

When examined using the zinc iodide-osmium (ZIO) reaction (Akert & Sandri, 1968), which makes special electron dense deposits on the nervous elements, especially on the synaptic vesicles, nerve endings and ganglion-like cells have special electron dense deposits (Figs. 18 and 19). This may indicate that the ganglion-like cells are neural elements other than islet cells. The ganglion-like cells, including the nerve endings with the large-cored vesicles, may possibly have neurosecretory activity as reported by Watari (1972). However, further observations are needed to ascertain the function of the cells and the nature of the large-cored vesicles.

MORPHOLOGICAL ALTERATIONS OF THE NERVOUS ELEMENTS IN SOME EXPERIMENTAL CONDITIONS

Recently Shorr *et al.* (1970) have reported changes of the nervous elements in the rat pancreatic islets after alloxan administration. They observed that the distrophic nerve endings contain abundant mitochondria, numerous vesicles (250–700 Å in diameter), and dispersely deposited glycogen particles. These structures most frequently come in contact with degranulated B cells, and they concluded that these nerve endings might result from neuronal hyperactivity for the purpose of secreting insulin from the post-alloxan insulin-deposited B cells.

Similar observations have been made in the carp pancreatic islets after the excess administration of glucose (Kurosumi *et al.* 1970). Some nerve endings were swollen and contained peculiar tubular structures (Fig. 12). The electron dense contents of the large-cored vesicles in these endings were depleted and left empty vesicles (Figs. 13 and 20). After the injection of reserpine, the content of the large-cored vesicles was not depleted (Fig. 14), but such nerve endings sometimes contained unusual myelinated structures (Fig. 14). This phenomenon may suggest that the content of the large-cored vesicles is not norepinephrine (noradrenaline) (Bloom & Barrnett, 1966), and is not affected by administration of reserpine.

REFERENCES

Akert, K. & Sandri, C. (1968). An electron-microscopic study of zinc iodide-osmium impregnation of neurons. I. Staining of synaptic vesicles at cholinergic junctions. *Brain Res.* **7**, 286–295.

Bargmann, W. (1937). Kolloidbindung im Inselgewebe des Pankreas von *Scorpaena porcus*. *Z. Zellforsch.* **27**, 450–454.

Bargmann, W. (1939). Die Langerhansschen Inseln des Pankreas. In *Handbuch der mikroskopischen Anatomie des Menschen*. Bd. VI/2, pp 196–288.

Bargmann, W., Lindner, E. & Andres, K. H. (1967). Über synapsen an endokrinen Epithelzellen und die Definition skeretorischer Neurone.

Untersuchungen am Zwischenlappen der Katzenhypophyse. *Z. Zellforsch.* **77**, 282–298.

Baumgarten, H. G. & Holstein, A.-F. (1968). Adrenerge Innervation im Hoden und Nebenhoden vom Schwan (*Cygnus olor*). *Z. Zellforsch.* **91**, 402–410.

Bencosme, S. A. (1959). Studies on the terminal autonomic nervous system with special reference to the pancreatic islets. *Lab. Invest.* **8**, 629–646.

Bensch, D. G., Gorden, G. B. & Miller, L. R. (1965). Studies on the bronchial counterpart of the Kultschitzky (argentaffin) cell and innervation of bronchial glands. *J. Ultrastruct. Res.* **12**, 668–686.

Bloom, F. E. & Barrnett, R. J. (1966). Fine structural localization of noradrenaline in vesicles of autonomic nerve endings. *Nature, Lond.* **210**, 599–601.

Coujard, R. & Theret, C. (1952). Les réactions insularies du pancréas par agression sur le sympathique périphérique. *C.R. Soc. Biol.* (Paris) **146**, 1883–1886.

Coupland, R. E. (1958). The innervation of pancreas of the rat, cat and rabbit as revealed by the cholinesterase technique. *J. Anat.* **92**, 143–149.

Creutzfeldt, W. (1949). Zur Histophysiologie des Inselapparates. *Z. Zellforsch.* **34**, 280–336.

Csillik, B. & Joó, F. (1967). Effect of hemicholinium on the number of synaptic vesicles. *Nature, Lond.* **213**, 508–509.

De Robertis, E. D. P. (1967). Ultrastructure and cytochemistry of the synaptic region. The macromolecular components involved in nerve transmission are being studied. *Science, N.Y.* **156**, 907–914.

De Robertis, E. D. P. & Bennett, H. S. (1955). Some features of the submicroscopic morphology of synapses in frog and earthworm. *J. biophys. biochem. Cytol.* **1**, 47–58.

Esterhuizen, A. C., Spriggs, T. L. B. & Lever, J. D. (1968). Nature of islet-cell innervation in the cat pancreas. *Diabetes* **17**, 33–36.

Ferner, H. (1952). Das Inselsystem des Pankreas. S. 104. Stuttgart: Georg Thieme.

Hagen, E. (1956). Über die Innervation der exkretorischen Drüsen und der Langerhansschen Inseln des Pankreas beim Hund. *Z. Zellforsch.* **43**, 486–500.

Hökfelt, T. (1967). On the ultrastructural localization of noradrenaline in the central nervous system of the rat. *Z. Zellforsch.* **79**, 110–117.

Honjin, R. (1956). The innervation of the pancreas of the mouse, with special reference to the structure of the peripheral extension of the vegetative nervous system. *J. comp. Neurol.* **104**, 331–371.

Ishii, S. (1967). Morphological studies on the distribution and properties of the granulated vesicles in the brain. *Arch. Histol. Jap.* **28**, 355–376.

Iversen, L. L. (1967). The catecholamines. *Nature, Lond.* **214**, 8–14.

Kanerva, L. (1972). Light and electron microscopic observations on the postnatal development of the rat paracervical (Frankenhäuser) ganglion. *Z. Anat. Entwickl. -Gesch.* **136**, 33–55.

Kern, H. F., Hofmann, H. V. & Kern, D. (1971). Light- und elektronenmikroskopische Untersuchungen der Langerhansschen Inseln von Nutria (*Myocastor coypus*), mit besonderer Berücksichtigung der neuroinsulären Komplexe. *Z. Zellforsch.* **111**, 216–229.

Klein, C. (1971). Innervation des cellules du pancréas endocrine du poisson téléostéen *Xiphophorus helleri H. Z. Zellforsch.* **113**, 564–580.

Kobayashi, S. & Fujita, T. (1969). Fine structure of mammalian and avian pancreatic islets with special reference to D cells and nervous elements. *Z. Zellforsch.* **100**, 340–363.

Kurosumi, K. (1967). Function and fine structure of the cell. *Heredity* (Tokyo) **21**, 63–71.

Kurosumi, K., Kobayashi, Y. & Watari, N. (1970). Ultrastructural morphology of the neuroendocrine integration. *Saishin-Igaku* (Japanese Journal) **25**, 2132–2145.

Legg, P. G. (1967). The fine structure and innervation of the beta and delta cells in the islet of Langerhans of the cat. *Z. Zellforsch.* **80**, 307–321.

Lever, J. D. & Esterhuizen, A. C. (1961). Fine structure of the arteriolar nerves in the guinea pig pancreas. *Nature, Lond.* **192**, 566–567.

Palade, G. E. (1956). Intracisternal granules in the exocrine cells of the pancreas. *J. biophys. biochem. Cytol.* **2**, 417–422.

Richardson, D. C. (1964). The fine structure of the rabbit iris with special reference to the identification of adrenergic and cholinergic nerves and nerve endings in its intrinsic muscles. *Am. J. Anat.* **114**, 173–205.

Ritter, O. (1946). Etude sur les nerfs du pancréas exocrine et endocrine. *Vjschr. Naturforsch. Ges.* (Zurich) **91**, 51–57.

Robertson, D. R. (1967). Ultimobranchial body in *Rana pipiens*. III. Sympathetic innervation of the secretory parenchyma. *Z. Zellforsch.* **78**, 328–340.

Sakagami, M. (1972). Ultrastructural changes of pancreatic B cells induced by electrical stimulation of the hypothalamus in rabbits. *Acta Anat. Nippon* **47**, 154–171.

Scott, B. L. & Pease, D. S. (1959). Electron microscopy of the salivary and lacrimal glands of the rat. *Am. J. Anat.* **104**, 115–161.

Shorr, S. S. & Bloom, F. E. (1970). Fine structure of islet cell innervation in the pancreas of normal and alloxan-treated rats. *Z. Zellforsch.* **103**, 12–25.

Simard, L. C. (1935). Les complexes neuro-insulaires du pancréas de l'homme adulte. *C. R. Soc. Biol.* (Paris) **119**, 27–28.

Simard, L. C. (1937). Les complexes neuro-insulaires du pancréas humain. *Arch. Anat. Micr. Morph. Exp.* **33**, 49–64.

Simard, L. C. (1942). Les complexes neuro-insulaires du pancréas chez les mannifères adultes. *Rev. Canad. Biol.* **1**, 1–49.

Stahl, M. (1963). Elektronenmikroskopische Untersuchungen über die vegetative Innervation der Bauchspeicheldrüse. *Z. mikrosk.-anat. Forsch.* **70**, 62–102.

Stoeckel, M.-E. & Porte, A. (1967). Sur l'ultrastructure des corps ultimobranchiaux du poussin. *Compt. Rend. Acad. Sci.*, Ser. D **265**, 2051-2053.

Uchizono, K. (1966). Excitatory and inhibitory synapses in vertebrate and invertebrate animals (Sixth Intern. Cong. for Electron Microsc., Kyoto), Electron Microsc. Vol. II, ed. R. Uyeda, pp. 431-432. Tokyo: Maruzen Co., Ltd.

Unsicker, K. (1969). Zur Innervation der Nebennierenrinde vom Goldenhamster. Eine Fluoreszenz- und elektronenmikroskopische Studie. *Z. Zellforsch.* **95**, 608–619.

Van Campenhout, E. (1925). Etude sur le développement et la signification morphologique des ilôts endocrines du pancréas chez l'embryon de mouton. *Arch. Biol.* (Liège) **35**, 45–88.

Van Campenhout, E. (1927). Contribution à l'étude de l'histogénèse du pancréas chez quelques mammifères. Les complexes sympathico-insulaires. *Arch. Biol.* (Liège) **37**, 121–171.

Watari, N. (1968). Fine structure of nervous elements in the pancreas of some vertebrates. *Z. Zellforsch.* **85**, 291–314.

Watari, N. (1970). The correlative light and electron microscopy of the islets of Langerhans in the pancreas of some vertebrates, with special reference to the synthesis, storage and extrusion of the islet hormones. *Gunma Symp. on Endocrinol.* **7**, 125–150.

Watari, N. (1971*a*). The absorption and secretion of the cell. *Igaku-no-Ayumi* (Japanese Journal) **76**, 445–454.

Watari, N. (1971*b*). Fine structure and function of the pancreatic islets. *Cell* (Japanese Journal) **3** (8), 2–24.

Watari, N. (1972). Ganglion-like cell occurring in the carp pancreatic islets examined by zinc iodide-osmium impregnation. Fourth Intern. Congr. Histochem. Cytochem., Kyoto. *Histochem. Cytochem.* ed. Takeuchi T., Ogawa K., Fujita S., pp. 507–508. Kyoto.

Watari, N., Torizawa, K., Saeki, S. & Takada, K. (1972). Ultrastructural alterations of the mouse pancreas after the prolonged and overdose administration of vitamin A. *J. Electron Microsc.* **21**, 40–54.

Watari, N. Tsukagoshi, N. & Honma, Y. (1970). The correlative light and electron microscopy of the islets of Langerhans in some lower vertebrates. *Arch. Histol. Jap.* **31**, 371–392.

Winborn, W. B. (1963). Light and electron microscopy of the islets of Langerhans of the Saimiri monkey pancreas. *Anat. Rec.* **147**, 65–94.

Wolfe, D. E., Axelrod, J., Potter, L. T. & Richerdson, K. C. (1962*a*). Localization of norepinephrine in adrenergic axons by light and electron-microscopic autoradiography. In *Electron Microsc.* Vol. 2, ed. Breese, S. S. Jr., p. U-1. New York: Academic Press.

Wolfe, D. E., Richerdson, K. C. & Axelrod, J. (1962*b*). Localizing tritiated norepinephrine in sympathetic axons by electron microscopic autoradiography. *Science, N.Y.* **138**, 440–442.

Wood, J. G. (1966). Electron microscopic localization of amines in central nervous tissue. *Nature, Lond.* **209**, 1131–1133.

Yamada, E. (1965). Some observations of the nerve terminal on the liver parenchymal cell of the mouse as revealed by electron microscopy. *Okajimas Folia Anat. Jap.* **40**, 663–677.

Yamamoto, T. (1966*a*). On the fine structure of the terminal portion of nasal gland in guinea pig, with special reference to the interrelationship between glandular cells and nerve endings. *Arch. Histol. Jap.* **27**, 311–325.

Yamamoto, T. (1966*b*). The fine structure of the neuro-muscular relationships in the smooth muscle tissue and the mixed tissue of smooth and striated muscles. *Symposia Cell Chem.* (Japanese Journal) **16**, 43–58.

Yokote, M. (1970). Sekoke disease, spontaneous diabetes in carp, *Cyprinus carpio*, found in fish farms. 1. Pathological study. *Bull. Freshwater Fish Res. Lab.* **20** (1), 39–72.

Young, B. A. & Harrison, R. J. (1969). Ultrastructure of light cells in the dolphin thyroid. *Z. Zellforsch.* **96**, 222–228.

EXPLANATION OF FIGURES

Fig. 1. Photomicrograph of ganglion cells (GC) occurring within the bat pancreatic tissue. The cells contain a large nucleus with a large nucleolus and pale cytoplasm with some pigment granules. They are accompanied by some satellite cells (S). V, Blood vessels; EX, Acinar cells. ×1,800 (from Watari, 1968).

Fig. 2. A part of a ganglion cell observed in the bat pancreas. The cytoplasm contains well-developed Golgi apparatus (G), abundant mitochondria (M), pigment granules (P), and rough-surfaced endoplasmic reticulum (ER). The arrow points to a large-cored vesicle. N, Nucleus. Fixed with osmic acid alone. ×15,000 (from Watari, 1968).

Fig. 3. An unmyelinated nerve fiber composed of a number of crosscut axons (AX) and a Schwann cell (SW). Some axons contain synaptic vesicles (Arrow 1) and a few large-cored vesicles (Arrow 2). Dog pancreas fixed with osmic acid alone. ×17,000.

Fig. 4. Two axons (AX) occurring within the hamster pancreatic islet. They are covered with Schwann cell cytoplasm (SW), but the portions in intimate contact with the B cell (B) are partly bared (Arrow). V, Blood capillary. Fixed with osmic acid alone. ×19,000 (from Kurosumi, Kobayashi & Watari, 1970).

Fig. 5. A nerve ending (NE) containing synaptic vesicles and some mitochondria is interposed between B_1 (B_1) and B_2 (B_2) cells. Yellow-tail fish fixed with osmic acid alone. ×22,000 (from Watari, Tsukagoshi & Honma, 1970).

Fig. 6. A nerve ending (NE) (Type 1-b, cholinergic) containing synaptic vesicles and a large-cored vesicle (Arrow) comes in contact with an A cell (A). AX, An axon; SW, Cytoplasmic processes of Schwann cell. Bat pancreas fixed with osmic acid alone. ×29,000 (from Watari, 1968).

Fig. 7. A nerve ending (NE) continuous to the axon (AX) contains some small-cored vesicles (Arrow 1) and a large-cored vesicle (Arrow 2) (Type 2-a). This nerve ending is seen close to the B cell (B) containing bar-shaped granules. Dog pancreas fixed with osmic acid alone. ×30,000 (from Watari, 1968).

Fig. 8. Some nerve endings (NE) containing small crystalline-cored vesicles (Arrow 1) are observed at the perivascular space. The nerve endings are covered with the attenuated cytoplasm of the Schwann cell (Arrow 2). Bat pancreas fixed with osmic acid alone. ×32,000 (from Watari, 1968).

Fig. 9. A survey picture of the pancreatic islet in the carp. A number of nerve endings containing large-cored vesicles are distributed among A (A) and B (B) cells (Arrows). Fixed with a mixture of glutaraldehyde and osmic acid. ×3,500.

Fig. 10. A closer view of the nerve ending (NE) containing many of the large-cored vesicles and a cluster of the synaptic vesicles (Arrow). T, Neurotubules; AX, Axon; A, A cell; B, B cell. Carp pancreatic islet fixed with a mixture of glutaraldehyde and osmic acid. ×14,000 (from Watari, 1970).

Fig. 11. A nerve ending (NE) containing small-cored vesicles (Arrow 1) and large-cored vesicles is seen in close contact with B cells (B). Arrow 2 indicates a thin cytoplasmic process of Schwann cell. Snake pancreas fixed with a mixture of glutaraldehyde and osmic acid. ×22,000.

Fig. 12. Degenerated nerve endings occurring 2 hr after the injection of excess glucose contain twisted tubule-like structures (DT). Normal nerve endings (NE) contain large-cored vesicles with variable density. Axons (AX) contain neural tubules and some large-cored vesicles (Arrow). Carp pancreatic islet fixed with a mixture of glutaraldehyde and osmic acid. ×22,000 (from Kurosumi, Kobayashi & Watari, 1970).

Fig. 13. Large nerve endings (NE) containing a majority of large empty vacuoles which may be the depleted large-cored vesicles interposed between B cells (B). They are continuous to the axon (AX). Carp pancreatic islet fixed with a mixture of glutaraldehyde and osmic acid. ×17,000.

Fig. 14. A portion of the carp islet after the injection of reserpine. Nerve endings (NE) contain electron dense large-cored vesicles which are not depleted. An arrow indicates the myelinated structure. A, A cell; B, B cell. Fixed with a mixture of glutaraldehyde and osmic acid. ×16,000.

Fig. 15. A part of the ganglion-like cell containing well-developed rough-surfaced endoplasmic reticulum (ER), a number of mitochondria (M), and Golgi apparatus (G) in which are seen some large-cored vesicles (Arrow). A, A cell. Carp pancreatic islet fixed with a mixture of glutaraldehyde and osmic acid. ×22,000.

Fig. 16. An activated ganglion-like cell occurring within the pancreatic islet of a carp with back muscle atrophy. The cell contains vesiculated rough-surfaced endoplasmic reticulum (ER) containing some flocculent substance (Arrow 4). Arrow 1, Large-cored vesicles; Arrow 2, Microtubules; Arrow 3, Enlarged perinuclear cisternae; M, Mitochondria; G, Golgi apparatus; N, Nucleus. Fixed with a mixture of glutaraldehyde and osmic acid. ×14,000.

Fig. 17. A ganglion-like cell (GL) in the carp 2 hr after the injection of excess glucose containing irregularly enlarged cisternae of the rough-surfaced endoplasmic reticulum (ER), in which is seen an electron dense colloidal substance. Some axons (AX) come in contact with the ganglion-like cell. D, D cell;

A, A cell; B, B cell; N, Nucleus. Fixed with a mixture of glutaraldehyde and osmic acid. ×5,000 (from Watari, 1972).

Fig. 18. A nerve ending (NE) observed between B cells (B) contains electron dense deposits of the zinc iodide-osmium impregnation (Arrow). Carp pancreatic islet postfixed with osmic acid. ×23,000 (from Watari, 1972).

Fig. 19. Ganglion-like cell (GL) occurring within the carp pancreatic islet. The ZIO deposits also are seen within the cytoplasm (Arrows). Postfixed with osmic acid. B, B cell; NE, nerve ending containing large-cored vesicles. ×30,000.

Fig. 20. Nerve endings (NE) containing synaptic vesicles and large-cored vesicles (Arrow 2) are observed within the carp pancreatic islet 2 hr after the injection of excess glucose. Some large-cored vesicles are depleted. Arrow 1 indicates membrane thickening between the nerve ending and B cell (B). AX, Axons. Fixed with a mixture of glutaraldehyde and osmic acid. ×38,000.

Fig. 21. A possible nerve ending (NE) occurring within the mouse pancreas, containing a number of vesicles with electron dense cores (Arrows), which are similar to the large-cored vesicles. B, B cells. Fixed with osmic acid alone. ×15,000.

Fig. 22. Ganglion-like cell (GL) projects an axon (AX) between B cells (B) (Thick arrow). The axon contains cross-cuttings of the neurotubules (Arrow 1) and synaptic vesicles (Arrow 2). Some large-cored vesicles (Arrow 3) are observed in the ganglion-like cell. Fixed with a mixture of glutaraldehyde and osmic acid. ×23,000.

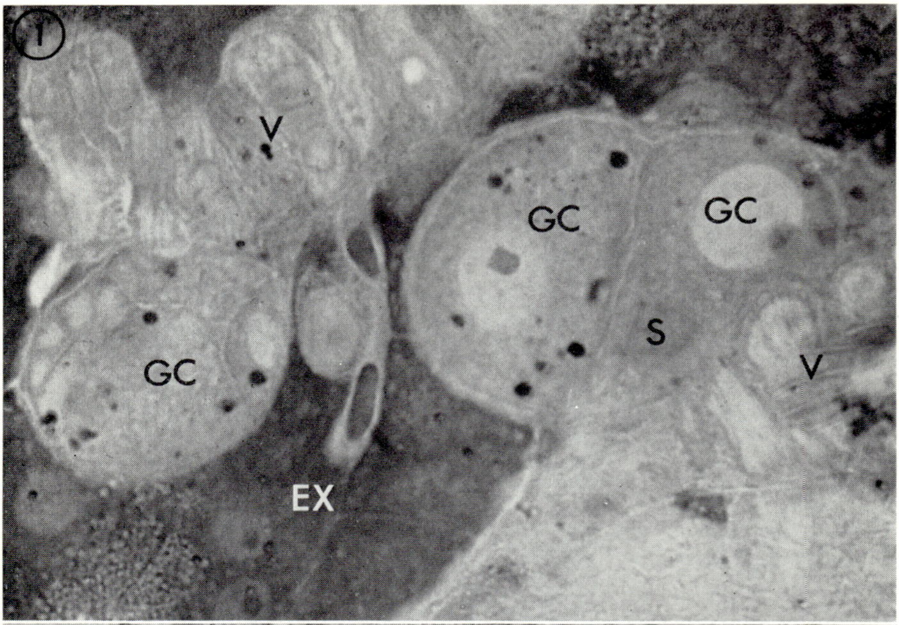
1
V
GC
GC
GC
S
V
EX

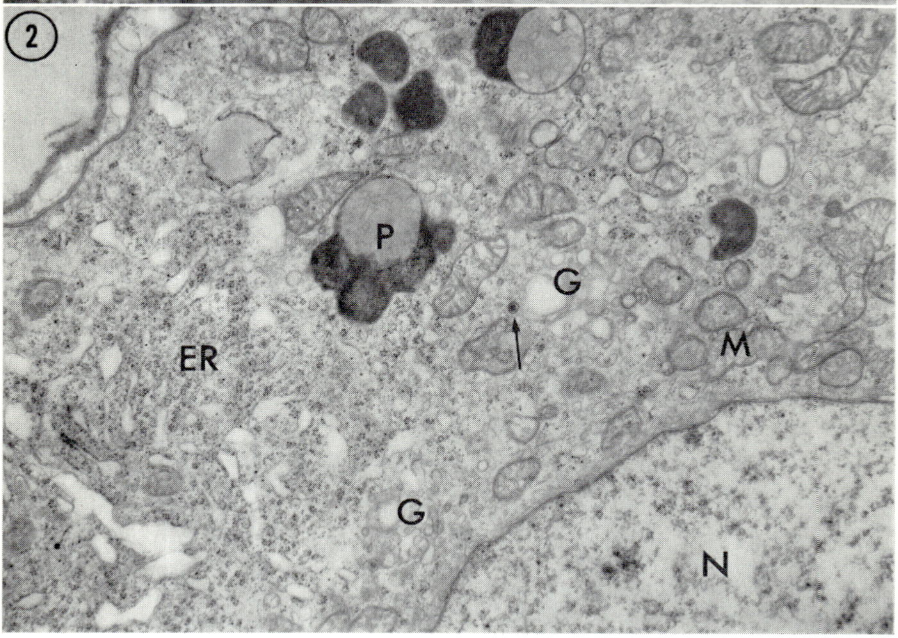
2
P
G
ER
M
G
N

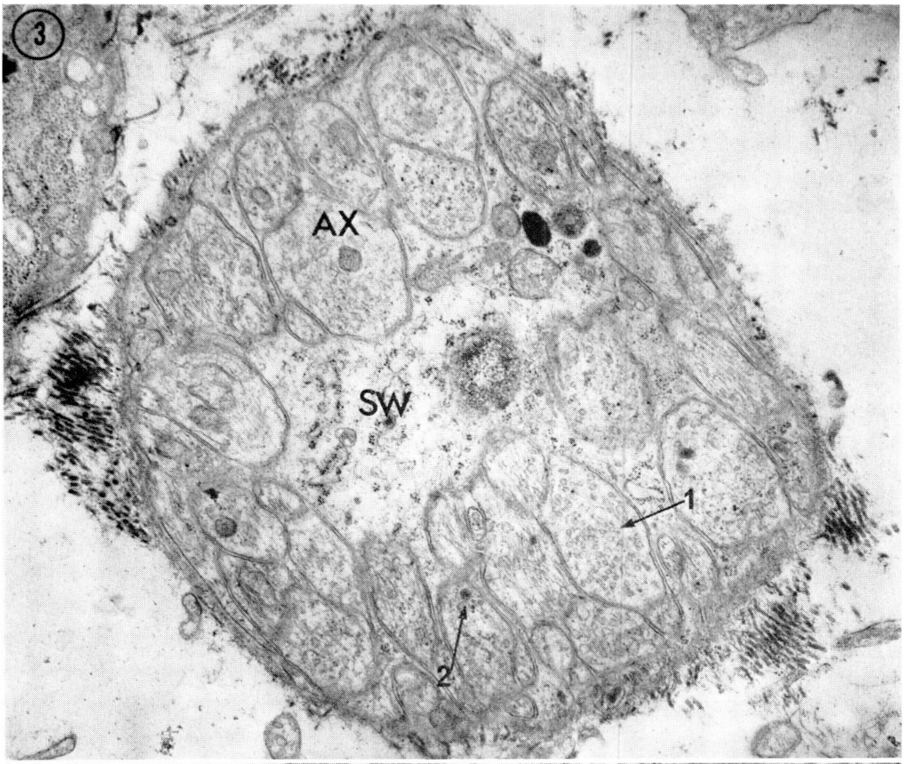
3
AX
SW
1
2

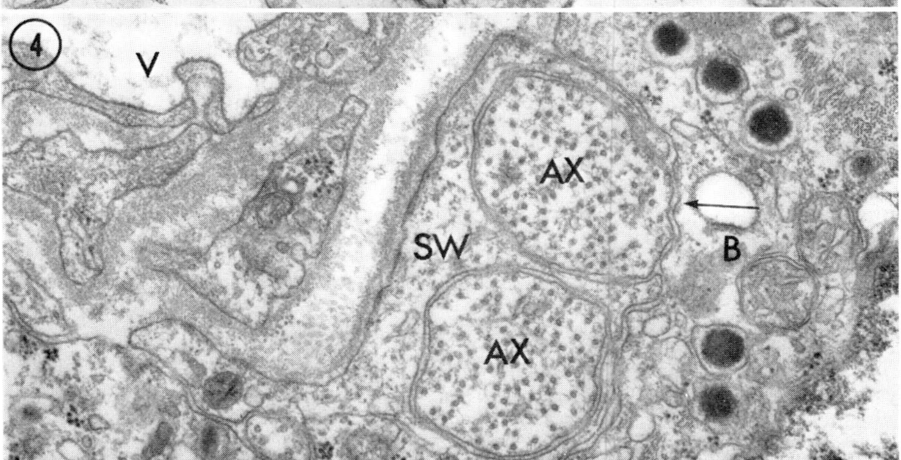
4
V
AX
SW
B
AX

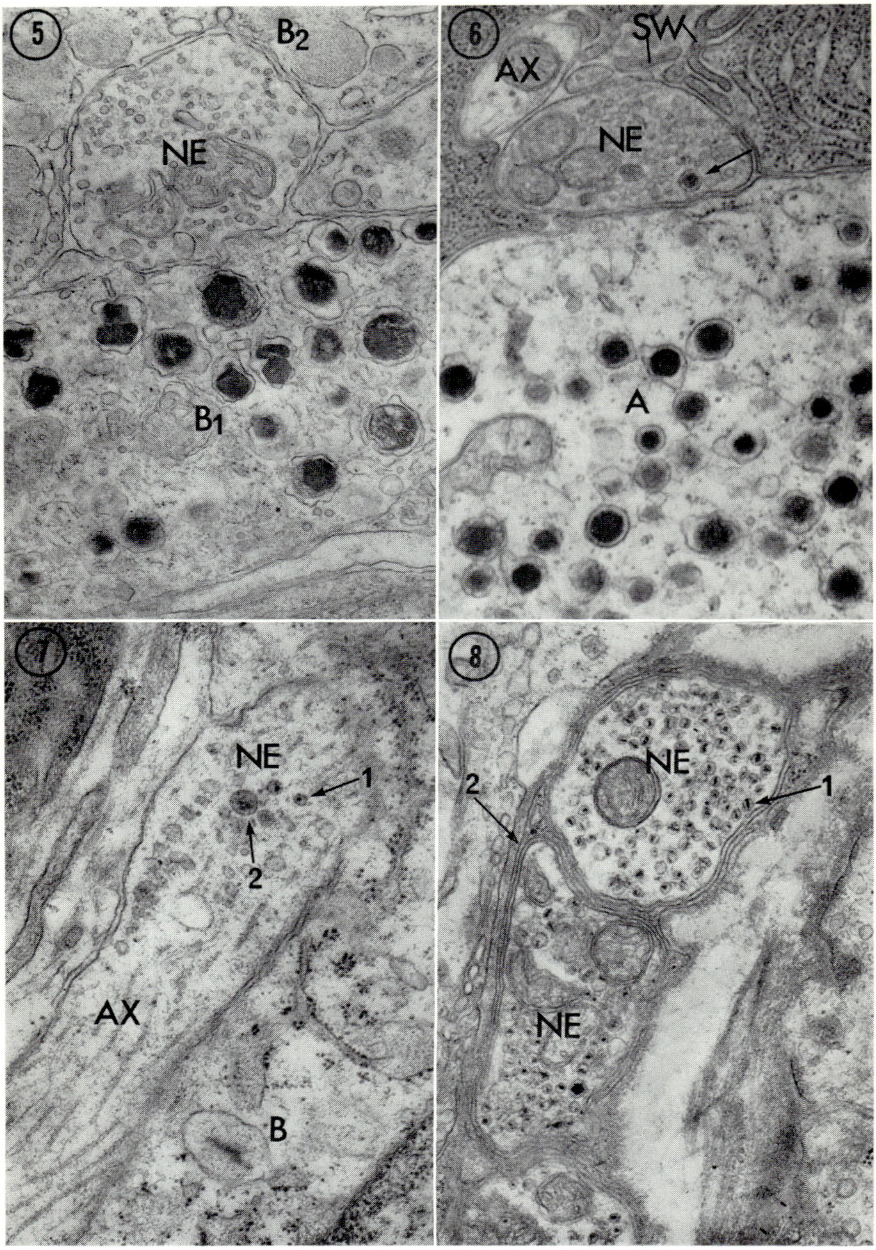
5
B2
NE
B1
6
AX
SW
NE
A
7
NE
1
2
AX
B
8
NE
2
1
NE

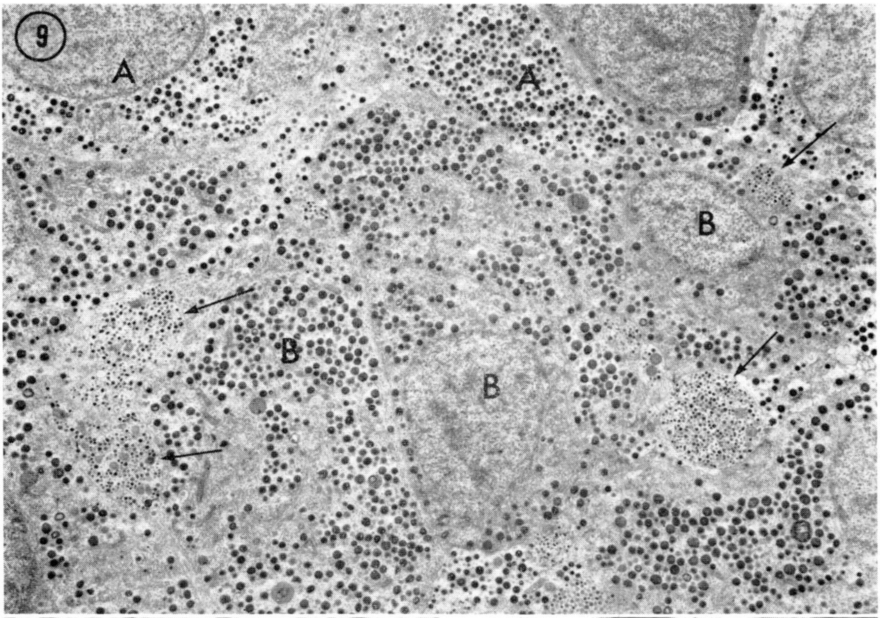
9
A
A
B
B
B

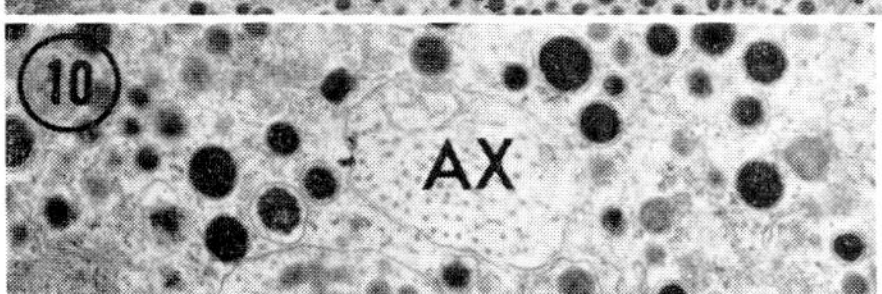
10
AX
NE
A
T
B

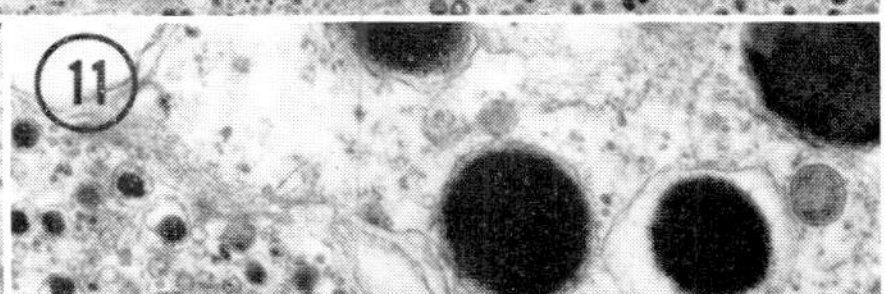
11
1
NE
B
2

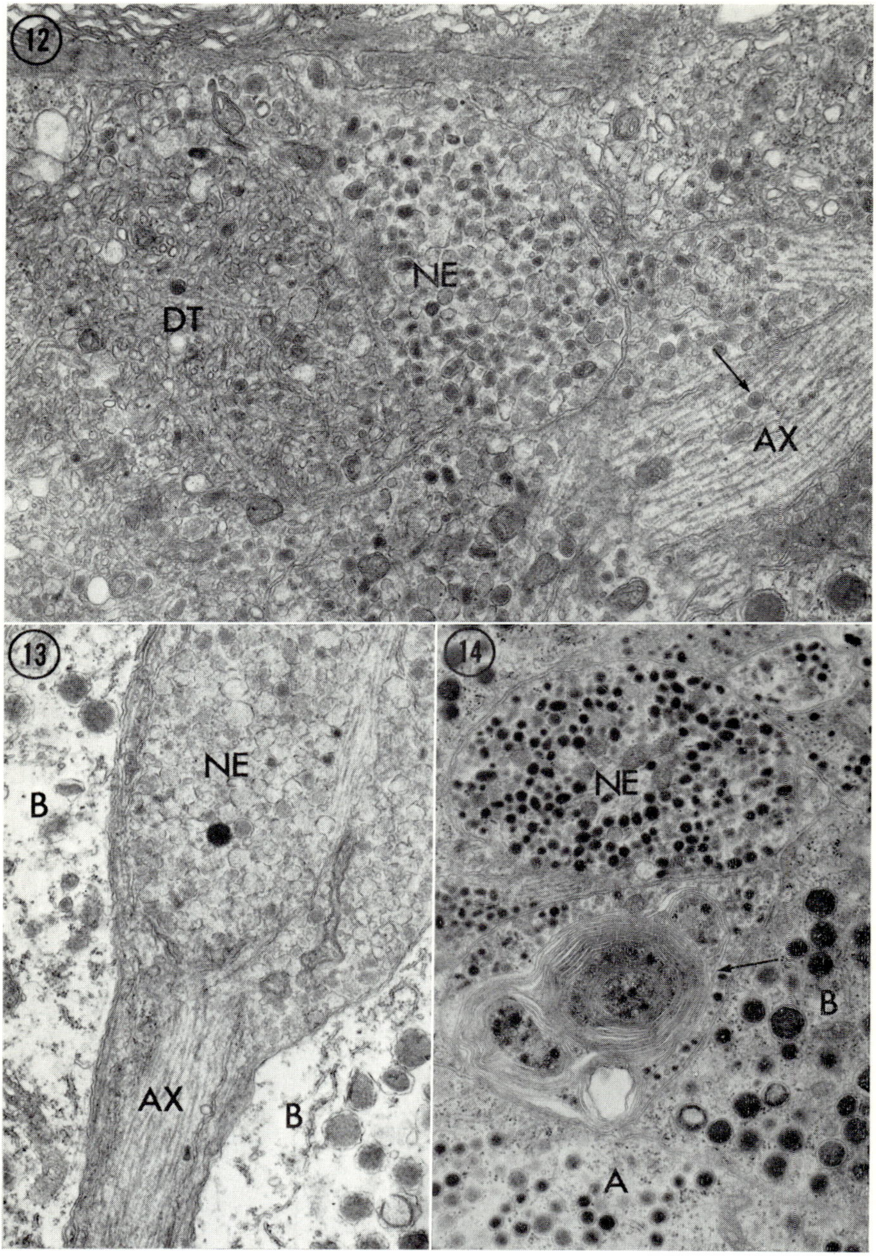
12
NE
DT
AX
13
NE
B
AX
B
14
NE
B
A

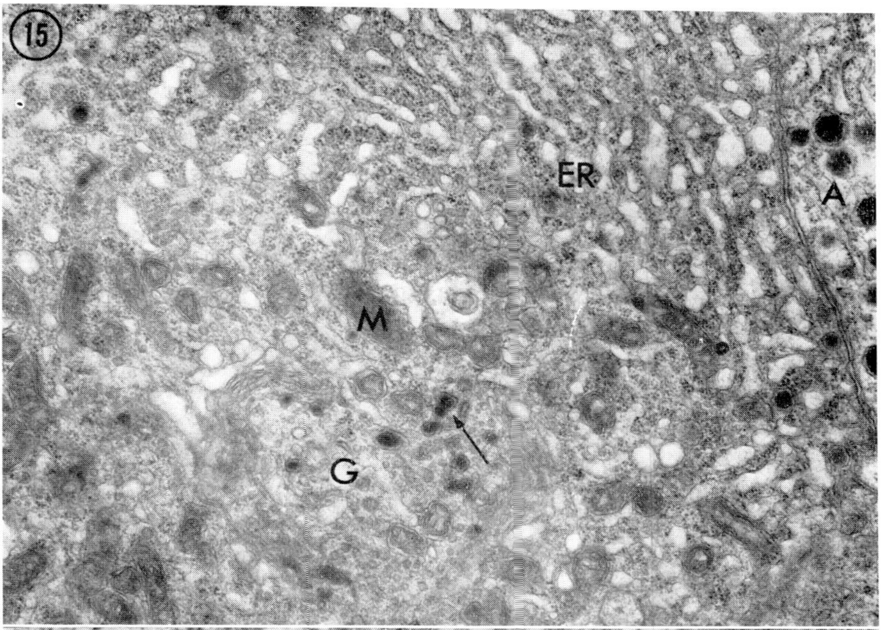
15
ER
A
M
G

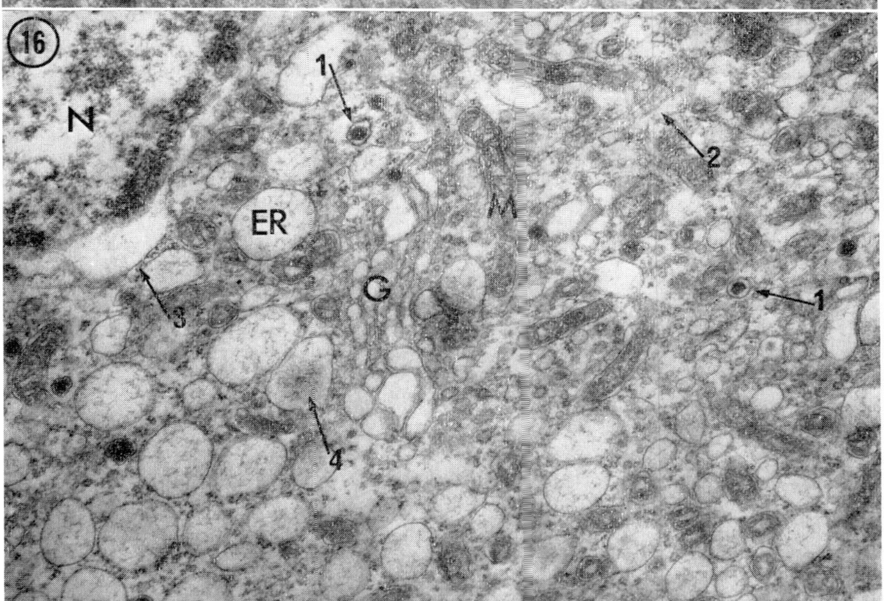
16
N
1
ER
M
2
G
3
1
4

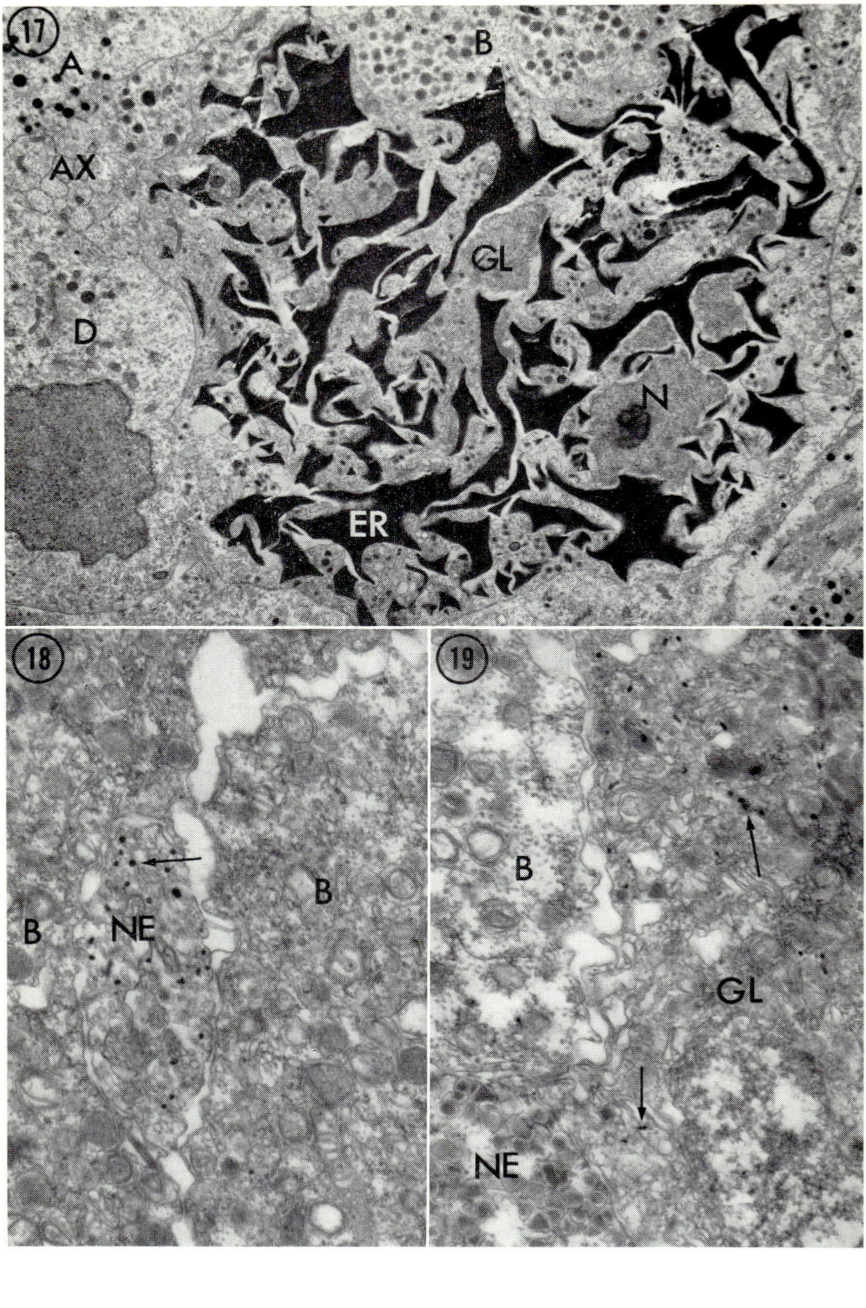
17
A
AX
D
B
GL
N
ER
18
B
NE
B
19
B
GL
NE

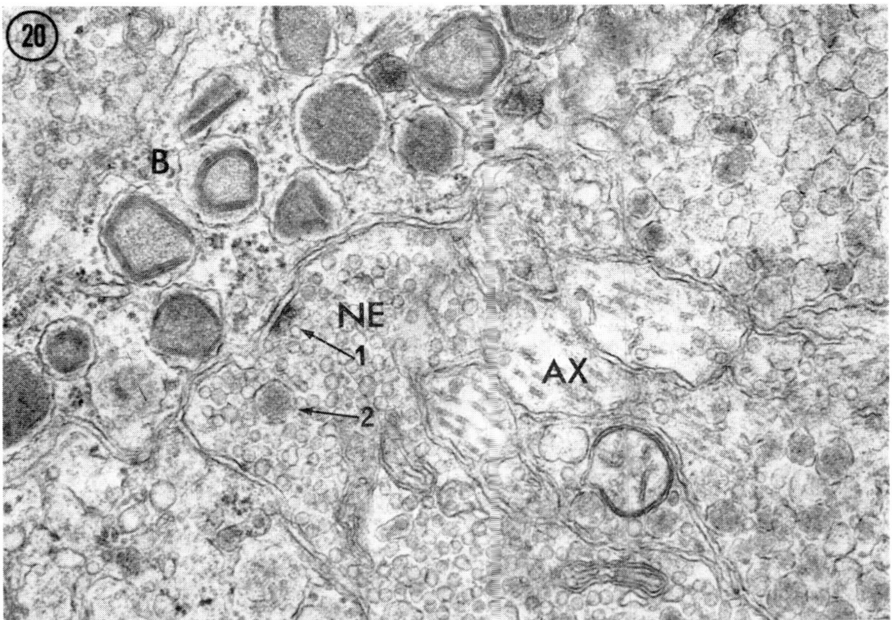
20
B
NE
1
2
AX

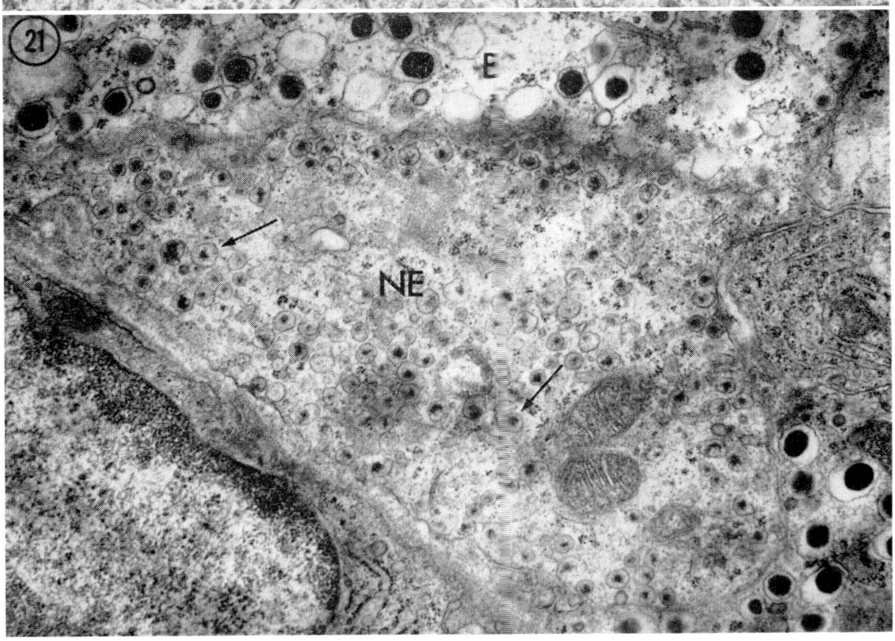
21
E
NE

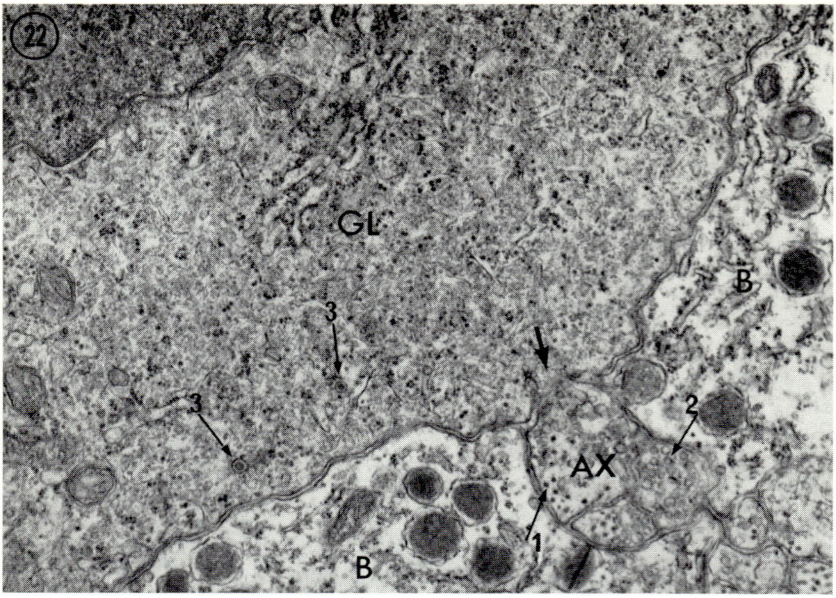
22
GL
3
3
B
B
AX
1
2

CONTROL OF VASOPRESSIN SECRETION

Sho Yoshida, Toshikazu Saito, and
Toshio Murase

The control of vasopressin secretion, which is concerned with osmoregulation and volume regulation of the body fluids, is reviewed. The receptors for ADH release in response to changes in these two variables are localized separately. The osmoreceptor is located in the hypothalamus. It has been reported that deafferentiation of the hypothalamus did not disturb antidiuretic responses caused by osmotic stimuli The increase of ADH titers by injection of a hypertonic saline was observed in dogs with diencephalic islands. It appears that the osmoreceptor is located in the hypothalamus. No other osmoreceptors are necessary for adequate control of the plasma osmolality. However, it is not possible to confine the osmosensitive elements to a restricted area in the hypothalamus.

The left atrium volume receptor theory proposed by Gauer and Henry (1963) is discussed. Though it has been repeatedly reported that changes in pressure of the left atrium produced alterations in ADH secretion, the experimental procedures which were designed to change the atrial pressure also caused changes in the hemodynamics of the peripheral arterial system. The graded hemorrhage in dogs performed in our laboratory did not cause any increase in plasma ADH titers unless a significant fall in the mean arterial pressure was observed.

It appears that acceptance of the left atrium volume receptor theory should be reserved until evidence shows that changes in pressure of the left atrium without any concomitant changes in hemodynamics in the arterial system caused alterations in ADH secretion.

INTRODUCTION

Vasopressin or antidiuretic hormone, ADH, is a hormone that regulates water excretion from the kidney. The hormone plays a very important role in the homeostatic regulation of water metabolism in terrestorial mammals. Animals keep the volume and tonicity of the extracellular fluids constant by controlling the amount of the hormone released from the neurohypophysis. It is implicit in the principle of the homeostatic state that there is not only an effector mechanism but also a receptor and possibly an integrative mechanism (Smith, 1957). As the control of vasopressin release is a typical homeostatic mechanism, much work has been done on this topic in the past decade.

Though release of ADH is influenced by many other factors, such as pain, emotional stress, and administrations of certain drugs, their physiological meaning is probably minor, and this review is confined to the field of osmoregulation and volume regulation.

OSMOREGULATION AND ADH

Verney (1947) is a pioneer who focused his attention on control of ADH release by the blood osmolality. He produced alterations in the osmotic pressure of the blood by the injection of hypertonic solutions into the common carotid arteries of unanesthetized dogs provided with carotid loops. The output of ADH was measured by the effect of injection upon a concurrently established water diuresis. By these means it was found that an increase of some 2% in the osmotic pressure of the blood in one carotid trunk caused a reduction of urine flow of some 90%. On the basis of these investigations, it was postulated that the neurohypophysis was functionally linked with sensory elements sensitive to change in the osmotic pressure of their vascular environment.

Such elements were termed osmoreceptors, and these were envisaged as being continually engaged in transmuting osmotic pressure changes in their environment into appropriately effective messages to the neurohypophysis. The ADH increases the amount of water reabsorbed at renal tubules, resulting in a decrease in the osmolality of the extracellular fluids. This effect in turn decreases the ADH output from the neurohypophysis according to signals discharged by the osmoreceptor. Thus, the release of ADH from the neurohypophysis is self-limiting, or it is under control of a negative feedback system. Verney's experiments were carefully designed and the results were clear cut, so that the osmoreceptor theory was accepted without any serious challenge.

After these experiments, research into the nature and exact locality of the osmoreceptor followed. By ligation of portions of the arterial tree, the cephalic field of distribution of the carotid arteries in the dog has been restricted to a defined region of the brain. Then, the response to intracarotid injection of the hypertonic solution was examined. By this means Jewell & Verney (1957) were able to confine the osmosensitive area to the restricted area of the anterior hypothalamus, which included the suprachiasmatic nucleus, the nucleus supraopticus diffusus, the ventromedial nucleus, and the supraoptic nucleus. Though Verney (1947) described, within the field of the supraoptic nucleus, vesicles which appear to be suited to an osmoreceptive function, he did not substantiate this hypothesis.

Sundsten & Sawyer (1961) and Woods, Bard & Bleier (1966) surgically ablated most parts of the forebrain of animals. The very restricted area of the hypothalamus remained intact without any direct neural connection. Responses to the osmotic stimuli were examined in these animals. The results generally confirmed Verney's concept that the receptors are located in the anterior hypothalamus. In these experiments release of ADH was inferred from an abrupt decrease in the urine flow of hydrated animals or by ejection of milk in lactating animals.

Since experiments with direct measurement of ADH release are desirable, we examined responses of dogs bearing diencephalic islands to administration of hypertonic saline solutions (Saito, Yoshida & Nakao, 1969). The levels of ADH in plasma were measured before

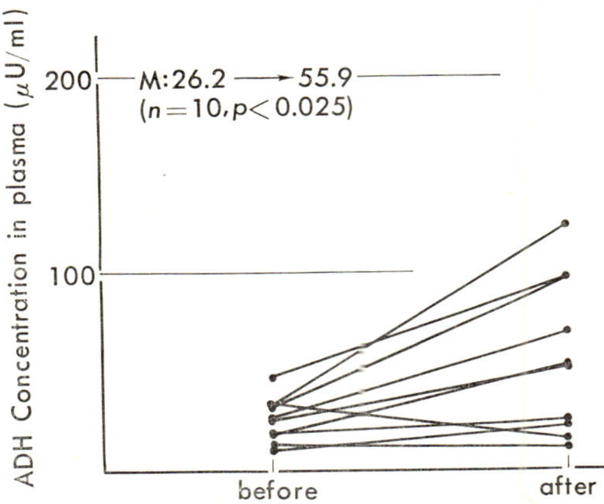

Fig. 1. Effect of infusion of hypertonic saline on release of ADH in dogs with diencephalic island (from Saito *et al.* 1969).

and after the injection of the hypertonic saline. The telencephalon of the dog was removed by suction, and transection of the midbrain was done at the level of the superior colliculus by removing a wedge of tissue by aspiration. The plasma ADH titers were measured using ethanol-anesthetized hydrated rats after extraction of ADH from plasma with trichloracetic acid and purification of the hormone by resin column chromatography (Yoshida, Motohashi, Ibayashi & Okinaka, 1963).

The intravenous infusion of the hypertonic saline in intact dogs resulted in an increase in ADH titers in jugular venous plasma from 34.4 ± 8.4 μU/ml (mean $\pm$SE) to 72.2 ± 12.3 μU/ml. In dogs with diencephalic islands, concentrations of ADH rose from a control level of 26.2 ± 3.6 to 55.9 ± 12.5 after the hypertonic saline infusion (Fig. 1). These increases in plasma ADH titers were statistically significant. There was no significant difference in the increments in plasma ADH titers between these two groups of dogs. These results support the concept that the osmoreceptor is located in the hypothalamus.

However, the possibility that the osmoreceptors are not confined to a particular cytoarchitectonically defined cell group or area but are distributed diffusely in the hypothalamus is not excluded by these ablation experiments. Indeed, many electrophysiological studies revealed the existence of osmosensitive cells spread widely throughout the anterior hypothalamus and in other central nervous structures (Clemente, Sustin & Silverstone, 1957; Cross & Green, 1959; Brooks, Ushiyama & Lange, 1962; Joynt, 1964; Hayward & Vincent, 1970). Euler (1953), in defining the receptors, said that they generate impulses with a frequency roughly proportional to the stimulus strength within certain limits and do not merely modulate pre-existing impulse traffic and that they do not have any other function, e.g., that of a synapse or an interneuron. According to Joynt (1964) the osmosensitive cells in the hypothalamus fulfill these conditions.

Hayward *et al.* (1970) recorded the activity of single neurons in the anterior hypothalamus while repeatedly injecting solutions of varying tonicity into a common carotid artery of unanesthetized monkeys. There were two types of osmosensitive cells, namely, specific and non-specific. They identified as specific osmosensitive neurons those that were responsive almost exclusively to intracarotid osmotic stimuli and poorly responsive to non-noxious arousing stimuli. These neurons were located in and around the supraoptic nuclei. Other nonspecific osmosensitive cells, responsive to both osmotic and arousing stimuli, were distributed throughout the anterolateral hypothalamus. They further subdivided the specific osmosensitive cells into the monophasic osmosensitive cells lying in the immediate perinuclear zone of the supraoptic nucleus and the biphasic cells in the supraoptic nucleus. They suggested the monophasic cells are the osmoreceptor of Verney. These electrophysiological methods are indeed very good tools in the search for osmosensitive elements. However, the osmosensitive cells cannot be identified as those concerned with ADH release by this technique. Evidence is limited to circumstantial.

It may be concluded that enough evidence that shows the presence of osmosensitive elements in the hypothalamus has been accumulated at present. But it is not clear whether these elements are in a discrete cell group or are diffusely dispersed in the structure. Nor is it impossible that osmoreceptors other than Verney's exist somewhere in the body. It has been noted that orally administered water was excreted more rapidly than the same amount of water parenterally administered (Ginsburg, 1912).

Haberich, Aziz & Nowacki (1965) reported the presence of an osmoreceptor located in the liver. They observed a diuretic response in rats when a small amount of water was infused via a portal vein. The same amount of water did not produce diuresis when it was administered via a peripheral vein. However, Schneider, Davis, Robb, Baumber, Johnson & Wright (1970) failed to confirm these results in the dog. Although Lydtin (1969) reported diuresis after portal infusion of a hypotonic saline in rats, this diuresis was not water diuresis but osmotic.

Besides ADH release, osmotic stimuli cause thirst sensation and drinking. The existence of a unitary control center for the joint activation of these responses is a very attractive hypothesis. One of the approaches to this problem is to induce drinking experimentally unaccompanied by antidiuresis. Drinking can be produced in animals by chemical or electrical stimulation of the lateral hypothalamic area. Andersson & McCann (1955) demonstrated in the goat that electrical stimulation of the rostral part of the drinking area produced an inhibition of water diuresis as well as polydipsia. They did not observe inhibition of water diuresis when they stimulated the most posterior part of the drinking area.

Recently Andersson and his colleagues (1967, 1969) examined this problem. They have shown that polydipsia and the inhibition of water diuresis may both be elicited by the infusion of certain hypertonic solutions into the third ventricle. The two centers reacted differently, however, to different solutions. They have concluded that there are probably two groups of osmoreceptors, partially overlapping in the hypothalamus, one group concerned with water intake and the other with the regulation of ADH secretion. According to Mogenson (1969), the drinking area in the rat is considerably more posterior than in the goat, so that osmoreceptors concerned with water intake and those concerned with ADH may not overlap in the rat. Mendelson, Feinberg & Berman (1971) were also able to produce drinking without antidiuresis in most of the rats they used. It can be concluded that the two effects seem to be mediated by anatomically distinct systems.

Very little information has been gathered in recent years on the nature of the osmoreceptor cells. According to Verney (1947) the osmoreceptors are stimulated by a rise in tonicity in the extracellular fluids, which reduces the volume of the receptor cells. This hypothesis was based on the finding that ADH release was not obtained by intracarotid injection of hypertonic urea solutions. The lack of response to urea was explained as due to the high diffusibility of the substance. This concept needs reinvestigation in light of the later information that urea enters the brain at about one-tenth the rate at which it enters muscle and that it causes marked dehydration of the brain. This suggests either that the permeability of the hypothalamic osmoreceptors to urea differs from that of central nervous system tissue in general, or that dehydration of these hypothalamic areas does not stimulate release of ADH (Kleeman & Cutler, 1963). Andersson & Westbye (1970) suggested that, rather than strictly osmotic factors, the intracellular Na concentration may determine the activity of these mechanisms.

VOLUME REGULATION AND ADH

Changes in the blood volume can be monitored by receptors located in portions of the vascular tree by sensing the transmural pressure or by sensing the degree of stretch of the vascular wall. Gauer *et al.* (1963) have divided the cardiovascular system into the low and high pressure systems. The low pressure system consists of the capacitance vessels of systemic circulation, the right side of the heart, the pulmonary vessels, and the left atrium. The high pressure system is the arterial system. According to them, receptors located in the low pressure system are more sensitive to a volume change than those that are located in the high pressure system. The degree of fullness of the bloodstream is reflected primarily in the tone of the capacitance vessels, and the status of the high pressure system is determined by the regulation of cardiac output and the tone of the resistance vessels. They called receptors located in the low pressure system volume receptors. The receptors located in the high pressure system are generally called baroreceptors.

The concept of volume receptors proposed by Henry and Gauer has promoted studies on control of ADH secretion by changes in the blood volume. The left atrium has been suggested as the volume receptor for ADH release. A great deal of evidence for and against the left atrium volume receptor concept has been accumulated. However, the controversy has not yet been settled. The author feels that the division of the circulatory system into the low and high pressure systems is somewhat arbitrary. It is very difficult to find any phenomenon or procedure that changes the degree of filling of the low pressure system without concomitant changes in the circulatory status in the high pressure system. Following is a review of studies on volume receptors for ADH release including experiments performed in the author's laboratory.

Hemorrhage is one of the phenomena that acutely reduces the circulating blood volume. Many investigators have reported a large increase of ADH titers in blood after hemorrhage (Ginsburg & Heller, 1953; Baratz & Ingraham, 1960; Weinstein, Berne & Sachs, 1960; Dyball, 1968; Rocha e Silva & Rosenberg, 1969). We measured plasma ADH titers before and after hemorrhage in dogs under Nembutal anesthesia (Saito *et al.* 1969; Saito & Yoshida, 1971; Murase & Yoshida, 1971). The hemorrhage was induced via a femoral artery until the mean arterial blood pressure fell to a level below 50% of the initial value and less than 50 mmHg. The plasma ADH concentrations rose from 41.4 ± 11.0 μU/ml to 360 ± 74.6 μU/ml in jugular venous blood (mean $\pm$SE) or from 19.4 ± 7.9 to 110 ± 37.7 μU/ml in femoral arterial blood.

The rise of ADH levels in blood after hemorrhage may be due to two factors. One is an increased release of ADH from the neurohypophysis, and the other is a decreased blood ADH clearance. Errington & Rocha e Silva (1971) estimated these two rates during the hypotensive hemorrhage in the dog. The results showed that the rapid increase of blood ADH after hemorrhage was mainly caused by an increased discharge of the hormone from neurohypophysis.

Weinstein *et al.* (1960) estimated the rate of release of the hormone from the neurohypophysis of two dogs after hemorrhage as 142 and 297 mU/min. Rocha *et al.* (1969) estimated that the secretion rate ranges from 0.8 to 6.0 mU/min/kg. The secretion rate of the hormone during hemorrhage seems to be rather superfluous. The infusion of the hormone at a rate of 50 μU/kg/min produces nearly maximal antidiuresis in the dog (Lauson, 1960). Recently McNeill, Stark & Greenway (1970) reported the contribution of vasopressin in intestinal vasoconstriction after hemorrhage. It is possible that the large amount of vasopressin in blood plays a role in the defensive mechanism against hemorrhagic shock.

According to Gauer *et al.* (1963), if a volume receptor located somewhere in the low pressure system is involved in vasopressin release during hemorrhage, the blood ADH titers should increase before changes appear in the hemodynamics of the high pressure system. In order to verify this possibility we performed the following experiments. Three mongrel dogs were used. The spleen of each dog was extirpated at least 4 weeks before the experiment in order to

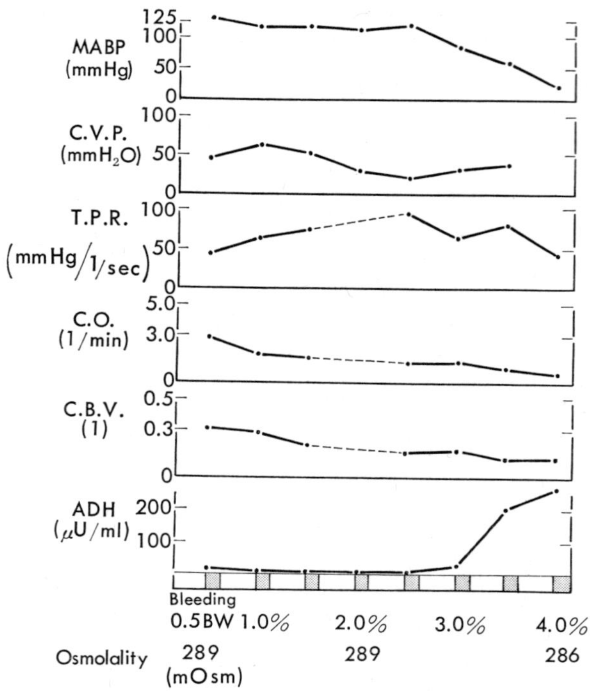

Fig. 2. Effect of hemorrhage on the level of plasma ADH and the cardiovascular system in the dog (Murase & Yoshida, unpublished observations).

eliminate discharge of the stored blood from the spleen in response to hemorrhage. The animals under chloralose-urethan anesthesia were subjected to graded hemorrhage. The mean arterial pressure (mABP), and the central venous pressure (CVP) were continuously recorded. The cardiac output (CO) was measured using the standard dye dilution technique with indocyan green and a Gilson cuvett densitometer. The total peripheral resistance (TPR) and the central blood volume (CBV) were calculated as TPR=mABP/CO and CBV=CO/60$\times\bar{t}$ ($\bar{t}$, mean circulation time). The plasma ADH was assayed by the method of Yoshida *et al.* (1963). The results of one of the experiments are shown in Fig. 2. We failed to observe a significant ADH increase before the decrease in cardiac output appeared.

Share (1967*a*) observed that the threshold for ADH release in response to hemorrhage was between 10 and 20% loss of blood volume. The hemodynamic changes observed during the hemorrhage were a slight reduction in arterial pulse pressure, an increase in heart rate, and a fall in central venous pressure. These results were at variance with our results. In our experiments, there were large differences between the dogs in the amount of blood withheld before the hemorrhage caused a significant increase in plasma ADH concentrations. The significant increase in the plasma ADH titers was always accompanied by a significant fall in the mean arterial blood pressure. These observations suggest that the role of the volume receptor in release of ADH in response to hemorrhage is minor.

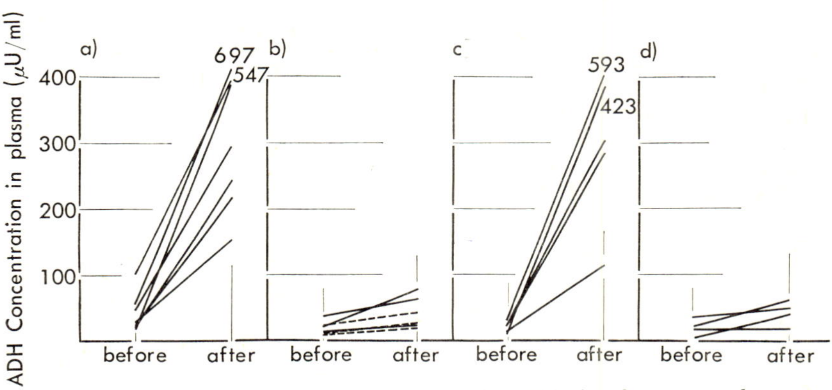

Fig. 3. Effect of hemorrhage on the level of ADH in jugular venous plasma. ——, 4 hr after operation; ----, 24 hr after operation. a) Intact dogs. b) Dogs with cervical vagus nerve sectioned. c) Dogs with abdominal vagus nerve sectioned. d) Dogs with sinus nerves divided (from Saito *et al.* 1969).

These experiments, however, did not exclude the possibilities that the anesthesia used in these experiments paralyzed the volume receptors or that the volume receptor stimulated a release of ADH in an amount that was not detectable by our experimental procedures. Besides these experiments, we induced graded hemorrhage in three intact dogs under Nembutal anesthesia. The results were the same as those observed in the experiments in which splenectomized dogs were used.

The effects of vagotomy and of denervation of the carotid sinus are shown in Fig. 3. The cervical vagotomy as well as the denervation of the carotid sinus attenuated ADH release in response to hemorrhage. Clark & Rocha e Silva (1967) and Share (1967*b*) also reported the existence of a baroreceptor for ADH release in the carotid sinus. We have reported that the hypotension produced by injection of nitroprussid soda solution caused an increase in plasma ADH titers in the dog (Murase *et al.* 1971). It is apparent that the hypotension without changes in the circulating blood volume causes ADH release from the neurohypophysis.

The controversy over the significance of the left atrium as a volume receptor has not been settled at present. Since the proposal of Henry, Gauer & Reeves (1956), this concept has been a subject of many investigations. Evidence for this concept has been derived from experiments utilizing such experimental methods as distension of the left atrium by a ballon (Henry *et al.* 1956; Share, 1965; Brennan, Malvin, Jochim & Roberts, 1971) and positive or negative pressure breathing (Gauer, Henry, Sieker & Wendt, 1954; Baratz *et al.*, 1960). Hemorrhage (Henry, Gupta, Meehan, Sinclair & Share, 1968) and fluid administration (Zehr, Johnson & Moore, 1969) were also used as tools to change the left atrial pressure. To a varying degree, however, these techniques also cause associated changes in arterial blood pressure, cardiac output, and other hemodynamic variables (Goetz, Hermreck, Slick & Starke, 1970*b*). It has been previously reported that the positive pressure breathing caused antidiuresis and an increase in the antidiuretic activity in blood (Baratz *et al.* 1960).

Recently Baratz, Philbin & Patterson (1970, 1971) re-examined this problem. The plasma ADH levels did not change significantly in the dog during intermittent positive pressure breathing with an end-expiratory pressure of zero (Baratz *et al.* 1970). According to them, this type of intermittent positive pressure breathing did not impose any stress on the circulatory system. Neither cardiac output nor peripheral arterial pressure changed significantly. In contrast, positive pressure breathing with positive end-expiratory pressure or continuous positive pressure breathing caused a significant increase in ADH concentrations. Vagotomy did not abolish this increase. Concomitantly with the increase in ADH levels in plasma, a decrease in the cardiac output was observed. It has been known that continuous positive pressure breathing causes severe circulatory stress. The peripheral arterial pressure decreases significantly, the central venous pressure rises, and the heart rate decreases (Baratz *et al.* 1960).

The distension of atrial chambers with a balloon, which causes diuresis and a decrease in plasma ADH levels, also causes associated changes in the hemodynamics. Brennan *et al.* (1971) reported an increase in the heart rate on distension of the left atrium by inflation of a balloon. This may increase the number of impulses per unit time arising from the baroreceptors, such as the sino-aortic receptor. And in turn, this increase may inhibit ADH release from the neurohypophysis.

Goetz and his associates (1970*a*, *b*) decreased the mean atrial transmural pressure in dogs by atrial tamponade instead of distension of the chamber by inflation of a balloon. There was no significant change from control values in the level of plasma ADH after 5 and 30 min of atrial tamponade. This procedure did not produce any alteretions in the mean arterial pressure, cardiac output, or heart rate. These recent observations cast some doubt on the left atrium volume receptor concept and favor the view that the baroreceptors play an important role in ADH release from the posterior pituitary.

It is interesting that the conditions that increase the plasma ADH concentration also produce an increase in the activity of the renin-angiotensin system in blood. Hemorrhage, postural changes, occlusion of the common carotids, and vagotomy were reported to stimulate the renin-angiotensin system (Brown, Davies, Lever, Robertson & Verniory, 1966*a*; Brown, Davies, Lever, McPherson & Robertson, 1966*b*; Conn, Rovner & Cohen, 1965; Hodge, Lowe & Vane, 1969). Bonjour & Malvin (1970) reported stimulation of ADH release by the renin-angiotensin system. They suggested that the rise of plasma ADH titers observed in conditions associated with increased plasma renin activities might be due to stimulation of the neurohypophysis by angiotensin. We have examined the response of the neurohypophysis to hemorrhage in dogs with diencephalic islands. The dogs were decorticated, and the connections between the diencephalon

TABLE 1. Effect of infusion of saline solutions and subsequent hemorrhage on the concentration of ADH in jugular vein plasma (from Saito *et al.* 1971)

Solutions	Number of experiments	ADH concentration (μU/ml)		
		Before infusion	After infusion	After hemorrhage
Normal saline	10	10.2±3.4	8.4±1.4	204.0± 60.4
10% saline	13	12.9±3.0	22.2±7.3	274.8± 70.8
30% saline	10	12.8±2.6	19.6±3.6	325.7±112.1

and the midbrain were divided. The animals failed to increase the level of plasma ADH in response to hemorrhage (Saito *et al.* 1969). This observation proved the importance of neural stimuli for ADH release from the neurohypophysis rather than humoral stimuli such as angiotensin.

During this series of experiments we have observed attenuation in ADH release in response to infusion of hypertonic saline as compared with hemorrhage as shown in Table 1 (Saito *et al.* 1971). While the mechanism of this attenuation remains to be studied further, it seems likely that hemorrhage is a more effective stimulus to ADH release than plasma hypertonicity. The receptors for release of ADH in response to these stimuli are known to be localized separately. It is possible that the amount of ADH released is limited by the nature of the stimuli. Moran (1971) recently discussed this phenomenon and called it a hierachy control of ADH release.

REFERENCES

Andersson, B., Dallman, M. F. & Olsson, K. (1969). Observation on central control of drinking and of the release of antidiuretic hormone (ADH). *Life Sci. Oxford* **8**, 425–432.

Andersson, B. & McCann, S. M. (1955). Drinking antidiuresis and milk ejection from electrical stimulation within the hypothalamus of the goat. *Acta physiol. scand.* **35**, 191–201.

Andersson, B., Olsson, K. & Warner, R. G. (1967). Dissimilarities between the central control of thirst and the release of antidiuretic hormone (ADH). *Acta physiol. scand.* **61**, 57–64.

Andersson, B. & Westbye, O. (1970). Synergistic action of sodium and angiotensin on brain mechanisms controlling fluid balance. *Life Sci. Oxford* **9**, 601–608.

Baratz, R. A. & Ingraham, R. C. (1960). Renal hemodynamics and antidiuretic hormone release associated with volume regulation. *Am. J. Physiol.* **198**, 565–570.

Baratz, R. A., Philbin, D. M. & Patterson, R. W. (1970). Urinary output and plasma levels of antidiuretic hormone during intermittent positive pressure breathing in the dog. *Anesthesiology* **32**, 17–21.

Baratz, R. A., Philbin, D. M. & Patterson, R. W. (1971). Plasma antidiuretic hormone and urinary output during continuous positive pressure breathing in dogs. *Anesthesiology* **34**, 510–513.

Bonjour, J. P. & Malvin, R. L. (1970). Stimulation of ADH release by the renin-angiotensin system. *Am. J. Physiol.* **218**, 1555–1559.

Brennan, L. A., Jr., Malvin, R. L., Jochim, K. E. & Roberts, D. E. (1971). Influence of right and left atrial receptors on plasma concentrations of ADH and renin. *Am. J. Physiol.* **221**, 273–278.

Brooks, C. McC., Ushiyama, J. & Lange, G. (1962). Reactions of neurones in or near the supraoptic nuclei. *Am. J. Physiol.* **202**, 478–490.

Brown, J. J., Davies, D. L., Lever, A. F., Robertson, J. I. S. & Verniory, A. (1966*a*). The effect of acute hemorrhage in the dog and man on plasma renin concentration. *J. Physiol.* **182**, 649–663.

Brown, J. J., Davies, D. L., Lever, A. F., McPherson, D. & Robertson, J. I. S. (1966*b*). Plasma renin concentration in relation to changes in posture. *Clin. Sci.* **30**, 278–284.

Bunag, R. D., Page, I. H. & McCubbin, J. W. (1966). Neural stimulation of release of renin. *Circulation Res.* **19**, 851–858.

Clark, B. J. & Rocha e Silva, M., Jr. (1967). An afferent pathway for the selective release of vasopressin in response to carotid occlusion and hemorrhage in the cat. *J. Physiol.* **191**, 529–542.

Clemente, C. D., Sustin, J. & Silverstone, J. T. (1957). Changes in electrical activity of the medulla on the intravenous injection of hypertonic solutions. *Am. J. Physiol.* **188**, 193–198.

Conn, J. W., Rovner, D. R. & Cohen, E. L. (1965). Normal and altered function of the renin-angiotensin-aldosterone system in man: Application in clinical and research medicine. *Ann. Intern. Med.* **63**, 266–284.

Cross, B. A. & Green, J. D. (1959). Activity of single neurones in the hypothalamus: Effect of osmotic and other stimuli. *J. Physiol.* **148**, 554–569.

Dyball, R. E. (1968). Stimuli for the release of neurohypophysial hormones. *Br. J. Pharmacol. Chemother.* **33**, 319–328.

Erringron, M. L. & Rocha e Silva, M., Jr. (1971). The secretion and clearance of vasopressin during the development of irreversible hemorrhagic shock. *J. Physiol.* **217**, 43–45

Euler, C. von (1953). A preliminary note on slow hypothalamic osmopotentials. *Acta Physiol. Scand.* **29**, 133–136.

Gauer, O. H. & Henry, J. P. (1963). Circulatory basis of fluid volume control. *Physiol. Rev.* **43**, 432–481.

Gauer, O. H., Henry, J. P., Sieker, H. O. & Wendt, W. E. (1954). The effect of negative pressure breathing on urine flow. *J. Clin. Invest.* **33**, 287–296.

Ginsburg, W. (1912). Diureseversuche. *Naunyn-Schmiedebergs Arch. Exp. Path. Pharmak.* **69**, 381–392.

Ginsburg, M. & Heller, H. (1953). Antidiuretic activity in blood obtained from various parts of the cardiovascular system. *J. Endocr.* **9**, 274–282.

Goetz, K. L., Bond, G. C., Hermreck, A. S. & Trank, J. W. (1970*a*). Plasma ADH levels following a decrease in mean atrial transmural pressure in dogs. *Am. J. Physiol.* **219**, 1424–1428.

Goetz, K. L., Hermreck, A. S., Slick, G. I. & Starke, H. S. (1970*b*). Atrial receptors and renal function in conscious dogs. *Am. J. Physiol.* **219**, 1417–1423.

Haberich, F. J., Aziz, O. & Nowacki, P. E. (1965). Über einen osmoreceptorisch tätigen mechanismus in der Leber. *Pflügers Arch.* **285**, 73–89.

Hayward, J. N. & Vincent, J. D. (1970). Osmosensitive single neurones in the hypothalamus of unanesthetized monkeys. *J. Physiol.* **210**, 947–972.

Henry, J. P., Gauer, O. H. & Reeves, J. L. (1956). Evidence of the atrial location of receptors influencing urine flow. *Circulation Res.* **4**, 85–90.

Henry, J. P., Gupta, P. D., Meehan, J. P., Sinclair, R. & Share, L. (1968). The role of afferents from the low pressure system in the release of antidiuretic hormone during nonhypotensive hemorrhage. *Can. J. Physiol. Pharmacol.* **46**, 287–295.

Hodge, R. L., Lowe, R. D. & Vane, J. R. (1969). Role of the vagus nerve in the control of concentration of angiotensin II in the circulation. *Nature, Lond.* **221**, 177–179.

Jewell, P. A. & Verney, E. B. (1957). An experimental attempt to determined the site of the neurohypophysial osmoreceptors in the dog. *Phil. Trans. B.* **240**, 197–324.

Joynt, R. J. (1964). Functional significance of osmosensitive units in the anterior hypothalamus. *Neurology, Minneap.* **14**, 584–590.

Kleeman, C. R. & Cutler, R. E. (1963). The neurohypophysis. *Ann. Rev. Physiol.* **25**, 385–432.

Lauson, H. D. (1960). Vasopressin and oxytocin in the plasma of man and other mammals. In *Hormones in Human Plasma*, ed. Antoniades, H. N., pp. 225–293. Boston: Little Brown.

Lydtin, H. (1969). Untersuchungen über mechanismen der Osmo- und Volumenregulation. II. Untersuchungen über den Einfluss intravenos, intraportal und oral zugefuhrter hypotoner Kochsalzlöungen auf die Diurese des Hundes. *Z. ges. exp. Med.* **149**, 193–210.

McNeill, J. R., Stark, R. D. & Greenway, C. V. (1970). Intestinal vasoconstiction after hemorrhage: Roles of vasopressin and angiotensin. *Am. J. Physiol.* **219**, 1342–1347.

Mendelson, J., Feinberg, L. E. & Berman, J. (1971). Dissociation of diencephalic mechanisms controlling water intake and water retention in the rat. *Am. J. Physiol.* **220**, 1768–1774.

Mogenson, F. J. (1969). Test of a possible influence of the drinking system of the lateral hypothalamus on the mechanisms controlling antidiuretic hormone. *Can. J. Physiol. Pharmacol.* **47**, 109–111.

Moran, W. H. (1971). CPPB and vasopressin secretion. *Anesthesiology* **34**, 501–504.

Murase, T. & Yoshida, S. (1971). Effect of hypotension on levels of antidiuretic hormone in plasma in dogs. *Endocr. Jap.* **18**, 215–219.

Philbin, D. M., Baratz, R. A. & Patterson, R. W. (1970). The effect of carbon dioxide on plasma antidiuretic hormone levels during intermittent positive pressure breathing. *Anesthesiology* **33**, 345–349.

Rocha e Silva, M., Jr. & Rosenberg, M. (1969). The release of vasopressin in response to hemorrhage and its role in the mechanism of blood pressure regulation. *J. Physiol.* **202**, 535–557.

Saito, T. & Yoshida, S. (1971). Levels of antidiuretic hormone in plasma after hemorrhage and infusion of hypertonic saline in dogs. *Endocrinology* **88**, 1511–1513.

Saito, T., Yoshida, S. & Nakao, K. (1969). Release of antidiuretic hormone from neurohypophysis in response to hemorrhage and infusion of hypertonic saline in dogs. *Endocrinology* **85**, 72–78.

Schneider, E. G., Davis, J. O., Robb, C. A., Baumber, J. S., Johnson, J. A. & Wright, R. S. (1970). Lack of evidence for hepatic osmoreceptor mechanism in conscious dogs. *Am. J. Physiol.* **218**, 42–45.

Share, L. (1965). Effects of carotid occlusion and left atrial distension of plasma vasopressin titer. *Am. J. Physiol.* **208**, 219–223.

Share L. (1967*a*). Vasopressin, its bioassay and the physiological control of its release. *Am. J. Med.* **42**, 701–712.

Share, L. (1967*b*). Role of peripheral receptors in the increased release of vasopressin in response to hemorrhage. *Endocrinology* **81**, 1140–1146.

Smith, M. W. (1957). Salt and water volume receptors. *Am. J. Med.* **23**, 623–652.

Sundsten, J. W. & Sawyer, C. H. (1961). Osmotic activation of neurohypophysial hormone release in rabbits with hypothalamic islands. *Expl Neurol.* **4**, 548–561.

Verney, E. B. (1947). The antidiuretic hormone and the factors which determine its release. *Proc. R. Soc.* B. **135**, 25–106.

Weinstein, H., Berne, R. M. & Sachs, H. (1960). Vasopressin in blood: Effect of hemorrhage. *Endocrinology* **66**, 712–718.

Woods, J. W., Bard, P. & Bleier, R. (1966). Functional capacity of the deafferented hypothalamus: Water balance and responses to osmotic stimuli in the decerebrate cat and rat. *J. Neurophysiol.* **29**, 751–768.

Yoshida, S., Motohashi, K., Ibayashi, H. & Okinaka, S. (1963). Method for the assay of antidiuretic hormone in plasma with a note on the antidiuretic titer of human plasma. *J. Lab. clin. Med.* **62**, 279–293.

Zehr, J. E., Johnson, J. A. & Moore, W. W. (1969). Left atrial pressure, plasma osmolality, and ADH levels in the unanesthetized ewe. *Am. J. Physiol.* **217**, 1672–1680.

ROLE OF BRAIN BIOGENIC AMINES IN THE REGULATION OF ACTH SECRETION

TSUTOMU HIROSHIGE and KAZUO ABE

The role of brain biogenic amines in the regulation of ACTH secretion under basal and stressful conditions was studied. First, the ACTH-releasing potency of biogenic amines present in the hypothalamus was examined by an intrapituitary injection technique. It was found that most amines, including norepinephrine, epinephrine, L-dopa, dopamine, serotonin, putrescine, and acetylcholine did not show significant ACTH-releasing activity, whereas histamine and spermidine exhibited an inherent potential to release ACTH in relatively large amounts.

Next, CRF content in the rat hypothalamus was estimated after depleting brain catecholamines by reserpine pretreatment. CRF activity was found to be reduced markedly in the female rat, but in the male rat the activity persisted, and a circadian rhythm was demonstrated. Furthermore, an incremental response of CRF to stress was observed in the male rat under reserpine. The increment was demonstrated only when the basal level was low in the morning.

Finally, various amounts of monoamines were injected into the brain ventricles of unanesthetized and Nembutal-anesthetized rats. Under our experimental conditions, most monoamines injected caused a marked stimulation of ACTH release. Furthermore, these amines failed to suppress the stress-induced increment of ACTH release in the anesthetized male rat. The significance of these findings is discussed in relation to possible roles of brain monoamines in the manifestation of circadian rhythm and stress-induced increase of CRF activity in the hypothalamus.

INTRODUCTION

The presence of catecholamines in the central nervous system was demonstrated in 1946 by von Euler. However, he ascribed the origin of the amines to brain vasomotor nerves. In 1954, Vogt made an extensive study on the distribution of sympathin in the dog as well as in the cat and rat brains. She found that the distribution of amines was not uniform as expected from the vascular supply, but was most abundant in the hypothalamus (1 μg/g wet weight), followed by the midbrain area. In the dog hypothalamus, a definite difference in their concentration was found between dorsal and ventral parts and also between medial and lateral. The norepinephrine content was evidently higher in the medial ventral portion of the hypothalamus. On the basis of these findings, she suggested that the role of catecholamines in the brain may lie not in vasomotor control, but in other functions such as neurotransmitters or humoral transmitters, which are supposedly to convey information from the hypothalamus to the anterior pituitary. The last possibility, in fact, was suggested by Markee, Sawyer & Hollinshead (1948) and Sawyer and his colleagues (1949, 1950) in their studies on the role of epinephrine in inducing ovulation in the rabbit.

Introduction of the histochemical fluorescence technique by Carlsson, Falck & Hillarp (1962) brought about decisive progress in the study of the localization of monoamines in the central nervous system. Fuxe (1964, 1965) and Fuxe & Hökfelt (1967, 1969) applied this technique in an excellent series of experiments on the hypothalamic histochemistry and disclosed several morphologic characteristics of amine distribution, especially in relation to hormonal states of experimental animals. These results support the view that the hypothalamic monoamines play an important role in the neuroendocrine regulation of the anterior pituitary function. Wurtman (1970) enumerated six different loci at which brain catecholamines may participate in the regulation of secretion from the anterior pituitary.

(1) As neurotransmitters in a tract that provides an input to hypophysiotropic cells in the hypothalamus,

(2) as the neurotransmitters that cause the hypothalamic cells to secrete the releasing hormone,

(3) within the hypophysiotropic cell to liberate the releasing hormone,

(4) as neurohormones acting directly on the anterior pituitary,

(5) catecholamines might be delivered to the pituitary through the general circulation and modify the secretion of the releasing hormone, and

(6) they might alter the sensory input to the brain, thus altering secondarily the secretion of the releasing hormone.

More than 20 years ago, Long's group in Yale (McDermott, Fry, Brobeck & Long, 1950*a*,*b*; Long, 1956) and Fortier (1951) examined the effects of direct injection of epinephrine and histamine into the grafted pituitaries. These were, in a sense, the first trials of *in vivo* direct application of biogenic amines on the study of the ACTH release mechanism. On the basis of their findings, Long proposed an epinephrine theory purporting that under stress epinephrine released from the adrenal medulla reaches the anterior pituitary through the general circulation and activates the gland. This theory thus corresponds to Wurtman's locus (5) mentioned above.

Vogt (1952), however, demonstrated that even in the adrenalectomized rat stress-induced ACTH release (as the adrenal ascorbic acid depletion) was demonstrable, and no elevation of circulating epinephrine level, simultaneously determined, was observed. It is less likely, therefore, that catecholamines act directly, if at all, through the general circulation on the anterior pituitary. The possibility of Wurtman's locus (6) does not appear to be relevant to the present discussion. Thus, among four possibilities left, the possibility that hypothalamic amines *per se* are releasing hormones should be examined first.

This can be done by placing biogenic amines in the anterior pituitary, by use of our intrapituitary injection technique, which was successfully applied to detect the activity of the corticotropin-releasing factor (CRF) in the rat median eminence (Hiroshige, Kunita, Yoshimura & Itoh, 1968*b*). Finally, we should discern among the three possibilities loci (1) to (3). Unfortunately, no decisive methods are available at the moment to differentiate them. But, some attempts were made in this study to discover the effect of catecholamine depletion on ACTH release. In a further attempt, various amounts of biogenic amines were injected into the brain ventricles of unanesthetized and Nembutal-anesthetized rats in order to study the mode of participation of the amines in the regulation of ACTH secretion.

ACTH-RELEASING POTENCY OF BRAIN BIOGENIC AMINES

Catecholamines

Since norepinephrine is the major catecholamine present in the hypothalamus, the inherent ACTH-releasing potency of this amine

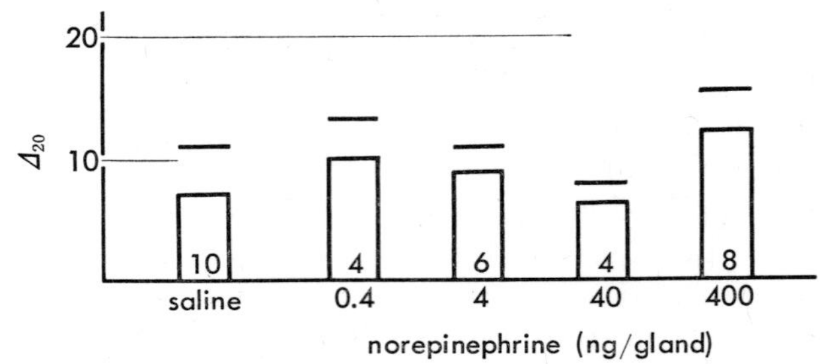

Fig. 1. ACTH-releasing activity of norepinephrine as assayed by direct intrapituitary injection into the anterior pituitary of dexamethasone (300 μg)-Nembutal blocked female rats. Increases of the plasma corticosterone level of assay rats 20 min after the microinjection in a volume of 0.4 μl were considered a measure of ACTH-releasing potency of test substances and are expressed as Δ_{20}. Numbers in the columns indicate the number of determinations. Bars on the top of the columns represent the standard errors (from Hiroshige & Itoh, 1968).

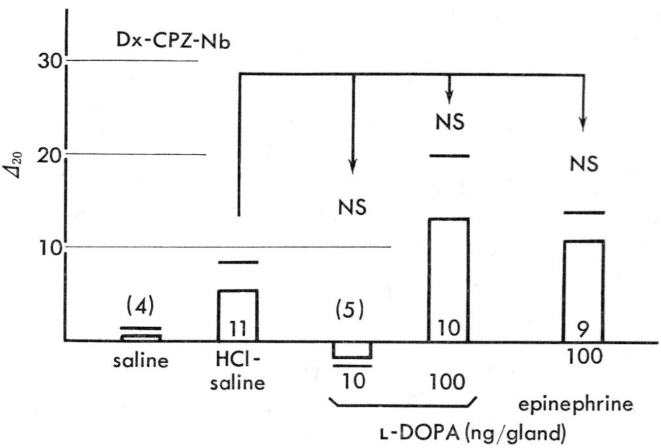

Fig. 2. Effect of intrapituitary injection of L-dopa and epinephrine on ACTH release from the anterior pituitary. Assay female rats were blocked with a combination of chlorpromazine along with dexamethasone (100 μg) and Nembutal. Since the catecholamines were supplied in hydrochloride form, 1.3 mM HCl-saline solution was also used as a control injection.

was examined by the intrapituitary injection method. Results are shown in Fig. 1. Norepinephrine at doses ranging from 0.4 to 400 ng per gland elicited no ACTH release, when assayed in dexamethasone (300 μg)-Nembutal blocked female rats (Hiroshige, Kunita, Yoshimura & Itoh, 1968). Next, the activity of L-dopa and epinephrine was examined, using dexamethasone (100 μg)-chloprromazine-Nembutal blocked female rats (Hiroshige 1973). Results in Fig. 2 indicate that 100 ng of both amines tended to cause an ACTH release, but because of its larger fluctuation the mean value was not significantly different from the HCl-saline injected control. Previously, we reported (Hiroshige, Kunita, Ogura & Itoh, 1698*a;* Hiroshige & Itoh, 1968) that epinephrine at doses of 10 and 100 ng caused a significant release of ACTH, the finding being in line with earlier observations of McDermott *et al.* (1950*a*) and Fortier (1951). But, under a more carefully controlled assay system with the additional use of chlorpromazine (Hiroshige, 1973), the increment did not attain significance. The hypothalamic content of epinephrine, however, is reported to be less than 10% of that of norepinephrine (McIlwain & Bachelard, 1971). In contrast, dopamine content in the hypothalamus is quite high and is comparable to that of norepinephrine. However, no inherent ACTH-releasing potency was found with dopamine as shown in Fig. 3.

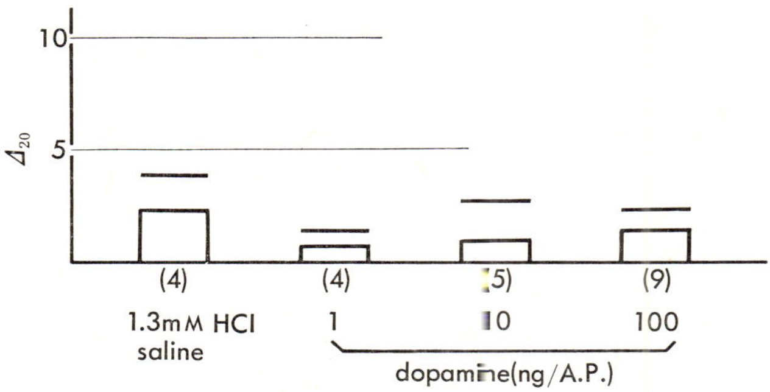

Fig. 3. Effect of intrapituitary injection of dopamine on ACTH release. Hypothalamic content of dopamine *ca.* 1 μg/g.

Serotonin

Serotonin is also known to be present in the hypothalamus in considerable amounts (McIlwain *et al.* 1971). However, serotonin in doses of 1 to 10 ng did not cause ACTH release, as shown in Fig. 4.

Histamine

As shown in Fig. 5, histamine at doses of 1 and 100 ng did not cause ACTH release, whereas 10 ng elicited a marked increment of ACTH secretion. In our previous studies with dexamethasone-Nembutal blocked assay rats, it was found that histamine at doses

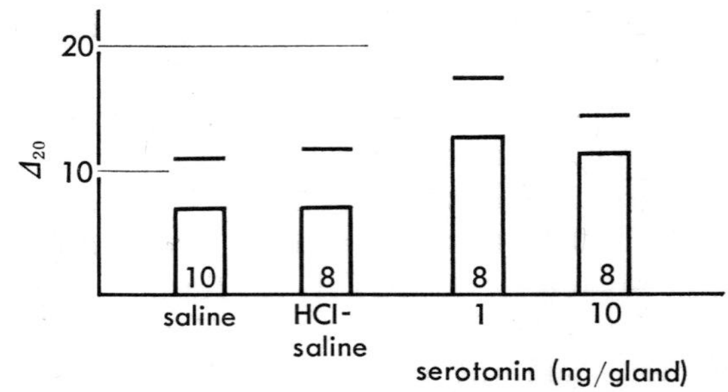

Fig. 4. Effect of intrapituitary injection of serotonin on ACTH release (from Hiroshige & Itoh, 1968).

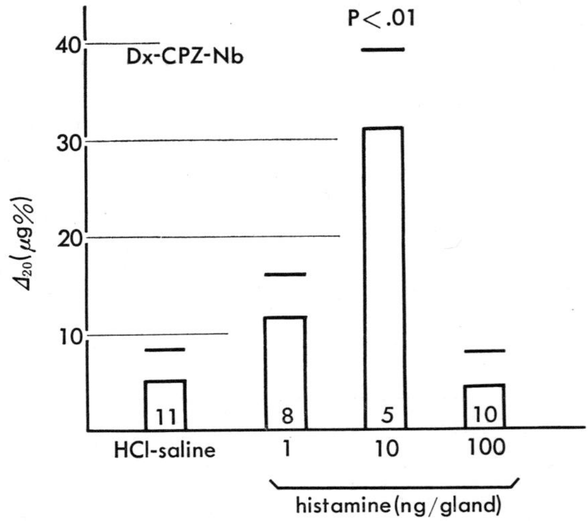

Fig. 5. ACTH-releasing activity of histamine. Histamine was injected directly into the anterior pituitary tissue of assay rats.

of 2 and 20 ng caused a significant ACTH release, while 0.2 ng did not (Hiroshige & Itoh, 1968; Hiroshige *et al.* 1968*a*). Thus, it appears that histamine is a potent ACTH releaser within a limited dose range.

Polyamines

Since polyamines are contained in the brain in considerable amounts (Pearce & Schanberg, 1969), the ACTH-releasing activity of polyamines was examined. Effects of intrapituitary injection of putrescine on ACTH release are shown in Fig. 6. Putrescine at the doses used exhibited no potency. On the other hand, spermidine

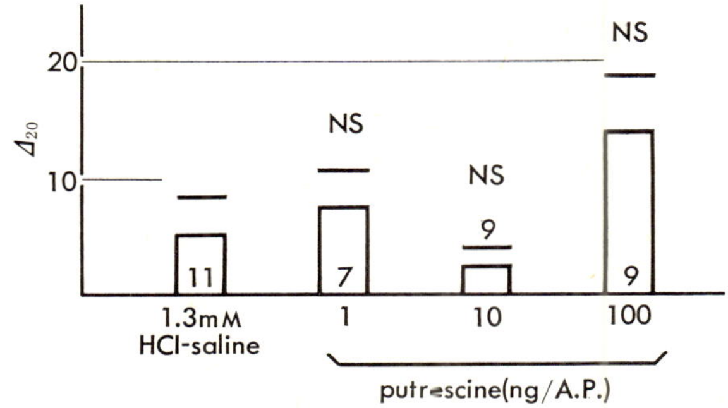

Fig. 6. Effect of intrapituitary injection of putrescine on ACTH release.

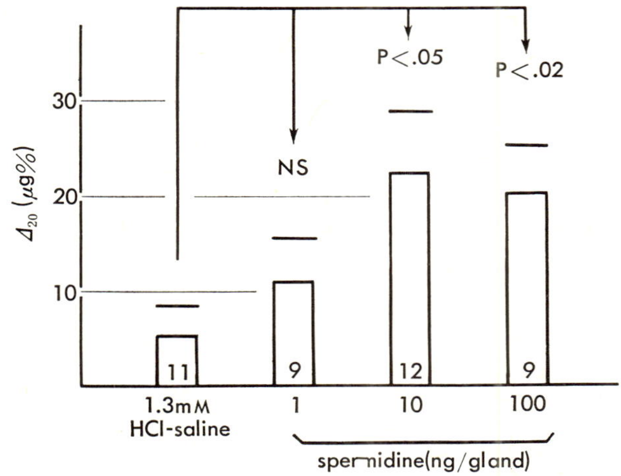

Fig. 7. ACTH-releasing potency of spermidine as assayed by intrapituitary injection technique.

in doses ranging from 10 to 100 ng was proved to be potent in eliciting ACTH release, as shown in Fig. 7, but a low dose (1 ng per gland) was without effect. Since the spermidine content in the rat brain was reported to be approximately 75 μg/g (Pearce *et al.* 1969), it is possible that this amine may act as a potent ACTH releaser in the crude median eminence extracts when the extracts are prepared from a larger tissue block.

Acetylcholine

Acetylcholine content in the brain is quite variable depending on the condition of the animal, but is reported to be about 1 μg/g. As shown in Fig. 8, it showed no potency within the doses used.

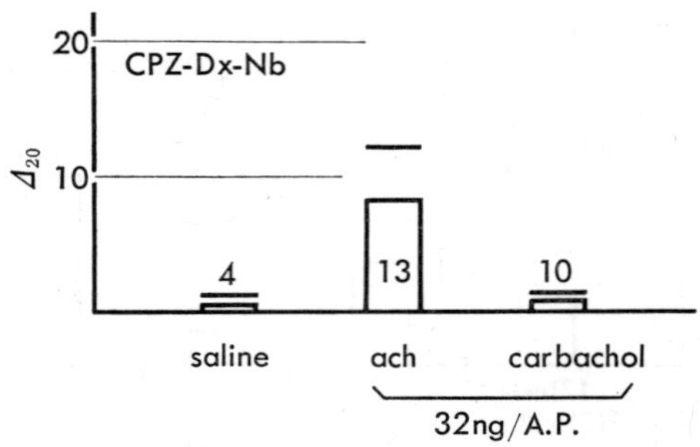

Fig. 8. Effect of intrapituitary injection of acetylcholine and carbachol on ACTH release. Final concentration 8×10^{-5}M.

Hormones

Several polypeptide hormones were tested for their ACTH-releasing potency for comparison. Bradykinin (Fig. 9) did not show any activity, while angiotensin (Fig. 10) exhibited marked activity at a dose of 75 ng per gland, injected directly into the pituitary of assay rats. It was reported (Hiroshige, 1971) that lysine vasopressin, similarly injected into the pituitary, caused no ACTH release at doses of less than 10 ng per gland, but at doses of more than 100 ng it elicited a marked release of ACTH.

A survey of substances tested so far revealed that most substances, such as norepinephrine, epinephrine, L-dopa, dopamine, serotonin, bradykinin, and acetylcholine, did not cause ACTH release at doses ranging from 1 to 100 ng. In contrast, vasopressin, histamine, spermidine, and angiotensin caused a significant release of ACTH at doses between 10 and 100 ng per gland. Although more extensive

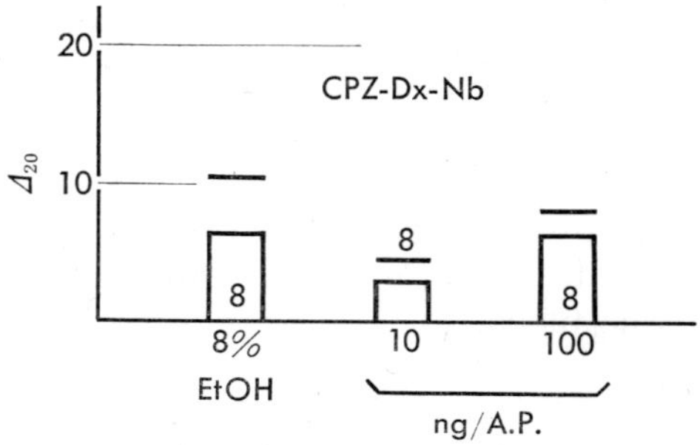

Fig. 9. Effect of intrapituitary injection of bradykinin on ACTH release from the anterior pituitary gland of dexamethasone-chlorpromazine-Nembutal blocked female rats.

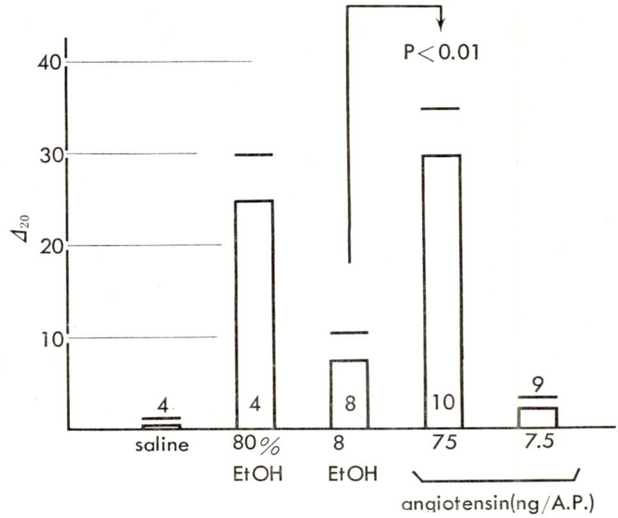

Fig. 10. ACTH-releasing actibity of angiotension as assayed by intrapituitary injection technique. Since the hormone was dissolved in 8% ethanol solution, this solution was also used as control.

studies using wider ranges of doses are needeed, it appears that these substances have an inherent ACTH-releasing potency in relatively large amounts. It is tempting therefore to separate corticotropin-releasing hormone (CRH) from corticotropin-releasing factors (CRFs), as proposed by Yates, Russell & Maran (1971), at least on the basis of results obtained with intrapituitary injection assays.

Into the category of CRFs, angiotensin and spermidine may be added and also histamine, if possible, in addition to vasopressin and still other substances which will be found to be potent releasers in the future. In fact, some of the vasoactive polypeptides were implicated in the ACTH release evoked by hypotension (Redgate, 1968). Nevertheless, it is less likely that these potent substances themselves act as releasers of ACTH in the crude median eminence extracts because of their limited quantity. The interesting possibility of a potentiation of CRF action by vasopressin was suggested by Yates *et al.* (1971). This possibility, corresponding to Wurtman's locus (4), appears to be unique and should be considered with other monoamines in the hypothalamus.

In the case of gonadotropin-releasing hormones, Schneider & McCann (1969, 1970) reported that dopamine exhibits neither LH-releasing activity nor a potentiating effect on LH-RH action, but does stimulate the release of LH-RH from the hypothalamic tissue *in vitro*. In contrast, Fuxe & Hökfelt (1969, 1970) claimed an inhibitory action of dopamine on the release of LH-RH on the basis of their histochemical studies on turnover rates of amines in the hypothalamus. Kamberi & McCann (1969) and Kamberi, Schneider & McCann (1970) reported also that dopamine induced a release of FSH-RH from the hypothalamic tissue. In any case, these results indicate that the role of hypothalamic monoamines may lie not in Wurtman's locus (4) but in loci (1) to (3). Elucidation of the loci of action thus awaits further exploration with the advent of new techniques.

EFFECT OF BRAIN AMINE DEPLETION ON ACTH RELEASE

One of the common ways to study the role of brain monoamines in the regulation of ACTH secretion is to examine the pattern of ACTH release after depleting the brain amine by drug administration. Among various drugs that are known to deplete brain amines, reserpine is most frequently used because of its easy availability, although its mode of action appears to be complicated (Bhattacharya & Marks, 1970; Hirsch & Moore, 1968).

Carr & Moore (1968) examined the effects of various noxious stimuli on ACTH release in the rat after reducing the brain amines by pretreatment with reserpine and α-methyltyrosine. They found that 24 hr after reserpine and 4 hr after α-methyltyrosine the brain contents of norepinephrine and dopamine was markedly reduced (1/8 and 1/6 of the control, respectively), while the plasma corticosterone levels were not significantly different from preinjection control values. Furthermore, all stressful procedures caused a significant increase of plasma corticosterone levels. No difference was noted between controls and drug-pretreated animals. Thus, they concluded that brain catecholamines do not play an essential role in stress-induced activation of the pituitary-adrenal system.

Smelik (1967, 1969) implanted reserpine pellets into the rat hypothalamus and detected no change in the stress-induced increments of plasma corticosterone. On the other hand, Bhattacharya & Marks (1969) found a marked depletion of CRF activity in the median eminence of female rats after reserpine or chlorpromazine administration. They interpreted this result as due to removal of an inhibi-

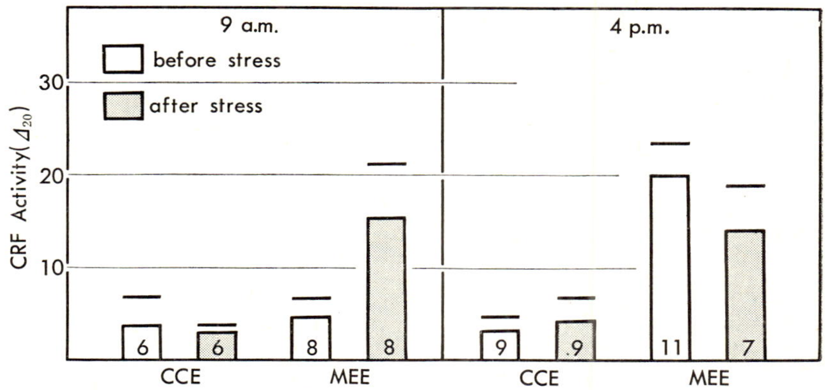

Plasma corticosterone level (μg/100ml)		9 a.m.	4 p.m.
	stress(−)	7.0+1 4(4)	35.4+13.9(4)
	stress(+)	18.2+2.7(4)	36.6+ 2.5(4)

Fig. 11. Stress-induced change and circadian variation of CRF activity in the hypothalamus of male rats after pretreatment with reserpine. Cerebral cortex extracts (CCE) were also assayed as tissue control. Blank columns indicate the activity before stress and dotted columns the activity after stress (ether-laparotomy stress of 2 min duration). MEE stands for median eminence extracts. The plasma corticosterone levels, simultaneously determined, are also shown in this figure (means ± standard errors).

tory action of the amines. We also noted, independently, a marked decrease of CRF content in the median eminence of female rats (Hiroshige & Ide, 1969, unpublished observations). In these experiments, animals were sacrificed at noon, 22 hr after a single injection (s.c.) of reserpine (5 mg/kg) and also 48 hr after the injection of a different dose of reserpine (2 mg/kg). In both cases, a marked depletion of CRF content was observed.

Since no other studies were done concerning this sort of change of CRF activity, we decided to follow the change of CRF activity after reserpinization in male rats. Male rats of the Wistar strain were injected with reserpine in a dose of 2 mg/kg s.c., and 48 hr later they were sacrificed by decapitation twice a day, at 9 a.m. and 4 p.m. In a preliminary experiment it was confirmed that brain norepinephrine content decreased to less than 1/10th of control under these conditions. Animals were housed under a lighting schedule of 12 hr light (from 6 a.m. to 6 p.m.) and 12 hr dark. CRF activity was estimated by our intrapituitary injection assay (Hiroshige, 1973).

In stress experiments animals were exposed to ether-laparotomy, and 2 min later they were decapitated for collecting median eminence block and blood. Frontal cortex was also collected and similarly extracted for assaying as tissue control (CCE). Results obtained are shown in Fig. 11. It is clear that even after reserpinization CRF content in the hypothalamus shows a marked circadian variation that is similar to the periodicity previously observed in intact male rats (Hiroshige & Sakakura, 1971). Circadian rhythmicity of the hypothalamo-pituitary-adrenal axis is also reflected in the level of plasma corticosterone (Fig. 11). Furthermore, a stress-induced increment of CRF activity was observed only when the basal level of the activity was low in the morning. Similar observations were reported in intact male rats (Hiroshige, Sato, Ohta & Itoh, 1969) as well as in adrenalectomized rats (Takebe & Sakakura, 1972). The cerebral cortex (CCE) exhibited no activity before or after stress.

It appears then that the dynamics of CRF activity in the hypothalamus is not essentially influenced after reserpine treatment in the male rat, even though this drug has evidently caused a marked depletion of hypothalamic catecholamine. It should be stressed here that a marked depletion of CRF content was observed in the female rat after reserpine. It is possible therefore that there is a sex difference in the effect of reserpine on CRF dynamics, since both the Columbus and Sapporo groups observed independently a CRF depletion in female rats. This should be explored further. Apart from this, the present results with male rats indicate that catecholamines in the brain appear to play no essential role in the regulation of ACTH release. We should, however, interpret the data obtained with psychopharmacological drugs with great caution, since, as Kopin (1966) emphasized, there could be a small functionally active store of catecholamines in the brain that is capable of maintaining noradrenergic or dopaminergic nerve function even after reserpinization.

EFFECTS OF INTRAVENTRICULAR INJECTION OF MONOAMINES

Participation of both adrenergic and cholinergic mechanisms in the regulation of the anterior pituitary function was suggested by Harris (1955), mainly on the basis of experiments concerning induction of ovulation. As to the ACTH release mechanism, Hall & Marks (1968) observed that carbachol injected intraventricularly was a powerful stimulator of ACTH release, and this effect was increased after reserpinization. Intraventricularly administered norepinephrine, on the other hand, was found to inhibit the ACTH secretion induced by reserpine. From these results, they concluded that excitatory cholinergic and inhibitory noradrenergic synaptic mechanisms are involved in the neural regulation of CRF secretion (Marks, Hall & Bhattacharya, 1970).

Hedge & Smelik (1968) reported an inhibition of corticosterone release by intrahypothalamic implantation of atropine pellets. Porter, Goldman & Wilber (1970) succeeded in suppressing a stress-induced ACTH release by perfusing the third ventricle of the rat with a high concentration of atropine solution. These results again support the view that cholinergic excitatory mechanisms are involved in the ACTH release mechanisms. On the other hand, Lorenzen & Ganong (1967) reported an inhibition of stress-induced ACTH release by α-ethyltryptamine (Monase) administration. Recently, van Loon, Hilger, King, Boryczka & Ganong (1971), van Loon, Scapagnini, Cohen & Ganong (1971) reported that the secretion rate of 17-OHCS into the adrenal venous blood in dogs under laparotomy stress was significantly reduced by pretreatment with L-dopa (50 mg/kg, i.v.). In order to confirm the central site of the action, several amines, such as L-norepinephrine, dopamine, and isoproterenol, were injected into the third ventricle of anesthetized dogs. They found again an inhibition of the stress-induced increment of ACTH release. Furthermore, van Loon, Scapagnini, Moberg & Ganong (1971*c*) and Ganong (1972) reported that the hypothalamic content of norepinephrine and dopamine is inversely correlated with the plasma corticosterone level in the rat. From these results they concluded that noradrenergic mechanisms involved are inhibitory to ACTH release under stress. Although these results are highly suggestive, it should be admitted that they were obtained under some peculiar experimental conditions, that is, these experiments were performed under almost maximum stress load and very large amounts of catecholamines were injected intraventricularly.

Attempts were made, therefore, in our laboratory to study the effect of intraventricular injections of much smaller amounts of monoamines on ACTH release. In the first series of experiments, polyethylene tubing was implanted stereotaxically into the third ventricle of male rats (about 300 g body weight) and fixed with a dental cement. Cannulated animals were accustomed to manual handling for 7 days prior to use. Various amounts of test substances, dissolved in 10 to 15 μl of saline solution, were injected through the

TABLE 1. Effects of injection of amines into the third ventricle of unanesthetized male rats on the plasma corticosterone levels

Series number	Injected substance	Plasma corticosterone levels at 4 p.m. (μg/100 ml)
I	No injection	23.1± 1.5 (9)*
II–1	Saline 10 μl	53.5± 6.9 (4)
–2	15 μl	48.2± 4.9 (9)
III–1	Norepinephrine 1.5 μg	53.0± 4.3 (5)
–2	15 μg	43.4± 4.7 (6)
IV	Dopamine 20 μg	49.8± 4.8 (4)
V–1	Serotonin 1 μg	62.3± 7.9 (3)
–2	11 μg	59.4± 4.3 (3)
VI–1	Carbachol 0.5 μg	52.2± 1.5 (3)
–2	5 μg	50.4± 2.1 (3)
VII–1	Atropine 14 μg	52.5± 9.4 (4)
–2	27 μg	51.1± 2.4 (2)
–3	41 μg	79.5± 7.3 (2)
VIII–1	Dexamethasone 5 μg	46.1± 2.8 (3)
–2	10 μg	34.9±16.1 (3)
–3	20 μg	26.8±13.0 (4)
–4	40 μg	3.1± 0.6 (4)

* Numbers in parentheses indicate the number of determinations. Means±standard errors.

cannula at noon by use of a microsyringe without anesthesia. Injected animals were decapitated at 4 p.m. Results obtained are summarized in Table 1. The plasma corticosterone level without injection was 23.1 μg/100 ml at 4 p.m., but it was 9.0±0.8 μg/100 ml at 8 a.m. (the value is not shown in the table). Therefore, it is clear that even in the chronically cannulated rats the plasma corticosterone level exhibits a typical circadian rhythm.

Then, various test substances were injected into conscious rats. All substances tested caused a marked increase of plasma corticosterone levels, which lasted up to 4 hr after the injection. It should be noted that even the injection of a saline solution caused a marked increase of ACTH release. This suggests that some unknown nonspecific mechanisms are involved in the injection-induced response. One of these mechanisms could be an elevation of ventricular hydrostatic pressure, since Edvinsson, Nielsen, Owman & West (1970) noticed in the rabbit that the implantation of the cannula into the lateral ventricle caused a fivefold increase in the intracranial pressure which persisted for about 30 hr. Nevertheless, it appears that the response induced by the injection was not suppressed at all by any of the monoamines injected simultaneously at doses given in Table 1. In contrast, dexamethasone in a dose of 40 μg/rat, injected intra ventricularly, caused a marked suppression under the same conditions.

Since saline injection *per se* was found to be a strong stimulus in the unanesthetized rats, similar experiments were repeated with Nembutal-anesthetized rats. In these experiments, the cannula was directed to the lateral ventricle. Animals were nembutalized at 2:30 p.m., and at 3:00 p.m. test solutions in a volume of 7.5 μl were injected through the cannula. At 4:00 p.m. they were decapitated, and trunk blood was collected. Results obtained are shown in Table 2. Under these conditions, saline injection alone did not cause a significant increase of ACTH release. But injections of

TABLE 2. Effect of injection of amines into the lateral ventricles of Nembutal-anesthetized male rats on the plasma corticosterone levels

Series number	Injected substances		Plasma corticosterone levels at 4 p.m. (μg/100 ml)	P versus I
I	Saline	7.5 μl	22.4±2.1 (10)*	
II	Norepinephrine			
-1		1 μg	29.7±4.0 (10)	NS
-2		10 μg	42.2±6.8 (4)	<0.005
-3		100 μg	62.3±3.6 (7)	<0.001
III	Dopamine			
-1		1 μg	34.5±8.3 (6)	NS
-2		10 μg	17.6±6.5 (4)	NS
-3		100 μg	46.8±7.9 (6)	<0.005
IV	Serotonin			
-1		1 μg	40.0±8.5 (10)	<0.05
-2		10 μg	57.8±8.7 (5)	<0.001
-3		100 μg	29.2±6.7 (7)	NS
V	Carbachol			
-1		0.5 μg	15.6±4.0 (10)	NS
-2		1 μg	11.7±1.4 (7)	<0.005
-3		5 μg	14.5±3.0 (10)	NS
-4		10 μg	21.7±6.0 (8)	NS

* Numbers in the parentheses indicate the number of animals used. Means±standard errors.

TABLE 3. Effects of intraventricular injection of biogenic amines on the stress-induced increase of plasma corticosterone levels in anesthetized rats
Ether laparotomy stress was applied 20 min prior to decapitation, and test substances were injected 60 min before sacrifice.

Series number	Injected substances and stress	Plasma corticosterone levels at 4 p.m. (μg/100 ml)
I	Saline 7.5 μl	22.4±2.1 (10)*
II	Saline 7.5 μl+stress	52.5±5.3 (4)
III	Norepinephrine 100 μg+stress	63.5±2.0 (5)
IV	Dopamine 100 μg+stress	63.1±2.3 (5)
V	Serotonin 100 μg+stress	55.4±2.8 (4)
VI	Carbachol 1 μg+stress	64.1±3.5 (4)

* Numbers in the parenthese indicate the number of animals used. Means±standard errors.

norepinephrine, dopamine, and serotonin caused a significant release of ACTH at doses more than 10 μg, except for serotonin where 100 μg was found ineffective.

The most striking feature in this experiment was that carbachol at doses ranging from 0.5 to 10 μg did not show any stimulatory effect. Carbachol at a dose of 1 μg even showed an inhibitory effect. These results are therefore a reciprocal of the roles of monoamines postulated by previous investigators. However, since these results were obtained in rats under Nembutal anesthesia, interpretation of the findings is hampered because rather complex interactions are expected to occur in the brain between Nembutal and injected monoamines. Finally, in a third series of experiments, suppressive effects, if any, of the monoamine injections on stress-induced ACTH release were examined in anesthetized rats. Test solutions were injected into the lateral ventricle 1 hr before sacrifice and ether laparotomy stress was applied 20 min prior to decapitation. Results obtained (Table 3) indicate that none of the amines tested proved to be effective in suppressing the stress-induced ACTH release under these experimental conditions.

POSSIBLE ROLE OF BIOGENIC AMINES IN ACTH RELEASE MECHANISMS

All of our efforts to prove an inhibitory action of monoamines were unsuccessful. This, of course, does not disprove that some monoamines in the brain act as inhibitory neurotransmitters at the nerve terminals in the hypothalamus But our failure to demonstrate this effect is not necessarily ascribable to technical errors in our approach, since other investigators also reported a stimulatory effect of biogenic amines when applied directly to the brain. For example, Krieger & Krieger (1970*a,b*) reported that pellets of substances such as carbachol, norepinephrine, serotonin, and gamma-aminobutyric acid, implanted into the cat hypothalamus, caused a significant increase of ACTH release.

Naumenko (1967, 1968) also reported in the guinea pig that implantation of epinephrine, acetylcholine, and serotonin into the hypothalamus elicited a marked rise of plasma corticosterone levels, the effect being abolished by midbrain transection, except that of serotonin. Thus, Naumenko emphasized a peculiar position of serotonin in ACTH release. Imura, Nakai, Kato, Yoshimoto & Moridera (1972) reported involvement of the α adrenergic receptor mechanism in enhanced ACTH secretion in man, the β receptor mechanism, in contrast, being inhibitory.

Dixit & Buckley (1969), on the other hand, succeeded in altering the brain serotonin level from 60 to 200% of control by pretreatment with various drugs. But these variations in the serotonin level were found not to affect the degree of stress-induced responses. These authors, however, concluded that the possibility remains that other parameters such as turnover rate may be of importance in determining the role of brain serotonin in the activation of the pituitary by stress.

In fact, Corrodi, Fuxe & Hökfelt (1968) reported an increased rate of amine depletion in the hypothalamus under the influence of immobilization stress. Kety, Javoy, Thierry, Julon & Glowinski (1967) also reported a persistent increase of norepinephrine turnover in various areas of the brain in response to electrical shock in the rat. Thierry, Blanc & Glowinski (1971) examined the turnover rate of norepinephrine in the rat brainstem and found an increased rate of discharge of dopamine and norepinephrine from the noradrenergic terminals following 15-min electrical stimulation. Since most of the monoaminergic fibers entering the basal hypothalamus as the medial forebrain bundle originate in the brainstem (Fuxe *et al.* 1969; Wurtman, 1971), it would be of interest to learn how the increment in the turnover rate of norepinephrine in the brainstem under stress is correlated with activation of CRF release in the hypothalamus.

In this connection, the role of the medial forebrain bundle is of special interest. In fact, Moore & Qavi (1971) reported an abolition of circadian rhythm of adrenal corticosterone levels after transection of this tract in the rat. Feldman, Conforti & Chowers (1971) reported a reduction of plasma corticosterone response to neurogenic stimuli after electrical lesioning of this bundle. They concluded that this bundle is the main final afferent pathway through which neural impulses to the hypothalamus are propagated. More detailed study in this direction is urgently needed, especially in relation to the dynamics of brain monoamines and CRF under various conditions.

Another aspect of interest is the possible participation of brain monoamines in the manifestation of the circadian rhythm of ACTH release. Krieger & Rizzo (1969) suggested that serotonin is a mediator of the circadian rhythm of basal ACTH secretion in the cat. Scapagnini, Moberg, van Loon, de Groot & Ganong (1971) found that serotonin content in the hippocampus and amygdala shows a circadian rhythm. Pretreatment with *p*-chlorophenylalanine resulted in the abolition not only of the serotonin rhythm but also of the plasma corticosterone rhythm. They concluded that serotoninergic neurons play a role in the diurnal fluctuation in the pituitary-adrenal function.

Agrawal, Glisson & Himwich (1966, 1968) examined the postnatal development of amine content in the rat and mouse brain. They found that serotonin content arrived at a maturation level at 25 days of age. Loizou (1971), using the histochemical fluorescence technique, demonstrated that the catecholamine content in the rat median eminence arrived at an adult level in the third week of postnatal life. Since the time of day when the animals were sacrificed was not described, the time of onset of the circadian rhythm of monoamine content was not clear. However, the period in which the monoamine content reached an adult level appears to be in line with the time when the circadian rhythm of CRF activity became manifest (Hiroshige & Sato, 1970). Okada (1971), in fact, examined the time of onset of the circadian rhythm of brain serotonin content and plasma corticosterone levels. But he found no close correlation between them, A detailed account of this problem is given by Yamashita, Moroji, Okada, Saito & Asano (1973) in this volume.

According to McIlwain *et al.* (1971), during the third week of postnatal life the chemical and enzymatic make-up of the rat brain reached almost an adult level. Therefore, manifestation of the circadian rhythm of CRF activity requires a higher degree of maturation of the central nervous system. It is possible that the achievement of maturation level of brain amines during this period is merely a general expression of the maturation process as a whole (see also Yamashita *et al.* 1973). Nevertheless, further studies are needed to clarify the relation, if any, between the postnatal development of the circadian rhythm of CRF activity and of brain monoamines, especially of serotonin.

In spite of these possible roles of biogenic amines in the control of ACTH secretion, it is our impression that these amines play a rather subsidiary role by modifying somehow the basic patterns of ACTH secretion under specific conditions. The exact role of biogenic amines as delicate modifiers of ACTH secretion remains to be elucidated by further exploration with new techniques.

REFERENCES

Agrawal, H. C., Glisson, S. N. & Himwich, W. A. (1966). Changes in monoamines of rat brain during postnatal ontogeny. *Biochim. Biophys. Acta* **130**, 511–513.

Agrawal, H. C., Glisson, S. N. & Himwich W. A. (1968). Developmental changes in monoamines of mouse brain. *Int. J. Neuropharmac.* **7**, 97–101.

Bhattacharya, A. N. & Marks, B. H. (1969). Reserpine- and chlorpromazine-induced changes in the hypothalamo-hypophyseal-adrenal system in rats in the presence and absence of hypothermia. *J. Pharmacol. exp. Ther.* **165**, 108–116.

Bhattacharya, A. N. & Marks, B. H. (1970). Effects of alpha methyl tyrosine and p-chlorophenylalanine on the regulation of ACTH secretion. *Neuroendocrinology* **6**, 49–55.

Carlsson, A., Falck, B. & Hillarp, N. Å. (1962). Cellular localization of monoamines. *Acta physiol. scand.* Suppl. 196, 1–28.

Carr, L. A. & Moore, K. E. (1968). Effects of reserpine and α-methyltyrosine on brain catecholamines and the pituitary-adrenal response to stress. *Neuroendocrinology* **3**, 285–302.

Corrodi, H., Fuxe, K. & Hökfelt, T. (1968). The effect of immobilization stress on the activity of central monoamine neurons. *Life Sci.* **7**, 107–112.

Dixit, B. N. & Buckley, J. P. (1969). Brain 5-hydroxytryptamine and anterior pituitary activation by stress. *Neuroendocrinology* **4**, 32–41.

Edvinsson, L., Nielsen, K. C., Owman, Ch. & West, K. A. (1970). Cited by Owman, Ch. In *The Pineal Gland* (1971), ed. Wolstenholme, G. E. W. & Knight, J., p. 292. London: Churchill Livingstone.

Feldman, S., Conforti, N. & Chowers, I. (1971). The role of the medial forebrain bundle in mediating adrenocortical responses to neurogenic stimuli. *J. Endocr.* **51**, 745–749.

Fortier, C. (1951). Dual control of adrenocorticotrophin release. *Endocrinology* **49**, 782–788.

Fuxe, K. (1964). Cellular localization of monoamines in the median eminence and the infundibular stem of some mammals. *Z. Zellforsch. mikrosk. Anat.* **61**, 710–724.

Fuxe, K. (1965). Evidence for the existence of monoamine neurons in the central nervous system. IV. Distribution of monoamine nerve terminals in the central nervous system. *Acta physiol. scand.* Suppl. 247, 39–85.

Fuxe, K. & Hökfelt, T. (1967). The influence of central catecholamine neurons on the hormone secretion from the anterior and posterior pituitary. In *Neurosecretion*, ed. Stutinsky, F., pp. 165–177. Berlin: Springer-Verlag.

Fuxe, K. & Hökfelt, T. (1969). Catecholamines in the hypothalamus and the pituitary gland. In *Frontiers in Neuroendocrinology* (1969), ed. Ganong, W. F. & Martini, L., pp. 47–96. New York: Oxford Univ. Press.

Fuxe, K. & Hökfelt, T. (1970). Participation of central monoamine neurons in the regulation of anterior pituitary function with special regard to the neuro-endocrine role of tubero-infundibular dopamine neurons. In *Aspects of Neuroendocrinology*, ed. Bargmann, W. & Scharrer, B., pp. 192–205. Berlin: Springer-Verlag.

Ganong, W. F. (1972). Evidence for a central noradrenergic system that inhibits ACTH secretion. In *Brain-Endocrine Interaction*, ed. Knigge, K. M., Scott, D. E. & Weindl, A., pp. 254–266. Basel: S. Karger.

Hall, M. M. & Marks, B. H. (1968). Neurotransmitter effects of biogenic amines in the regulation of pituitary ACTH secretion. *Pharmacologist* **10**, 225.

Harris, G. W. (1955). Neural Control of the Pituitary Gland. Monographs of the Physiol. Soc., No. 3. London: Arnold.

Hedge, G. A. & Smelik, P. G. (1968). Corticosterone release: Inhibittion by intrahypothalamic implantation of atropine. *Science, N.Y.* **159**, 891–892.

Hiroshige, T. (1971). Role of vasopressin in the regulation of ACTH secretion: Studies with intrapituitary injection technique. *Med. J. Osaka Univ.* **21**, 161–180.

Hiroshige, T. (1973). CRF assay by intrapituitary injection through the parapharyngeal approach and its physiological validation. In *Brain-Pituitary-Adrenal Interrelationships*, ed. Brodish, A. & Redgate, E. S. Basel: S. Karger (in press).

Hiroshige, T. & Itoh, S. (1968). Studies on the release of corticotropin by means of intrapituitary microinjection. In *Integrative Mechanism of Neuroendocrine System*, vol. 1, ed. Itoh, S., pp. 21–40. Sapporo : Hokkaido Univ. Sch. Med.

Hiroshige, T., Kunita, H., Ogura, C. & Itoh, S. (1968). Effects on ACTH release of intrapituitary injections of posterior pituitary hormones and several amines in the hypothalamus. *Jap. J. Physiol.* **18**, 609–619.

Hiroshige, T., Kunita, H., Yoshimura, K. & Itoh, S. (1968). An assay method for corticotropin-releasing activity by intrapituitary microinjection in the rat. *Jap. J. Physiol.* **18**, 179–189.

Hiroshige, T. & Sakakura, M. (1971). Circadian rhythm of corticotropin-releasing activity in the hypothalamus of normal and adrenalectomized rats. *Neuroendocrinology* **7**, 25–36.

Hiroshige, T. & Sato, T. (1970). Postnatal development of circadian rhythm of corticotropin-releasing activity in the rat hypothalamus. *Endocr. jap.* **17**, 331–336.

Hiroshige, T., Sato, T., Ohta, R. & Itoh, S. (1969). Increase of corticotropin-releasing activety in the rat hypothalamus following noxious stimuli. *Jap. J. Physiol.* **19**, 866–875.

Hirsch, G. H. & Moore, K. E. (1968). Brain catecholamines and the reserpine-induced stimulation of the pituitary-adrenal system. *Neuroendocrinology* **3**, 398–405.

Imura, H., Nakai, Y., Kato, Y., Yoshimoto, Y. & Moridera, K. (1972). Biogenic amines in the hypothalamus and pituitary function. *Taisha* **9**, 447–457 (in Japanese).

Kamberi, I. A. & McGann, S. M. (1969). Effect of biogenic amines, FSH-releasing factor (FRF) and other substances on the release of FSH by pituitaries incubated *in vitro*. *Endocrinology* **85**, 815–824.

Kamberi, I. A., Schneider, H. P. G. & McCann, S. M. (1970). Action of dopamine to induce release of FSH-releasing factor (FRF) from hypothalamic tissue *in vitro*. *Endocrinology* **86**, 278–284.

Kety, S. S., Javoy, F., Thierry, A. M., Julon, J. & Glowinski, J. (1967). A sustained effect of electroconvulsive shock on the turnover of norepinephrine in the central nervous system of the rat. *Proc. natn. Acad. Sci. U.S.A.* **58**, 1249–1254.

Kopin, I. J. (1966). Metabolism and disposition of catecholamines in the central and peripheral nervous system. In *Endocrines and the Central Nervous System*, ed. Levine, R., pp. 343–351. Baltimore : Williams & Wilkins.

Krieger, D. T. & Krieger, H. P. (1970*a*). Effect of dexamethasone on pituitary-adrenal activation following intrahypothalamic implantation of ' Neurotransmitters.' *Endocrinology* **87**, 179–182.

Krieger, H. P. & Krieger, D. T. (1970*b*). Chemical stimulation of the brain : Effect on adrenal corticoid release. *Am. J. Physiol.* **218**, 1632–1641.

Krieger, D. T. & Rizzo, F. (1969). Serotonin mediation of circadian periodicity of plasma 17-hydroxycorticosteroids. *Am. J. Physiol.* **217**, 1703–1707.

Loizou, L. A. (1971). The postnatal development of monoamine-containing structures in the hypothalamo-hypophyseal system of the albino rat. *Z. Zellforsch.* **114**, 234–252.

Long, C. N. H. (1956). Pituitary-adrenal relationships. *A. Rev. Physiol.* **18**, 409–432.

Lorenzen, L. C. & Ganong, W. F. (1967). Effect of drugs related to α-ethyltryptamine on stress-induced ACTH secretion in the dog. *Endocrinology* **80**, 889–892.

Markee, J. E., Sawyer, C. H. & Hollinshead, W. H. (1948). Adrenergic control of the release of luteinizing hormone from the hypophysis of the rabbit. *Recent Progr. Horm. Res.* **2**, 117–131.

Marks, B. H., Hall, M. & Bhattacharya, A. N. (1970). Psychopharmacological effects and pituitary-adrenal activity. In *Pituitary, Adrenal and the Brain*, ed. de Wied, D. & Weijnen, J. A. W. M., pp. 57–69. Amsterdam: Elsevier.

McDermott, W. V., Fry, E. G., Brobeck, J. R. & Long, C. N. H. (1950*a*). Release of adrenocorticotrophic hormone by direct application of epinephrine to pituitary grafts. *Proc. Soc. exp. Biol. Med.* **73**, 609–610.

McDermott, W. V., Fry, E. G., Brobeck, J. R. & Long, C. N. H. (1950*b*). Mechanism of control of adrenocorticotrophic hormone. *Yale J. Biol. Med.* **23**, 52–66.

McIlwain, H. & Bachelard, H. S. (1971). Biochemistry and the Central Nervous System, 4th ed. London: Churchill.

Moore, R. Y. & Qavi, H. B. (1971). Circadian rhythm in adrenal adenyl cyclase and corticosterone abolished by medial forebrain bundle transection in the rat. *Experientia* **27**, 249–250.

Moroji, T., Yamazaki, K. Okada, F., Saito, Y. & Asano, Y. (1973). Circadian rhythm of glucocorticoid secretion. In *Neuroendocrine Control*, ed. Yagi, K. & Yoshida, S., pp. 57–81. Tokyo: University of Tokyo Press.

Naumenko, E. V. (1967). Role of adrenergic and cholinergic structures in the control of the pituitary-adrenal system. *Endocrinology* **80**, 69–76.

Naumenko, E. V. (1968). Hypothalamic chemoreactive structures and the regulation of pituitary-adrenal function. Effects of local injections of norepinephrine, carbachol and serotonin into the brain of guinea-pigs with intact brains and after mesencephalic transection. *Brain Res.* **11**, 1–10.

Okada, F. (1971). The maturation of the circadian rhythm of brain 5-HT in the rat, with special reference to its correlation with the sleep-wakefulness pattern and the circadian rhythm of plasma corticosterone. *Life Sci.* **10**, 77–86.

Pearce, L. A. & Schanberg, S. M. (1969). Histamine and spermidine content in brain during development. *Science, N.Y.* **166**, 1301–1303.

Porter, J. C., Goldman, B. D. & Wilber, J. F. (1970). Hypophysiotropic hormones in portal vessel blood. In *Hypophysiotropic Hormones of the Hypothalamus: Assay and Chemistry*, ed. Meites, J., pp. 282–293. Baltimore: Williams & Wilkins.

Redgate, E. S. (1968). Role of the baroreceptor reflexes and vasoactive polypeptides in the corticotropin release evoked by hypotension. *Endocrinology* **82**, 704–720.

Sawyer, C. H., Markee, J. E. & Everett, J. W. (1950). Activation of the adenohypophysis by intravenous injections of epinephrine in the atropinized rabbit. *Endocrinology* **46**, 536–543.

Sawyer, C. H., Markee, J. E. & Townsend, B. F. (1949). Cholinergic and adrenergic components in the neurohumoral control of the release of LH in the rabbit. *Endocrinology* **44**, 18–37.

Scapagnini, U., Moberg, G. P., van Loon, G. R., de Groot, J. & Ganong, W. F. (1971). Relation of brain 5-hydroxytryptamine content to the diurnal variation in plasma corticosterone in the rat. *Neuroendocrinology* **7**, 90–96.

Schneider, H. P. G. & McCann, S. M. (1969). Possible role of dopamine as transmitter to promote discharge of LH-releasing factor. *Endocrinology* **85**, 121–132.

Schneider, H. P. G. & McCann, S. M. (1970). Mono- and indolamines and control of LH secretion. *Endocrinology* **86**, 1127–1133.

Smelik, P. G. (1967). ACTH secretion after depletion of hypothalamic monoamines by reserpine implants. *Neuroendocrinology* **2**, 247–254.

Smelik, P. G. (1969). Role of hypothalamic monoamines in the control of pituitary secretions. *J. Neuro-Visc. Rel.* Suppl. IX, 277–282.

Takebe, K. & Sakakura, M. (1972). Circadian rhythm in pituitary-adrenocortical function. In *Advances in Climatic Physiology*, ed. Itoh, S., Ogata, K. & Yoshimura, H., pp. 367–380. Tokyo: Igaku-Shoin.

Thierry, A. M., Blanc, G. & Glowinski, J. (1971). Effect of stress on the disposition of catecholamines localized in various intraneuronal storage forms in the brain stem of the rat. *J. Neurochem.* **18**, 449–461.

van Loon, G. R., Hilger, L., King, A. B., Boryczka, A. T. & Ganong, W. F. (1971). Inhibitory effect of L-dihydroxyphenylalanine on the adrenal venous 17-hydroxycorticosteroid response to surgical stress in dogs. *Endocrinology* **88**, 1404–1414.

van Loon, G. R., Scapagnini, U., Cohen, R. & Ganong, W. F. (1971). Effect of intraventricular administration of adrenergic drugs on the adrenal venous 17-hydroxycorticosteroid response to surgical stress in the dog. *Neuroendocrinology* **8**, 257–272.

van Loon, G. R., Scapagnini, U., Moberg, G. P. & Ganong, W. F. (1971). Evidence for central adrenergic neural inhibition of ACTH secretion in the rat. *Endocrinology* **89**, 1464–1469.

Vogt, M. (1952). The role of adrenaline in the response of the adrenal cortex to stress of various kinds, including emotional stress. *Ciba Found. Colloq. Endocrinol.* **4**, 154–160.

Vogt, M. (1954). The concentration of sympathin in different parts of the central nervous system under normal conditions and after the administration of drugs. *J. Physiol.* **123**, 451–481.

von Euler, U. S. (1946). A specific sympathomimetic ergone in adrenergic nerve fibres (sympathin) and its relations to adrenaline and noradrenaline. *Acta physiol. scand.* **12**, 73–97.

Wurtman, R. J. (1970). Brain catecholamines and the control of secretion from the anterior pituitary gland. In *Hypophysiotropic Hormones of the Hypothalamus: Assay and Chemistry*, ed. Meites, J., pp. 184–189. Baltimore: Williams & Wilkins.

Wurtman, R. J. (1971). Brain monoamines and endocrine function. *Neurosci. Res. Progr. Bull.* **9**, 172–297.

Yates, F. E., Russell, S. M., Dallman, M. F., Hedge, G. A., McCann, S. M. & Dhariwal, A. P. S. (1971). Potentiation by vasopressin of corticotropin release induced by corticotropin-releasing factor. *Endocrinology* **88**, 3–15.

Yates, F. E., Russell, S. M. & Maran, J. W. (1971). Brain-adenohypophysical communication in mammals. *A. Rev. Physiol.* **33**, 393–444.

NATURE OF THE HIPPOCAMPAL FUNCTION IN RELATION TO GONADOTROPIN SECRETION

Masazumi Kawakami, Katsuo Seto, Fukuko Kimura, and Ei Terasawa

We attempt to report findings obtained in our laboratory and offer some hypotheses concerning gonadotropin regulation by the hippocampus, reviewing a large number of experiments in the same field.

There appear to be two types of hippocampal modulatory mechanism for gonadotropin secretion. One acts phasically in relation to the changes in the humoral environment. The other acts in an inhibitory manner independent of the humoral factors.

The hippocampus reveals its phasic modulatory influence especially on FSH regulation. At the first stage of the estrous cycle (in mature) and of puberty (in immature), it enhances the release of FSH, and just before the discharge of ovulating hormone (in mature and immature) it inhibits the release. Such function may be based on the hippocampal sensitivity to sex steroids and gonadotropins, because its activity is influenced by these hormones of exogenous as well as endogenous origin. Furthermore it is probable that the hippocampal impulses influencing gonadotropin secretion change their route to the hypothalamus in relation to the humoral environment.

INTRODUCTION

More than 90 years ago, Broca (1878) drew attention to the existence of a phylogenetically ancient portion of the cerebral cortex (the great limbic lobe) in several vertebrate species. In subsequent comparative neuroanatomical studies, the limbic cortical areas and related nuclear masses were assumed to be a prevailingly olfactory structure; hence, the term rhinencephalon was generally used to denote these prosencephalic structures.

However, modern anatomical investigations have clarified some of the components and circuitry of this area of the brain (Adey, Dunlop & Sunderland, 1958; Powell, Guillery & Cowan, 1957), and in addition, neurophysiological and behavioral studies have emphasized the nonolfactory functional aspects in higher forms (Bargmann & Schade, 1963; Green, 1960; Sawyer, 1960).

At present, in view of these more recent findings, this portion of the brain is generally called the limbic system (MacLean, 1949), including phylogenetically ancient portions of the cerebral cortex and related subcortical derivatives, as well as intrinsic fiber connections and extramural pathways linking these cortical components with the diencephalon and other areas in the brain stem. These portions of the brain largely correspond to the telencephalic system (Gloor, 1956).

Among them, the hippocampus, which is included in the rhiencephalon, has been studied from two somewhat different points of view. First, because of its simple stratification of cortical composition compared with the neocortex, it offers special experimental opportunities. It has been especially suitable for the study of field potentials and unit activity. Second, because no clear role has been attributed to so large a part of the brain, it has been the subject of an increasing number of investigations in various fields.

In fact, since Papez's (1937) theoretical speculation that these structures bear an important relationship to emotional behavior and the almost simultaneous striking demonstration by Klüver & Bucy (1937) that these structures are involved in the balancing, integration, and elaboration of critically profound motivational-emotional behavior patterns, a host of important subsequent neurophysiological experiments have been carried out to elucidate the function of these structures. However, no completely satisfactory theory concerning the role of the limbic system has yet emerged, although a possible role in self-preservation and preservation of the species has been suggested (MacLean, 1959).

This is true, also, for the hippocampus; that is, the hippocampus does not participate in the somatomotor or autonomic activities (Kaada, 1951), but is involved in emotional behavior (MacLean & Delgado, 1953) and in memory function (Penfield & Milner, 1958). Furthermore, it has been proposed that it participates in higher nervous activity only indirectly, *e.g.*, by inhibition of the oriental cortex (Grastyan, 1959) or by hippocampo-reticular influence modulating the activity of the cerebral cortex (Green, 1960). Despite these assumptions, however, the specific function was considered to be another matter (Green, 1960; Brady, 1960). Indeed, Gloor (1956) proposed that the role of the limbic system is a modulation of functional patterns integrated at the level of the hypothalamus and the brain stem tegmentum.

In these earlier experimental efforts, attention of the neuroendocrinologists was focused on the modulatory influence of the limbic system, particularly of the hippocampus, amygdala, and septal complex, on the hypothalamic centers controlling endocrine systems. Investigating neural and hormonal factors pertaining to extra-hypothalamic influence on functions of the hypothalamo-hypophyseal system, they have noted that these areas appear to be involved in modulating the release of gonadotropin, as well as thyrotropin or adrenocorticotropin. Schreiner & Kling (1953) showed first that lesions of the amygdala lead to hypersexualism in the male, and subsequent work has confirmed the involvement of the amygdala in ovulation in rabbits (Hayward, Hilliard & Sawyer, 1964), rats (Bunn & Everett, 1957), and cats (Koikegami, Yamada & Usui, 1954).

On the other hand, the function of the hippocampus is confused, *i.e.*, Riss, Burstein & Johnson (1963) reported that lesions of the hippocampus induced delayed puberty, whereas Velasco & Taleisnik (1969) proposed that this area inhibits gonadotropin release. These reports are somewhat diverse and even conflicting, and in the experiments on adrenocorticotropin regulation this is also true. It seems that an enigma arises in the function of the hippocampus.

The diversity of the experimental results is considered to be, in part, due to difficulties both in methodology of experiments and interpretation of results. For instance, electrical stimulation of the hippocampus easily elicits paroxysmal afterdischarges spreading to other parts of the brain, and lesions may induce epileptic foci in the remaining hippocampal tissue. Green (1960) pointed out that effects of ablation or stimulation of the hippocampus may thus be due not only to impairment of hippocampal function but also to remote effects of this paroxysmal activity.

It seems necessary requisite to review a large number of experiments dealing with neuroendocrine regulation of the brain from these points of view. The present paper reports on knowledge obtained in our laboratory and gives some hypotheses on function of the hippocampus.

HIPPOCAMPAL CONNECTIONS WITH THE HYPOTHALAMUS

The individual components of the limbic system receive neocortical and neo-striatal afferent connections, which suggests that they act as intermediaries between the newer part of the endbrain and the older part of the diencephalon—the hypothalamus and epithalamus—to which these new parts have no direct access otherwise (Gloor, 1956).

These old cortical and striatal formations funnel down their activity to the hypothalamus through the following main system: the hippocampo-hypothalamic system discharging over the fornix, the amygdalo-hypothalamic system running through the stria terminalis and other shorter amygdalo-hypothalamic connections, and the orbito-hypothalamic system. Some of the fibers originating in the hippocampus and the amygdala are interrupted in their downward course at the level of the septal nuclei acting as relay stations in these cortico- and amygdalo-hypothalamic pathways. At the hypothalamic level all of these systems enter the medial forebrain bundle (Crosby & Woodburne, 1940; Brodal, 1947). Besides this common characteristic of the main connections there are, however, patterns more specific for each individual system.

In higher forms, the hippocampal formation with its fiber pathways is thought to be divided into two main parts: (1) the hippocampus proper (Ammon's horn or Cornu Ammonis) and (2) the dentate fascia (gyrus dentatus) and its continuation over the corpus callosum (gyrus supracallosus). The term dorsal hippocampus is used to imply the dorsal subcallosal hippocampus. A system of commissural fibers develops between the rostral portion of the hippocampus proper (hippocampal commissure).

It is now generally accepted that the efferent connections of the hippocampus are composed of pyramidal cell axons, which are gathered together in the alveus and distributed through two pathways, the fimbria and the dorsal fornix. There is some evidence to suggest that the fimbria and dorsal fornix differ in their distribution and in their origin from the hippocampus. The bulk of the fibers of the fimbria diverges in two bundles, the postcommissural and precommissural fornix. These three well-defined fornix fibers terminate mainly in the mammillary body, in the medial mammillary nucleus. A great number of fornix fibers, however, leave the main course of the bundle from the septum onward and end in various septal, thalamic, and hypothalamic nuclei and areas.

Nauta (1956) reported that all of the septal cell groups receive hippocampal fibers through either the precommissural or postcommissural fornix. Fornix terminals in the tuber cinereum close to the infundibulum of the pituitary gland were described first by Ramóny Cajal (1901) and later by Gurdjian (1927) as tractus cortico-hypothalamicus medialis. In our laboratory, the existence of this hippocampo-hypothalamic system is indicated by the short latency (4–6 ms) evoked potential recorded in the arcuate nucleus or in the supra-optic nucleus following hippocampal stimulation (Kawakami, Seto, Kimura, Ishida, Yoshida & Kubo, 1968). Through these efferent pathways, the hippocampal activity can readily affect that of the hypothalamus (Kawakami, Terasawa, Kimura & Kubo, 1971; Kawakami & Kubo, unpublished data).

CHARACTERISTICS OF HIPPOCAMPAL ELECTRICAL ACTIVITY

The hippocampal arousal pattern (Green & Arduini, 1954) is characterized by the clear-cut 4–7 Hz activity of almost sinusoidal shape, the " theta " rhythm, occurring simultaneously with desynchronization in the neocortex. It was noted that the brain of cat, rabbit, rat and guinea-pig also showed the same theta rhythm following a variety of stimuli. It is now clear from studies by Petsche & Stumpf (1962) that the theta wave has several interesting properties: (1) It tends to travel from the region of the septum along longitudinal axes of the two hippocampi, but at a very slow rate; (2) it may be recorded above and below the hippocampus, but a phase reversal occurs near its dorsal aspect; (3) the frequency of the wave is partly determined by the intensity of the peripheral stimulus. This rhythm could also be induced by stimulating the reticular activating system of Moruzzi & Magoun (1949). No theta rhythm, however, was induced by stimulation of the hippocampus directly.

Evidence suggests that evoked potential is generated in area CAl following stimulation of the dorsal fornix or a homotopic point of the contralateral hippocampus. This evoked potential commonly increases in size with repetitive stimulation and further repetitive stimulation leads to distortions of the evoked potentials which herald the onset of seizure discharges. Furthermore, in cats electrolytic lesions in the hippocampus induce seizures that show most of the

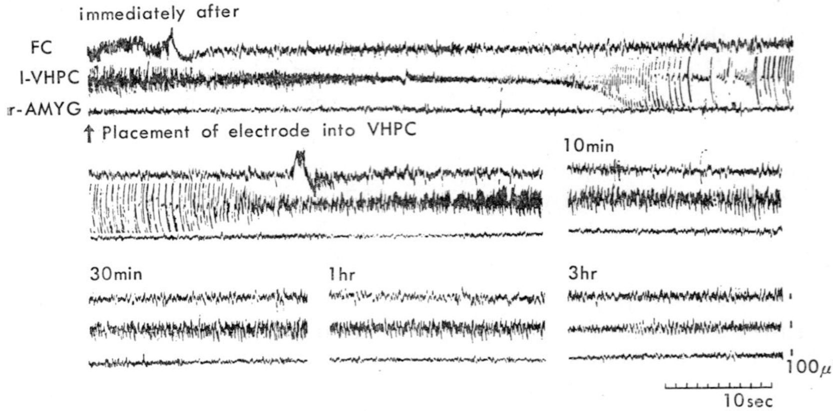

Fig. 1. An example of EEG seizure pattern induced by insertion of an electrode into the ventral hippocampus (VHPC) spreading to the frontal cortex (FC) and the contralateral amygdala (AMYG). The experiment was performed under ether anesthesia (Kawakami, Kimura & Wakabayashi, 1972).

characteristics of psychomotor epilepsy, and secondary injury to the hippocampus when the region of primary destruction is elsewhere, particularly in the amygdala, also induces it (Green, 1960).

The hippocampus has the lowest threshold for induction of seizure discharges of any region in the brain, and due to such an exceedingly low threshold it could induce paroxysmal electrical discharges simply by the insertion of electrodes (Fig. 1). Seizure discharges propagate from the hippocampus in all directions, but they are not dependent necessarily upon external connections. They propagate both by axonal pathways and by routes that are not strictly anatomical pathways but simply regions of contiguity (Green, 1960).

THE ROLE OF THE HIPPOCAMPUS IN CONTROLLING GONADOTROPIN SECRETION IN THE FEMALE

Stimulation experiments

Involvement of the hippocampus in the pituitary-ovarian function was first indirectly indicated by an experiment recording EEG activity after copulation of estrous rabbits (Sawyer & Kawakami, 1959). In these animals, a characteristic hyperarousal record with 8 Hz sinosoidal waves (EEG afterreaction) appeared in several rhinencephalic areas related to the hippocampus; it was hypothesized that the hyperarousal was due to the action of the released ovulating hormone on the brain. This hypothesis led to the idea that the hippocampus may have some correlation with the regulation of gonadotropin secretion.

First attempts (Kawakami, Seto & Yoshida, 1966; Kawakami, Seto, Terasawa & Yoshida, 1967; Kawakami, Seto & Yoshida, 1968) were made using mature female New Zealand white rabbits. Concentric bipolar electrodes were implanted chronically. After recovery from the operative procedure, the rabbit was primed with estrogen for 2 days, and stimulation consisting of monophasic square wave pulses of 0.1 ms, 60 Hz, and 190 to 210 μA was applied for 30 min at 30-sec periods alternating with 30-sec rest periods under freely moving conditions. Usually this parameter of stimulation did not induce seizure discharges in the hippocampal and cortical

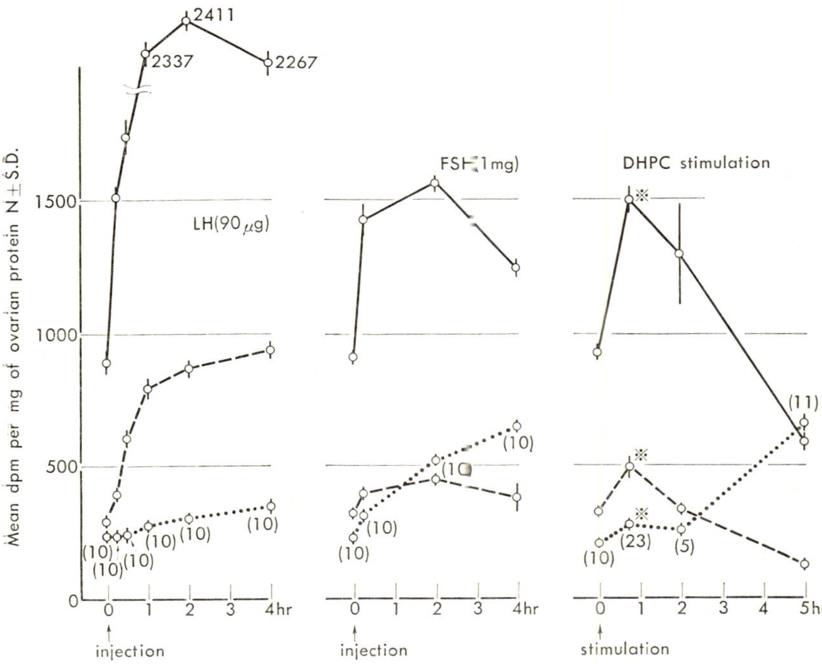

Fig. 2. Time course of the *in vitro* transfer rates of ^{14}C from ^{14}C–1-acetate into progesterone, 20α–OH–P, and estrogen following intravenous injection of LH, FSH, and hippocampal electrical stimulation. The number in parenthesis represents the number of animals in each group (Kawakami, Kimura & Seto, unpublished data). ※ From Kawakami Seto & Yoshida (1968). ——, Progesteron; - - - -, 20α–OH-P; ······, Estrogen.

TABLE 1. Effect of electrical stimulation of the hippocampus in mature rats (Kawakami,

Experimental animals		Site of stimulation	Anterior pituitary weight (mg)	Pituitary
				LH (μg/mg)
Mature female	DI	Control	8.9±0.5† (13)	0.55±0.02
		d-HPC	9.3±0.9 (6)	0.53±0.04
	DII	Control	8.8±0.4 (11)	0.67±0.03
		d-HPC	7.8±0.7 (7)	0.75±0.05
	P	Control	9.9±0.4 (13)	0.76±0.04
		d-HPC	9.1±0.4 (7)	0.79±0.10
	E	Control	9.6±0.6 (15)	0.41±0.03
		d-HPC	8.5±0.5 (9)	0.46±0.03
Androgenized female		Control	11.7±1.0 (10)	0.45±0.06
		d-HPC	12.2±0.8 (7)	0.55±0.04
Mature male		Control	8.2±0.5 (7)	1.41±0.08
		d-HPC	7.6±0.6 (4)	1.21±0.08

† Mean±S.E.; * $p<0.05$ *vs.* control. Number of animals in parentheses. Abbre-
hippocampus.

EEG. Stimulated rabbits were decapitated immediately after the stimulation to remove the ovaries. As a parameter of gonadotropin release, the *in vitro* transfer rates of ^{14}C from ^{14}C-1-acetate into progesterone, 20α-hydroxy-pregn-4-en-3-one (20α-OH-P), and estrogen (estradiol and estrone) was determined in the ovarian homogenates incubated for 1 hr.

Stimulation of the dorsal hippocampus, particularly the alveus, caused a remarkable increase in transfer rates of ^{14}C-1-acetate into progesterone and 20α-OH-P and a slight increase into estrogen (Fig. 2). These results indicate that hippocampal stimulation elicits the release of some gonadotropin from the pituitary. In the subsequent experiments, we observed that this hippocampal effect on ovarian biosynthetic activity differs in relation the period after the stimulation. That is, we observed an increase of transfer rates of ^{14}C into progesterone, no change into 20α-OH-P, and a slight increase into estrogen by 2 hr. By 5 hr, we observed a decrease into progesterone and 20α-OH-P, but a remarkable increase into estrogen. These findings indicate that, following electrical stimulation of the hippocampus, the transfer rates of ^{14}C into progesterone and 20α-OH-P significantly increase during the first 2 hr, and then they decrease precipitously, and that of estrogen consistently increases.

Kimura & Wakabayashi 1972, Kawakami, Terasawa & Kimura, unpublished data).

concentration of		Serum concentration of		
FSH (μg/mg)	Prolactin (mIU/mg)	LH (ng/ml)	FSH (ng/ml)	Prolactin (μIU/ml)
5.58±0.35	25.4±4.3	0.8 ±0.1	969±162	506± 95
5.42±0.58	17.5±1.4	0.8 ±0.1	888±135	411± 91
5.04±0.35	20.8±1.5	0.8 ±0.1	1238±215	527±117
5.70±0.54	15.7±2.4*	0.6 ±0.1	511±108*	420± 50
4.22±0.19	17.9±1.5	0.7 ±0.1	915±108	1495±181
4.41±0.71	13.6±2.0	0.7 ±0.1	1184±377	733±138*
3.67±0.16	22.3±2.2	0.7 ±0.1	727±135	1593±216
4.92±0.31*	16.5±1.6*	0.5 ±0.1	1561±242*	1125±241*
3.22±0.36	35.0±4.3	0.5 ±0.1	634± 39	1314±227
3.22±0.22	32.6±1.4	1.7 ±0.4*	844± 78*	1347±314
39.9±3.6	8.5±0.9	1.45±0.07	1855±149	699± 82
26.7±1.6*	14.0±1.7*	1.50±0.16	1601±158	339± 74*

viations: DI, diestrus I; DII, diestrus II; P, proestrus; E, estrus; *d*-HPC, dorsal

These characteristic patterns of biosynthetic activity in the time course after the stimulation are quite similar to those observed after a single injection of FSH (1 mg, i.v., Fig. 2). Thus, it was indicated that in the rabbit hippocampal stimulation induces the release of FSH, rather than LH.

On the basis of these findings, electrical stimulation of the rat hippocampus was performed. Previously, it had been observed that electrical stimulation of the hippocampus, particularly the ventral one, at any stage of the estrous cycle, could induce prolonged cornification of the vaginal smears following the expected vaginal cornification on the day of estrus. The ovaries of these animals, showing constant vaginal cornification, contained a large number of mature follicles, which is indicative of a dominant effect of FSH on the ovary (Kawakami & Kimura, unpublished data). However, this also indicated an inhibition of the ovulating surge of LH.

In recent work using the radioimmunoassay method (Kawakami, Kimura & Wakabayashi, 1972), the effect of hippocampal stimulation was more precisely clarified. Mature wistar rats with a 4-day cycle were stimulated through chronically implanted bipolar electrodes between 10:00 and 14:00 during the estrous cycle. The stimulation parameters were as follows; monophasic square wave, pulse duration of 0.1 ms, with a frequency of 100 Hz, and current

TABLE 2. Effect of electrical stimulation of the hippocampus in

	Experimental animals	Site of stimulation
Immature female	27-day-old rats	Control
		d-HPC sham
		d-HPC
	Estrogen primed 27-day-old rats	Control
		Estrogen
		Estrogen+sham
		Estrogen+d-HPC
	34-day-old rats	Control
		d-HPC
Immature male	27-day-old rats	Control
		d-HPC sham
		d-HPC
	Testosterone primed 27-day-old rats	Testosterone+d-HPC sham
		Testosterone+d-HPC
	34-day-old rats	Control
		d-HPC

† Mean±S.E.; * $p<0.05$ *vs.* sham; ** $p<0.05$ *vs.* control. Number of animals

of 60 μA. It was observed that electrical stimulation of the hippocampus did not induce any changes in serum and pituitary concentrations of LH during the estrous cycle. On the other hand,both serum and pituitary concentrations of FSH were elevated by the stimulation on the day of estrus, and the serum concentrations of FSH was reduced on the day of diestrus II (Table 1).

These results indicate that the hippocampus profoundly participates in the regulation of release and synthesis of FSH, but not in that of LH. In addition, this hippocampal function seems to be phasic, since with the humoral environment the stimulation effect altered, at the onset of the cycle facilitatory and before ovulation inhibitory. It was further confirmed by bioassay that stimulation on the day of estrus also increases the FSH-RF activity in the stalk median eminence and not LH-RFactivity, whiich is indicative of the separate effect of hippocampal stimulation on LH-RF and FSH-RF activities in the hypothalamus, and followingly the existence of two gonadotropin-releasing factors: LH-RF and FSH-RF (Kawakami, Kimura immature rats (Kawakami, Terasawa & Kimura, unpublished data).

Anterior pituitary weight (mg)	Pituitary concentration of		Serum concentration of	
	LH (μg/mg)	FSH (μg/mg)	LH (ng/ml)	FSH (ng/ml)
1.9±0.1†(11)	1.42±0.08	19.3± 1.6	0.6 ±0.03	400± 74
2.3±0.5 (6)	1.27±0.06	18.1± 1.8	0.8 ±0.13	323± 63
1.8±0.1 (7)	1.95±0.11*	39.5± 5.5*	0.5 ±0.06	794± 91*
2.4±0.3 (6)	0.80±0.10	19.3± 3.4	0.6 ±0.03	500± 67
2.0±0.1 (5)	1.47±0.20**	30.6± 2.5**	0.7 ±0.03	531±108
2.1±0.1 (6)	1.60±0.02	27.8± 6.7	0.7 ±0.05	423±156
2.0±0.1 (5)	1.56±0.21	28.9± 3.8	0.6 ±0.06	392± 86
2.6±0.1 (7)	1.83±0.18	48.5± 3.6	0.29±0.19	1572±213
2.1±0.4 (6)	1.37±0.16	48.1± 5.8	ND (0.5)	792± 64*
2.5±0.1 (5)	—	73.9± 1.7	—	3185±214
2.0±0.1 (5)	—	137.4±13.9*	—	2019±186*
2.1±0.2 (6)	—	121.3±21.3	—	1365±163**
1.6±0.2 (5)	—	161.9±21.3	—	1179±180
2.1±0.1 (5)	—	171.0±10.4	—	476±128**
2.5±0.1 (11)	2.02±0.12	74.6± 4.1	0.52±0.23	2481±326
2.8±0.3 (7)	2.58±0.24	79.3± 9.6	0.55±0.20	1571±280*

in parentheses. Abbreviations: *d*-HPC, dorsal hippocampus

& Wakabayashi, unpublished data). Simultaneously, the ovaries of rats stimulated in the hippocampus showed an increase in the transfer rate into estrogen throughout the estrous cycle (Kawakami, Seto, Terasawa & Kimura, 1972). As for prolactin, the hippocampal stimulation decreased the serum content, which is indicative of the existence of inhibitory mechanisms in the limbic system (amygdalar stimulation also decreased the serum content).

In an effort to investigate the role of the hippocampus in the onset of puberty, immature female rats were stimulated through a bipolar electrode implanted a day before (Kawakami & Terasawa, 1972). The stimulation parameters were the same as for mature rats. In 28-day-old rats, electrical stimulation of all examined areas, the arcuate nucleus, medial preoptic area, medial amygdala, and dorsal hippocampus, induced an increase of the pituitary concentrations of LH and FSH (Table 2). Only the dorsal hippocampus and the arcuate nucleus, however, could induce an increase of serum FSH, and no areas induced an increase of LH.

However, when animals were pretreated with estrogen (2.5 μg, s.c.), the serum concentration of FSH was increased by stimulation of the medial preoptic area, medial amygdala, and arcuate nucleus, but not by stimulation of the dorsal hippocampus. In addition, hippocampal stimulation prior to vaginal opening (at 34 days of age) decreased the serum content of FSH. In summary, the dorsal hippocampus seems to start stimulating the release of FSH at the first stage of puberty, which in turn causes the release of estrogen from the ovary.

In fact, in a correlated experiment, only hippocampal stimulation could facilitate the transfer rates of ^{14}C into estrogen in ovarian homogenates (Kawakami, Seto, Terasawa & Kimura, 1972). After the increase of circulatory estrogen, the amygdala and/or the medial preoptic area, in turn, may accelerate the further release of FSH to cause vaginal opening and first ovulation. However, the inhibitory influence of the hippocampus on FSH as seen in 34-day-old rats may have some role in first ovulation, because this behavior of the hippocampus resembles that of mature rats on the day of diestrus II.

Riss *et al.* (1963) noted that hippocampal or piriform lobe damage in infancy caused diminished gonadal development and Zarrow, Naqvi & Denenberg (1969) reported that lesions to the anterior ventro lateral hippocampus prevented androgen-induced precocious puberty. These findings also point out that the hippocampus in immature female rats is responsible for the advancement of maturation, though it was reported that there was no retardation in the onset of puberty after lesions in the hippocampus (Kling, 1964).

At any rate, the hippocampus both in mature and immature females appears to change its modulatory influence on gonadotropin release due to the hormonal environment.

The manner of hippocampal involvement in gonadotropin regulation was quite altered in androgenized persistent estrous rats (Table 1). These animals were obtained by testosterone propionate injection (50 μg, s.c.) at 5 days of age, and the stimulation experiment was performed chronically. In such animals, electrical stimulation of the hippocampus, with parameters similar to those mentioned above, increased both serum and pituitary concentrations of LH slightly and serum FSH significantly. The medial part of the amygdala also showed a different effect; *i.e.*, the stimulation induced no change in serum LH, decreased pituitary LH and decreased serum FSH, inducing no change in pituitary FSH. Stimulation of the medial preoptic area, however, induced an increase of serum LH but not pituitary LH. Both serum and pituitary FSH did not change. These results are indicative of a difference of the brain in androgenized rats from that of normal females, mainly of the limbic system rather than of the hypothalamus as proposed by Barraclough & Gorski (1961).

Recently, Velasco *et al.* (1969) proposed that the hippocampus exerts an inhibitory influence on gonadotropin secretion. They stimulated the rat hippocampus by electrolytic deposition of iron in acute preparation and noticed the inhibition of both spontaneous and stimulation-induced (amygdala or medial preoptic area) ovulation. This study recalled the earlier findings that the hippocampus exerted an inhibitory influence on the stress-induced ACTH release when it was stimulated during immobilization stress, while under unrestrained and unanesthetized conditions the stimulation facilitated ACTH release (Kawakami, Seto, Terasawa, Yoshida, Miyamoto, Sekiguchi & Hattori, 1968; Kawakami, Seto, Kimura & Yanase, 1971): Thus, we hypothetized that the hippocampus must change its influence on gonadotropin secretion as it does on ACTH, according to the experimental conditions.

At first, stimulation was applied to rats in proestrus just before the critical period through the chronically implanted electrode, and it was observed that ovulation was not inhibited by this stimulation (Table 3). Measurement of gonadotropin confirmed the results (Kawakami, Kimura & Wakabayashi 1972). On the other hand,

TABLE 3. Effect of electrical stimulation of the dorsal hippocampus on spontaneous ovulation in rats in proestrus (Kawakami, Kimura & Wakabayashi 1972)

Conditions	Ovulation		
	Number of animals*	%	Number ova/rat
Control	10/10	100.0	11.0±1.4**
Chronic experiment unanesthetized and unrestrained	7/7	100.0	10.1±2.0
Chronic experiment under ether anesthesia	5/5	100.0	12.0±1.3
Acute experiment under ether anesthesia	4/8	50.0	8.8±1.3

Stimulation parameters: square wave pulses of 0.1 ms duration, 100 Hz, 60 μA.
* Number of ovulating rats/number of stimulated rats. ** Mean±SE.

when dorsal-hippocampal stimulation was applied under acute preparation, stereotaxically restrained and anesthetized with ether, ovulation was blocked in 50% of the rats, and the gonadotropin surge was inhibited as well. Interesting enough was the fact that secretion of FSH and prolactin, which generally coincides with LH, was also inhibited.

What induced the different results in these two experiments? Probably (1) the so-called stress, such as ether or stereotaxic restraint, may alter the hippocampal function, and (2) the acute insertion of an electrode may alter the excitability of the hippocampus itself and that of the neural connections to the hypothalamus. For instance, changes of excitability of the hippocampus during the application of stress were indicated by the seizure threshold lowering in 6 hr of immobilization stress (Kawakami, unpublished) and in addition, insertion of an electrode into the hippocampus in many cases resulted in seizure discharges around this area, spreading to the other brain regions, such as the neocortex and the amygdala (Fig. 1, see Section I).

Accordingly, it seems relevant to mention that the function of the hippocampus changes due to the condition of the hippocampus, and in conditions that result in hippocampal seizure discharges, the hippocampus exerts an inhibitory influence on the hypothalamus. This assumption is supported by the electrophysiological findings that the multiple unit activity in the arcuate nucleus is inhibited by acute hippocampal stimulation, but is not affected by chronic stimulation. In relation to this concept, MacLean and his colleagues (1961) found that stimulation of the hippocampus and its related structures elicits penile erection in rats if that stimulation is just of sufficient intensity to provoke afterdischarges, and furthermore, some stimulations in the anterior thalamus result in aversive manifestations followed by the appearance of erection only after the termination of stimulation and concomitantly with the appearance of hippocampal afterdischarges. The neural mechanisms that induce such phenomena are still unclear, but one possibility is release of the inhibitory mechanism for penile erection by hippocampal seizure discharges.

To investigate more precisely the nature of hippocampal inhibition on gonadotropin secretion, combined stimulations in the dorsal hippocampus and medial preoptic area were applied acutely to rats made anovulatory by pentobarbital administration at 13:45 on the day of proestrus. After simultaneous stimulation of these two areas there was no increase in the serum and pituitary contents of LH, which had

TABLE 4. Effect of electrical stimulation of the dorsal hippocampus simultaneously or prior to medial preoptic stimulation in mature female rats (Kawakami, Terasawa, Kimura & Wakabayashi, 1972)

Day of cycle	Site of stimulation	Anterior pituitary weight (mg)	Pituitary concentration of LH (μg/mg)	Pituitary concentration of FSH (μg/mg)	Pituitary concentration of Prolactin (mIU/mg)	Serum concentration of LH (ng/ml)	Serum concentration of FSH (ng/ml)	Serum concentration of Prolactin (μIU/ml)
DI	Control	9.9±0.6† (6)	0.63±0.06	3.64±0.34	30.9±2.4	0.2±0.2	3522±934	862± 84
	MPO	9.4±0.9 (6)	0.96±0.09*	4.48±0.39	22.3±2.7	4.3±1.3*	3838±512	492± 62*
	HPC+MPO[a]	10.5±0.3 (7)	0.59±0.03**	5.80±0.20*	14.5±3.1	0.8±0.3**	3293±257	662±116
DII	Control	8.8±0.7 (6)	0.65±0.09	4.00±0.22	32.2±3.6	0.5±0.2	3106±632	720± 76
	MPO	9.3±0.6 (6)	1.08±0.09*	5.21±0.40	18.7±0.6*	4.9±1.2*	4631±308	394± 11*
	HPC+MPO[a]	9.6±0.6 (6)	0.69±0.08**	7.77±0.72*	21.6±5.9	0.5±0.4**	2458±551**	628± 87**
	HPC+MPO[b]	6.7±0.4 (9)	0.54±0.09**	5.91±0.63	14.1±2.3	1.8±0.3**	3952±1525	433± 59
	HPC+MPO[c]	7.8±0.5 (10)	0.67±0.03	5.78±0.76	20.6±2.7	0.7±0.3**	2899±375	440± 34
P	Control	9.6±0.6 (6)	0.78±0.04	3.06±0.30	28.1±3.4	1.1±0.2	3358±674	1857±404
	MPO	9.8±0.3 (9)	1.10±0.15*	5.60±0.40*	11.1±0.5	15.1±4.5*	5628±539*	607± 75*
	HPC+MPO[a]	9.9±0.5 (6)	0.60±0.05**	7.75±0.33*	25.5±3.7**	0.5±0.3**	2982±389**	413± 54
	HPC+MPO[b]	7.3±0.2 (7)	0.81±0.07**	6.60±0.62	20.5±3.9	1.0±0.3**	3820±701	456± 27
	HPC+MPO[c]	8.3±0.6 (8)	0.74±0.09**	4.72±0.58	23.5±3.1	1.6±0.6**	3965±402	521± 26
E	Control	9.7±0.5 (6)	0.48±0.05	3.27±0.39	41.4±2.5	0.2±0.1	2870±759	761±168
	MPO	9.7±0.8 (6)	0.61±0.08*	3.75±0.37	16.7±3.1*	16.7±0.5*	2805±408	502± 71
	HPC+MPO[a]	9.7±0.9 (6)	0.47±0.04**	6.06±0.61*	23.5±5.3	0.4±0.2**	4195±955	571± 93

[a] HPC stimulation and MPO stimulation were applied simultaneously. [b] HPC stimulation was applied 2 hr prior to MPO stimulation. [c] HPC stimulation was applied 4 hr prior to MPO stimulation. † Mean±SE. * $P<0.05$ *vs.* control; ** $P<0.05$ *vs.* MPO only. Number of animals in parentheses. Abbreviations: DI, diestrus I; DII, diestrus II; P, proestrus; E, estrus; d-HPC, dorsal hippocampus; MPO, medial preoptic area.

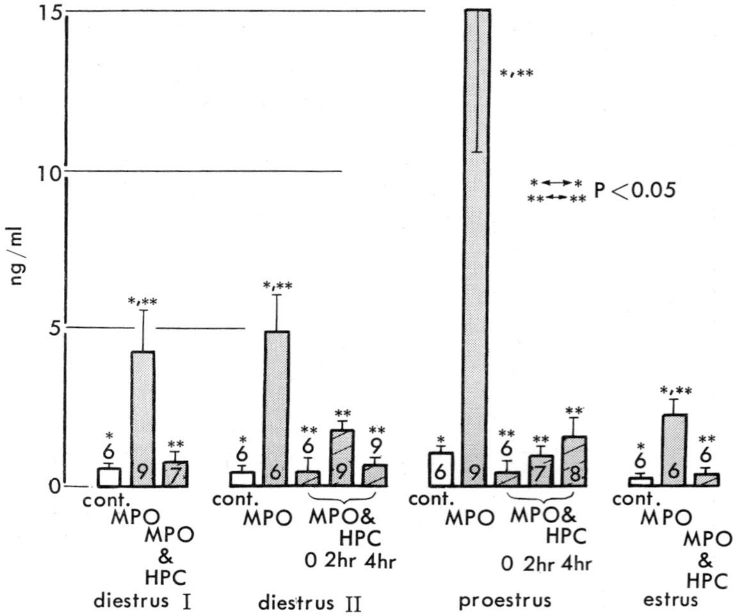

Fig. 3. Effects of electrical stimulation of the dorsal hippocampus (HPC) applied simultaneously (0) or preceding (2 and 4 hr) medial preoptic (MPO) stimulation on serum LH (Kawakami, Terasawa, Kimura & Wakabayashi, 1972).

been expected by stimulation of the medial preoptic area throughout the estrous cycle. Hippocampal stimulation applied even 2 to 4 hr prior to the preoptic stimulation could inhibit the LH increase (Kawakami, Terasawa, Kimura & Wakabayashi, 1972, Fig. 3 and Table 4).

An increase in the serum concentration of FSH, which should be induced by preoptic stimulation on the day of proestrus, was also inhibited by simultaneous or preceding stimulation of the hippocampus. However, it is of interest to note that the pituitary concentration of FSH increased from simultaneous stimulation of these areas at any stage during the estrous cycle. The simultaneous stimulation of the hippocampus and medial preoptic area induced no change in serum concentration of prolactin, in spite of the decrease by respective stimulation except on the day of proestrus, and pituitary concentration decreased as induced by preoptic stimulation alone.

Taking into consideration the results that acute hippocampal stimulation inhibits spontaneous discharge of gonadotropin or stress-induced ACTH release, it seems that such inhibitory action may occur only when some neural activation takes place to release or decrease pituitary hormone, however natural or artificial it may be. Furthermore, this inhibitory function of the hippocampus lasted for 2 to 4 hr and appeared to be independent of the estrous cycle.

In this form, the phenomenon seems to be identical to that described by Grastyan (1959). He noticed hippocampal inhibition in acute experiments in connection with movements either elicited from the motor cortex or induced reflexly. However, he proposed that those inhibitory effects are not simply due to epileptic manifestation, because of the existence of inhibition even in cases where afterdischarges failed to appear in the contralateral hippocampus or in the neocortex, either during or after stimulation.

Lesion experiments

The results of the above-mentioned experiments demonstrate that, in rats and rabbits, the hippocampus is capable of affecting pituitary-ovarian function in response to electrical stimulation. However, the physiological significance of the hippocampal function in the maintenance of functional capacity of the ovary or the maintenance of normal estrous cyclicity has not been evaluated, and this problem may be approached by the elimination of the hippocampal influence on the hypothalamus. For this purpose, the rabbit fornix was sectioned along the coronal plane by a 4-mm-wide knife, and after 3 to 4 months

TABLE 5. Effects of the fornix section on the transfer rates of ^{14}C into ovarian steroids (Kawakami, Seto & Kimura, 1972)

Group	Number of animals	Ovarian weight (mg)	^{14}C incorporated		
			Progesterone	20α-OH-P[a]	Estrogen[b]
Intact					
Control	10	209± 2[c]	878±15[d]	293± 3	225± 4
Estrogen primed*	10	—	912± 4	306± 2	200± 2
Estrogen primed and mated*	10	—	2763±25	1102±16	288± 8
Fornix sectioned					
Control	5	218±43	435±70	142±18	100± 9
Estrogen primed	5	215±47	446±51	176±54	78±10
Estrogen primed and mated	5	239±56	860±71	319±20	258±20

[a] 20α-hydroxy-pregn-4-en-3-one. [b] Estradiol and estrone. [c] Mean±standard error. [d] Mean dpm per mg of ovarian protein N±standard error. * Data from Kawakami, Seto & Yoshida (1968).

the animal was decapitated to measure the ovarian biosynthetic activity. The results were dramatic; the transfer rates of ^{14}C into progesterone, 20α-OH-P, and estrogen were decreased by approximately 50% when compared to intact rabbits (Kawakami, Seto & Kimura, 1972, Table 5). Combined with the results from stimulation experiments, it may be said that the hippocampus plays an important role in the maintenance of ovarian functional capacity.

In addition, the ovaries in sectioned rabbits failed to ovulate following copulation, although the stimulative effect of the ovulating hormone was indicated by the increase of transfer rates of ^{14}C into progesterone and 20α-OH-P. These results indicate that the matter is in the ovarian condition and not in the gonadotropin discharge mechanism. Previously, Sawyer (1959) reported that lesions in the female rabbit rhinencephalon did not inhibit reproductive behavior or block copulation-induced ovulation. A difference between his experiment and ours may be the time after lesion, ours being longer than his and inducing a greater reduction in the ovary. A number of investigations have shown that in rabbits the most important nervous pathways concerned with LH release are those that are activated directly or indirectly by copulation, and on the other hand the electrical stimulation of the hypothalamus can induce ovulation (Harris, 1937; Haterius & Derbyshire, 1937). The latter facts and above-mentioned finding by Sawyer indicate that, at least in rabbits, normal pituitary-gonadal function can be maintained only by the hypothalamus. However, the authors' experiment showed the importance of areas other than the hypothalamus, such as the hippocampus.

Another attempt was made to study the role of the hippocampus in the release of gonadotropin responsible for the events of proestrus in 4-day-cycle rats (Kawakami, Terasawa, Kimura, Shinohara & Hiroto, 1972). The fornix or medial corticohypothalamic tract was transected by a small L-shaped knife at various times during the estrous cycle. These lesions only partially prevented the events of the proestrous day. However, when the fornical transection was conducted including the septal complex above the anterior commissure with a larger knife on the day before diestrus II, it prevented the events of proestrus and estrus, indicating the blockage of gonadotropin secretion responsible for such events.

However, since all the transected rats resumed their vaginal 4-day cycle by about 2 weeks from the operation, the inside of this area, the preoptic-anterior basal hypothalamus, may also have the ability to release ovulating hormone. This assumption is supported by the study of Köves & Halász (1970) in which they observed a nearly normal vaginal cycle in rats with neural interruption of the bilateral, anterior, and superior connections to the anterior hypothalamic area. In the fornix-transected rats, however, both pituitary and serum contents of FSH on the day of diestrus II were higher (about 1.4 times) than those of the normal controls on the same day. This fact indicates that the transection of major efferent fibers of the hippocampus may result in a disinhibition to the hypothalamus, and therefore, neural transected animals differ from the normals, even though the estrous cycle was resumed after transection.

Further study was performed by lesions in the large part of hippocampus. Rats with the lesions in the hippocampus resumed vaginal 4-day-cycle. However, serum concentrations of LH and FSH on the day of proestrus were significantly higer than those in normal rats in proestrus, and those on the day of estrus were not different. The result seems to suggest that the hippocampus exertes inhibitory influence on the secretion of LH and FSH on the day of proestrus, as well as on the day of diestrus II.

THE ROLE OF THE HIPPOCAMPUS IN CONTROLLING GONADOTROPIN SECRETION IN THE MALE

As early as 1912, observations were made relating the brain to the control of reproductive function in male dogs (Aschner, 1912). However, since then no review specifically devoted to problems of gonadotropin regulation in males has appeared. Especially the extrahypothalamic afferents affecting gonadotropin secretion have received little attention, except by Riss *et al.* (1963), who tried hippocampal lesions in 1-week-old males and observed a decrease of testicular weight at 75–80 days of age.

In our recent work, immature and mature male rats were stimulated in the dorsal hippocampus through implanted electrodes with the same parameters used in the females. In mature males, the stimulation did not induce any change in either serum or pituitary contents of LH, as in the females (Table 2). The pituitary concentration of FSH decreased, and its serum content showed a tendency to decrease. In addition, pituitary and serum concentrations of prolactin decreased due to the stimulation. Following stimulation of the hippocampus in 27- and 35-day-old rats, pituitary and serum concentrations of LH did not change, but those of FSH decreased. Prolactin concentrations were not affected. This inhibitory effect on

FSH was observed when the 27-day-old rats were pretreated with testosterone (1 mg).

These results indicate that the male hippocampus does not participate in LH regulation, unlike the female, and involvement in FSH regulation is consistently inhibitory. The latter phenomenon may result from the constant and dominant circulation of one gonadal hormone, testosterone. A noticeable finding in the present experiment is that, in 35-day-old rats, the hippocampal stimulation did not inhibit prolactin release. Neither was prolactin inhibition seen following amygdalar or preoptic stimulation. This may indicate the lack of inhibitory mechanisms for prolactin in immature rats.

ELECTROPHYSIOLOGICAL ASPECTS OF THE HIPPOCAMPAL FUNCTION

Hohlweg and Junkmann suggested in 1932 that the gonadal hormones might influence gonadotropic function through a hypothetical " sexual center " located somewhere in the brain. In the light of experimental data, it is probable that the feedback effects of sexual hormones act to modify gonadotropin secretion at three levels capable of modifying the activity of the hypothalamus, the pituitary, hypothalamus, and extrahypothalamic area.

As mentioned in Section IV, the hippocampus drew our attention because of the probability that it possesses some neuronal component sensitive to gonadal hormones.

Electrical activity during the reproductive process

In chronically prepared female rabbits, characteristic sinusoidal waves (4–8 and 8–13 Hz) decreased significantly with the progress of the estrous state, and after copulation these EEG patterns were markedly dominated by rhythmic sinusoidal waves, particularly during the first 2–3 hr. Similar changes were observed during pregnancy (Kawakami *et al.* 1967). In addition, the evoked potentials recorded in the arcuate nucleus produced by single-shock stimuli of the dorsal hippocampus decreased in amplitude at estrus and increased at postestrus (Kawakami, Seto, Kimura, Ishida, Yoshida & Kubo, 1968).

From the above, it may be said that the electrical activity of the hippocampus itself and/or the neural pathway to the hypothalamus diminishes at the state of estrogen dominance over progesterone and increases at the state of progesterone dominance over estrogen. However, interpretation of these results may be very difficult because

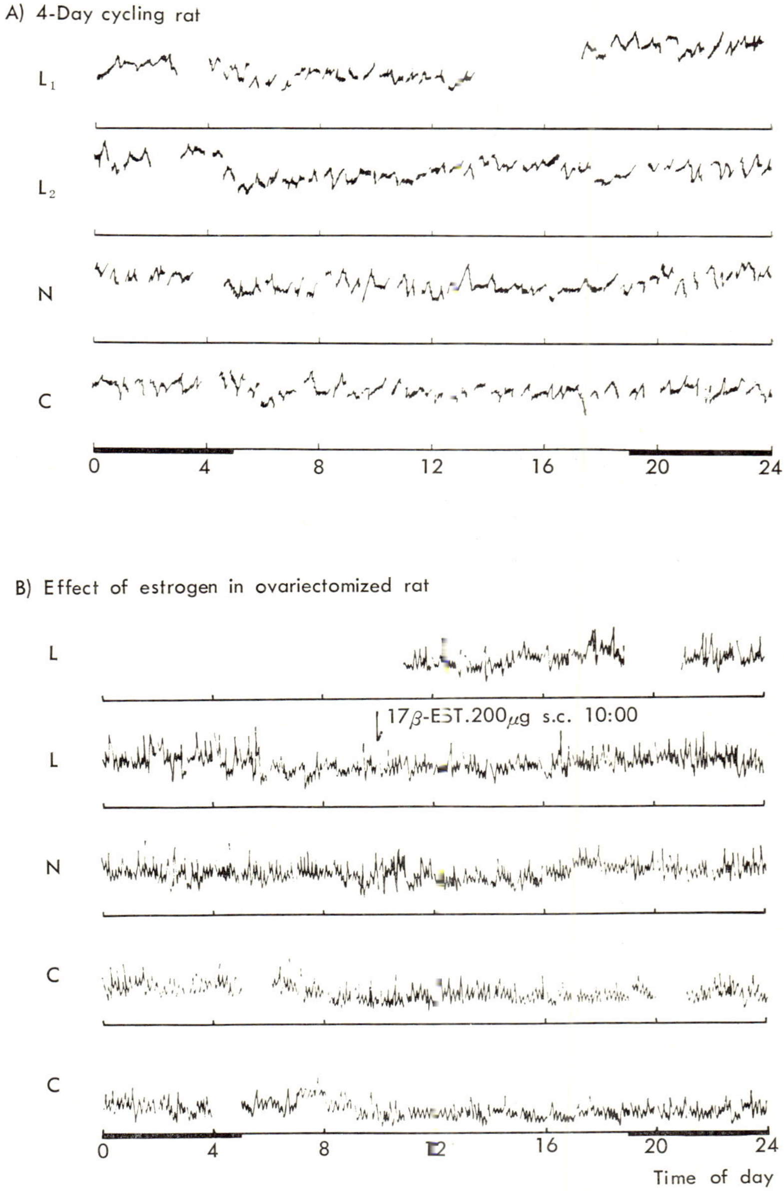

Fig. 4. Changes of integrated multiple unit activity in the dorsal hippocampus during the estrous cycle (A) and effects of 17 β-estradiol (200 μg) injection. L, N, and C represent diestrous, proestrous, and estrous smears, respectively (Kawakami & Manaka, unpublished data).

Grastyan & Karmos (1962) proposed that theta rhythm does not reflect an active functional state but represents an inhibited or resting state of the hippocampus.

In this respect, a study by Terasawa & Timiras (1968) is interesting. They investigated in a chronic experiment the changes of localized seizure threshold during the rat estrous cycle and observed that the hippocampal seizure threshold decreased from the morning of proestrus through the morning of estrus, with the lowest value at the night of proestrus. Considering that changes in seizure threshold reflect changes in irritability, they concluded that the hippocampal electrical activity increased in response to estrogen being dominant at that time.

On the other hand, most recently our investigation using the multiple unit recording technique (Sawyer, Kawakami, Meyerson & Whitmoyer, 1968) showed that integrated multiple unit activity in the dorsal hippocampus of both cycling and ovariectomized rats has only a diurnal variation, which is relatively higher during the day (5.00–17.00) than at night and not a variation according to the estrous cycle (Fig. 4).

Influence of sex steroids and gonadotropin

It is inferred from the changes of electrical activity during the reproductive process that the activity may change in response to exogenously administered steroid hormones or gonadotropin as well. In fact, progesterone (5 mg) administration to the ovariectomized and chronically prepared rabbits caused the hippocampal theta wave to increase its amplitude and frequency. Similarly, the response to LH (0.3 U.S.P unit) was a shift from irregular slow waves mixed with low amplitude sinusoidal waves of 3–4 Hz to a dominance of 4–8 and 8–13 Hz sinusoidal waves (Kawakami *et al.* 1967). These EEG changes correspond well to those after copulation (Section VI, A), and thus, the latter seems to reflect mainly the effects of LH and/or progesterone (Hilliard, Hayward & Sawyer, 1964), rather than the nervous activity that initiates the pituitary activation. The hippocampal-evoked potentials in the arcuate nucleus also changed in the same manner, increasing in amplitude due to progesterone or LH and decreasing due to FSH (Kawakami, Seto, Kimura, Ishida, Yoshida & Kubo, 1968).

Summarizing the results obtained by recording EEG and evoked potentials under different endocrine conditions, it may be said that the activities of the hippocampus and the synaptic chains conducting the hippocampal impulses to the arcuate nucleus were inhibited at estrus or by estrogen and were facilitated at postestrus or by progesterone. This assumption agrees with the facts that integrated multiple unit activity of ovariectomized rats significantly decreased after injection of 17β-estradiol (200 μg, s.c.) (Fig. 4).

Thus, it may be suggested that the hippocampus and its connections to the hypothalamus have a significant sensitivity to sex hormones and gonadotropin, and their excitability changes in response to the hormonal conditions, although the functional significance for the regulation of gonadotropin is not defined. In addition, it was confirmed that in rabbits the implantation of a minute amount of solid progesterone into the hippocampus resulted in an increase of gonadotropin release, which was indicated by the ovarian biosynthetic activity, and implantation of estrogen induced no change (Kawakami, Seto, Yoshida & Miyamoto, 1969).

CONCLUSION

There appear to be two types of hippocampal modulatory mechanism for gonadotropin secretion. One acts phasically in accordance with changes of the humoral environment. The other acts in an inhibitory manner in abnormal external and internal environments and is not specific for gonadotropin secretion.

The hippocampus reveals its phasic modulatory influence especially on FSH regulation. At the first stage of the estrous cycle (in mature animals) and of puberty (in immature animals), it enhances the release of FSH, and just before the discharge of ovulating hormone (in mature and immature animals) it inhibits the release. Such a function seems to be based on the hippocampal sensitivity to sex steroids and gonadotropin, because its electrical activity is easily changed by these hormones of endogenous or exogenous origin. Furthermore, it is probable that the hippocampal impulses influencing gonadotropin secretion change routes to the hypothalamus in relation to the humoral environment.

One point that should be emphasized is that, generally speaking, the amygdala seems to modulate gonadotropin secretion in a manner reciprocal to the hippocampus. This has been confirmed repeatedly in electrophysiological and biochemical studies (Kawakami *et al.* 1967; Kawakami & Kubo, 1971). In addition, a possible role of the septal complex as a modulator, as well as a relay station of hippocampal and amygdalar influence to the hypothalamic center is indicated (Kawakami, Terasawa, Kimura, Shinohara & Hiroto, 1972). Though this hypothalamic center alone can maintain almost normal gonadotropin secretion, for absolutely normal secretion the modulation of the limbic system may be required.

REFERENCES

Adey, W. R., Dunlop, C. W. & Sunderland, S. (1958). A survey of rhiencephalic interconnections with the brain stem. *J. comp. Neurol.* **110**, 173–203.

Aschner, B. (1912). Über die Funktion der Hypophyse. *Arch. ges. Physiol.* **146**, 1–146.

Bargmann, W. & Schade, J. P., eds. (1963). The Rhiencephalon and Related Structures. In *Progress in Brain Research*, Vol. 3, pp. 253. Amsterdam: Elsevier.

Barraclough, C. H. & Gorski, R. A. (1961). Evidence that hypothalamus is responsible for androgen-induced sterility in the female rat. *Endocrinology* **68**, 68–79.

Brady, J. V. (1960). Emotional behavior. In *Handbook of Physiology*, (Amer. Physiol. Soc., ed. Field, J.) Section I, Vol. III, pp. 1529–1552. Baltimore, Maryland: Williams & Wilkins.

Broca, P. (1878). Anatomie comparée des circonvolutions cérébrales: Le grand lobe limbique et la série des mammifères. *Rev. Anthropol.* (Paris) (**2**) **1**, 385–498.

Brodal, A. (1947). The hippocampus and the sense of smell. *Brain* **70**, 179–222.

Bunn, J. P. & Everett, J. W. (1957). Ovulation in persitentestrous rats after electrical stimulation of the brain. *Proc. Soc. exp. Biol. Med.* **96**, 369–371.

Crosby, E. C. & Woodburne, R. T. (1940). The comparative anatomy of the preoptic area and the hypothalamus. *Res. Publs Ass. Res. nerv. mental Dis.* **20**, 52–169.

Gloor, P. (1956). Telencephalic influences upon the hypothalamus. In *Hypothalamic-Hypophysial Interrelationships*, ed. Fields, W. S., pp. 74–113. Springfield: Thomas.

Grastyan, E. (1959). The hippocampus and higher nervous activity. In *The Central Nervous System and Behavior*, ed. Brazier, M. A. B., pp. 119–205. Madison, N. J.: Madison Printing Company.

Grastyan, E. & Karmos, G. (1962). The influence of hippocampal lesions on simple and delayed instrumental conditioned reflexes. In *Physiologie de L'hippocampe*, Éditions Du Centre National de la Recherche Scientifique. pp. 225–226, Paris.

Green, J. D. (1960). The Hippocampus. In *Handbook of Physiology*

(Amer. Physiol. Soc., ed. Field, J.) Section I, Vol. II, pp. 1373–1389. Baltimore, Maryland: Williams & Wilkins.

Green, J. D. & Arduini, A. A. (1954). Hippocampal electrical activity in arousal. *J. Neurophysiol.* **17**, 533–557.

Gurdjian, E. S. (1927). The diencephalon of the albino rat. *J. comp. Neurol.* **43**, 1–114.

Harris, G. W. (1937). The induction of ovulation in the rabbit by electrical stimulation of the hypothalamo-hypophyseal mechanism. *Proc. R. Soc.* B **122**, 374–394.

Haterius, H. O. & Derbyshire, A. J., Jr. (1937). Ovulation in the rabbit following upon stimulation of the hypothalamus. *Am. J. Physiol.* **119**, 329–330.

Hayward, J. N., Hilliard, J. & Sawyer, C. H. (1964). Time of release of pituitary gonadotropin induced by electrical stimulation of the rabbit brain. *Endocrinology* **74**, 108–113.

Hilliard, J., Hayward, J. N. & Sawyer, C. H. (1964). Postcoital patterns of secretion of pituitary gonadotropin and ovarian progesterone in the rabbit. *Endocrinology* **75**, 957–963.

Hohlweg, W. & Junkmann, K. (1932). Über die Wirkung intrazerebral verabreichten Dienoestroldiacetats bei Ratten. *Endokrinologie* **38** 46–51.

Kaada, B. R. (1951). Somatomotor, autonomic and electrocorticographic responses to electrical stimulation of " rhiencephalic " and other structures in primates, cat and dog. *Acta physiol. scand.* **24**, Suppl. 183.

Kawakami, M., Kimura, F. & Wakabayashi, K. (1972). Electrical stimulation of the hippocampus under the chronic preparation and changes of LH, FSH and prolactin levels in serum and pituitary. *Endocr. jap.* **19**, 85–96.

Kawakami, M. & Kubo, K. (1971). Neuro-correlate of limbic-hypothalamic-pituitary-gonadal axis in the rat: change in limbic-hypothalamic unit activity induced by vaginal and electrical stimulation. *Neuroendocrinology* **7**, 65–89.

Kawakami, M., Seto, K. & Kimura, F. (1972). Influence of the fornix section on the biosynthesis of ovarian steroids in the rabbit. *Endocr. jap.* **19**, 429–433.

Kawakami, M., Seto, K., Kimura, F., Ishida, S., Yoshida, K. & Kubo, K. (1968). The role of extrahypothalamic impulses to the hypothalamus in the control of gonadotropin release with reference to the limbic system as a locus of neuroendocrine interaction. In *Progress in endocrinology*, ed. Gaul, C., pp. 503–507. Amsterdam: Excerpta Medica Foundation.

Kawakami, M., Seto, K., Kimura, F. & Yanase, M. (1971). Difference in the buffer action between the Limbic Structures and the hippocampus to the immobilization stress in rabbits. In *Inflence of Hormones on the Nervous System*. Proc. Int. Soc. Psychoneuroendocrinology, Brooklyn, 1970. pp. 107–120. Basel: Karger.

Kawakami, M., Seto, K., Terasawa, E. & Kimura, F. (1972). Role of the limbic-hypothalamic system in relation to ovulation and ovarian steroidgenesis in the rats. *Med. J. Osaka Univ.* **23**, 57–75.

Kawakami, M., Seto, K., Terasawa, E. & Yoshida, K. (1967). Mechanisms in the limbic system controlling reproductive functions of the ovary with special reference to the positive feedback of progestin to the hippocampus. In *Progress in Brain Research*, ed. Adey, W. R. & Tokizane, T., Vol. 27, pp. 69–102. Amsterdam: Elsevier Publishing Company.

Kawakami, M., Seto, K., Terasawa, E., Yoshida, K., Miyamoto, T., Sekiguchi, M. & Hattori, Y. (1968). Influence of electrical stimulation and lesion in limbic structure upon biosynthesis of adrenocorticoids in the rabbit. *Neuroendocrinology* **3**, 337–348.

Kawakami, M., Seto, K. & Yoshida, K. (1966). Influence of the limbic system on ovulation and on progesterone and estrogen formation in rabbit's ovary. *Jap. J. Physiol.* **16**, 254–273.

Kawakami, M., Seto, K. & Yoshida, K. (1968). Influence of the limbic structure on biosynthesis of ovarian steroids in rabbits. *Jap. J. Physiol.* **18**, 356–372.

Kawakami, M., Seto, K., Yoshida, K. & Miyamoto, T. (1969). Biosynthesis of ovarian steroids in the rabbit: Influence of progesterone or estradiol implantation into the hypothalamus and limbic structures. *Neuroendocrinology* **5**, 303–321.

Kawakami, M. & Terasawa, E. (1972). Electrical stimulation of the brain on gonadotropin secretion in the female prepuberal rat. *Endocr. jap.* **19**, 335–347.

Kawakami, M., Terasawa, E., Kimura, F. & Kubo, K. (1971). Correlated changes in gonadotropin release and electrical activity of the hypothalamus induced by electrical stimulation of the hippocampus in immature and mature rats. *Proc. II Int. Soc. Psychoneuroendocrinology*, Budapest, 1971 (in press).

Kawakami, M., Terasawa, E., Kimura, F., Shinohara, Y. & Hiroto, S. (1972). A role of the septal complex in gonadotropin release in the rat. Proc. *IV International Congress of Endocrinology*, Washington D.C. 1972 (in press).

Kawakami, M., Terasawa, E., Kimura, F. & Wakabayashi, K. (1972). Modulating function of the limbic structure on gonadotropin release. *Neuroendocrinology* (in press).

Kling, A. (1964). Effects of rhiencephalic lesions on endocrine and somatic development in the rat. *Am. J. Physiol.* **206**, 1395–1400.

Klüver, H. & Bucy, P. C. (1937). Psychic blindness and other symptoms following bilateral temporal lobotomy on rhesus monkeys. *Am. J. Physiol.* **119**, 352–353.

Koikegami, H., Yamada, T. & Usui, K. (1954). Stimulation of the amygdaloid nuclei and periamygdaloid cortex with special reference to its effects on uterine movement and ovulation. *Folia Psychiat. Neurol. Japan.* **8**, 7–31.

Köves, T. & Halász, B. (1970). Location of the neural structures triggering ovulation in the rat. *Neuroendocrinology* **6**, 180–193.

MacLean, P. D. (1949). Psychosomatic disease and the " visceral brain." *Psychosom. Med.* **11**, 338–353.

MacLean, P. D. (1959). The limbic system with respect to two basic life principles. In *The central nervous system and behavior*, ed. Brazier, M. A. B., pp. 31–118. Madison, N. J.: Madison Printing Company.

MacLean, P. D. & Delgado, J. M. R. (1953). Electrical and chemical stimulation of frontotemporal portion of limbic system in the waking animal. *EEG Clin. Neurophysiol.* **5**, 91–100.

MacLean, P. D., Denniston, R H., Dua, S. & Ploog, D. W. (1961). Hippocampal changes with brain stimulation eliciting penile erection. In *Physiologie de L'hippocampe.* ed. Éditions Du Centre National de la Recherche Scientifique, pp. 491–510. Paris.

Moruzzi, G. & Magoun, W. H. (1949). Brain stem reticular formation and activation of EEG. *Electroencephalog: Clin. Neuro-Physiol.* **1**, 455–473.

Nauta, W. J. H. (1956). An experimental study of the fornix system in the rat. *J. comp. Neurol.* **104**, 247–273.

Papez, J. W. (1937). A proposed mechanism of emotion. *Arch. Neurol. Psychiat.* **38**, 725–743.

Penfield, W. & Milner, B. (1958). Memory deficit produced by bilateral lesions in the hippocampal zone. *Arch. Neurol. Psychiat.* **79**, 475–497.

Pestche, H. & Stumpf, Ch. (1962). Hippocampal arousal and seizure activity in rabbits ; toposcopical and microelectrode aspects. In *Physiologie de L'hippocampe.* ed. Du Centre National de la Recherche Scientifique, pp. 121–134. Paris.

Powell, T. P. S., Guillery, R. W. & Cowan, W. H. (1957). A quantitative study of the fornix-mammillo-thalamic system. *J. Anat.* **91**, 419–437.

Ramón y Cajal, S. (1901). Studies on the Cerebral Cortex (translated from the Spanish by L.M. Kraft in 1955). Chicago, Illinois: Year Book Publ.

Riss, W., Burstein, S. D. & Johnson, R. W. (1963). Hippocampal or pyriform lobe damage in infancy and endocrine development of rats. *Am. J. Physiol.* **204**, 861–866.

Sawyer, C. H. (1959). Effects of brain lesions on estrous behavior and reflexogenous ovulation in the rabbit. *J. exp. Zool.* **142**, 227–246.

Sawyer, C. H. (1960). Reproductive behavior. In *Handbook of Physiology.* (Amer. Physiol. Soc., ed. Field, J.), Section I, Vol. II, pp. 1225–1240. Baltimore, Maryland: Williams & Wilkins.

Sawyer, C. H. & Kawakami, M. (1959). Characteristics of behavioral and electroencephalographic after-reactions to copulation and vaginal stimulation in the female rabbit. *Endocrinology* **65**, 622–630.

Sawyer, C. H., Kawakami, M., Meyerson, B. & Whitmoyer, D. I. (1968). Effects of ACTH, dexamethasone and asphyxia on electrical activity of the hypothalamus. *Brain Res.* **10**, 213–226.

Schreiner, L. & Kling, A. (1953). Behavioral changes following rhiencephalic injury in cat. *J. Neurophysiol.* **16**, 643–659.

Terasawa, E. & Timiras, P. S. (1968). Electrical activity during the estrous cycle of the rat: cyclic changes in limbic structures. *Endocrinology* **83**, 207–216.

Velasco, M. E. & Taleisnik, S. (1969). Effect of hippocampal stimulation on release of gonadotropin. *Endocrinology* **85**, 1154–1159.

Zarrow, M. X., Naqvi, R. H. & Denenberg, V. H. (1969). Androgen-induced precocious puberty in the female rat and its inhibition by hippocampal lesions. *Endocrinology* **84**, 14–19.

FEED-BACK EFFECTS OF PERIPHERAL HORMONES ON THE ACTIVITY OF HYPOTHALAMO-HYPOPHYSIAL SYSTEMS

NEURAL CONTROL OF HUMAN FETAL ADENOHYPOPHYSIS

Takashi Tamura, Hiroshi Minaguchi, Takuro Kobayashi, and Shoichi Sakamoto

The pituitary of the human fetus in the second trimester of pregnancy contains a considerable amount of LH and HGH. The average LH and HGH contents of the pituitary were found to be 3.4 IU/gland and 2.3 μg/gland, respectively. The pituitary content and concentration of HGH increased significantly as the fetal age advanced, although there were no remarkable changes in the pituitary content of LH. The pituitary incubated *in vitro* secreted LH, FSH, and HGH into the medium. The addition of rat hypothalamic extract into the medium increased the release of LH and HGH by the pituitary. Synthetic LH–RH also enhanced the pituitary release of LH and FSH *in vitro*. However, in a pituitary of a fetus in the 16th week of gestation the secretion of LH and FSH *in vitro* was not stimulated by the addition of synthetic LH–RH. The results demonstrate that the anterior pituitary of the human fetus is responsive to hypothalamic releasing factors *in vitro* at least by the 19th week of fetal age.

INTRODUCTION

Information concerning the secretion of fetal pituitary hormones during intrauterine life is limited, especially for humans. By the 8th week of gestation, the anterior portion of the human pituitary is recognized as a cystic structure separated from the oral epithelium and lying in apposition to the diencephalic infundibulum (Antwell, 1926). Cellular differentiation in the anterior pituitary, which occurs primarily after 8 to 11 weeks' gestation (Palvlova, Pronina & Skebelskaya, 1968), is accompanied by the sequential appearance of several hormones within the gland (Levina, 1968). However, the presence of these hormones in the fetal pituitary does not represent actual synthesis and secretion. In addition, little is known of the physiological effect of the human fetal pituitary hormones on their target organs.

It is generally recognized that in the adult the anterior pituitary secretes various tropic hormones in response to hypothalamic releasing factors. The present study was performed to determine whether the human fetal pituitary is able to respond to hypothalamic releasing factors *in vitro* or not.

MATERIALS AND METHODS

Fetal pituitaries

Human fetal pituitaries studied were delivered from women in the 16th to 26th weeks of gestation. All fetuses were alive when they were born. The pituitaries were removed immediately after fetal death and placed in an incubation medium for *in vitro* study. Pituitaries for incubation Nos. FPI 6 and 7 were obtained from fetuses of single-ovum twins in the 26th week of gestation. For measurement of pituitary hormone contents, each pituitary was dried in aceton, weighed, and stored at 4°C. The pituitary was homogenized in saline with a glass homogenizer and assayed for human growth hormone (HGH) and luteiniging hormone (LH).

Preparation of acid extracts of rat hypothalamus

Mature female rats of the Wistar strain were sacrificed by decapitation. The hypothalami (including pituitary stalk and median eminence) were rapidly removed, homogenized in ice-cold 0.1 N HCl (0.2 ml/hypothalamus), and centrifuged at 10,000 rpm for 30 min at 4°C. The supernatants were stored at −20°C. Just prior to use, the supernatants were placed in protein-free medium 199, and the pH was adjusted to 7.4 by adding a drop of 1 N NaOH. The final volume was made up so that each milliliter of medium contained extract equivalent to 5 rat hypothalami.

In vitro incubation of fetal pituitaries

Each pituitary was divided into two equal parts by sagittal median section (Fig. 1). One half was placed into a control flask and the other half into an experimental flask, each containing 2 ml of medium 199. *In vitro* incubations were carried out in a Dubnoff metabolic shaker at 37°C. Media were renewed every 1 hr. Three to 5 hr after the initiation of incubation, the neutralized hypothalamic extracts, equivalent to 10 rat hypothalami, or synthetic LH-releasing

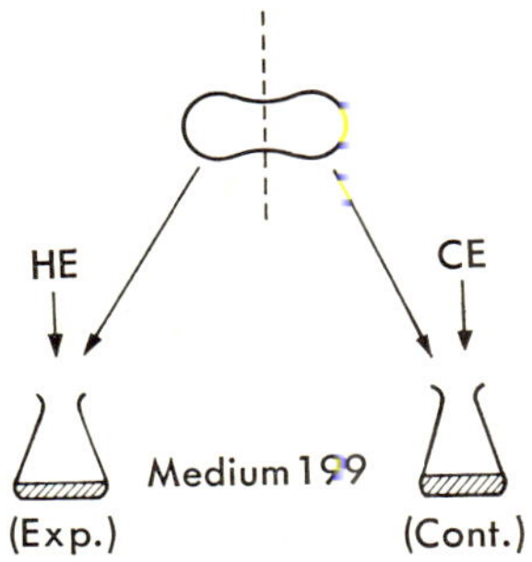

Fig. 1. Fetal pituitary. 1. Preincubate 3 to 5 hr at 37°C; 2. Add neutralized rat hypothalamic extract (HE) or synthetic LH–RH to the medium (experimental); 3. Add neutralized rat cerebral cortical extract (CE) or saline solution (control); 4. Each medium is renewed every 1 hr; 5. Radioimmunoassay of HGH, LH, and FSH in the medium.

hormone (LH-RH) was added to one of the paired flasks and neutralized cerebral cortical extracts or saline solution to the other control flask. The media were stored at −20°C until assayed later.

Assays of HGH, LH, and FSH

HGH, LH, and FSH in the fetal pituitaries and the media were measured by the radioimmunoassay technique using a double-antibody system. The standard for LH and FSH assays was human pituitary extract reference preparation (LER 907) kindly provided by the National Pituitary Agency, U.S.A., which had 48 mU of LH and 20 mU of FSH per microgram (bioassay, 2nd IRP-HMG standard). The radioimmunoassay kit for human GH was obtained from the Dainabot Radioisotope Lab., Tokyo. The standard for assay of HGH was human growth hormone preparation (Wilhelmi, Lot NIH-GH-HS 1147).

RESULTS

Pituitary contents of LH and HGH in human fetuses (Table 1)

Relatively high pituitary contents of both LH and HGH were observed in all of the 21 fetuses in the 16th to 26th weeks of gestation. The average LH content of fetal pituitaries was 3.4 IU/gland. There were no remarkable variations in the pituitary LH content of fetuses in the second trimester of pregnancy.

The average HGH content of these fetal pituitaries was 2.3 μg/gland. Both the pituitary content and concentration of HGH

TABLE 1. LH and HGH content in pituitaries of human fetuses

Fetal (months)	Age (weeks)	Fetal sex	Body weight (g)	LH IU/gland	LH IU/mg dry weight	HGH μg/gland	HGH μg/mg dry weight
V	16–6	male	100	2.5	5.7	0.50	1.1
	18–6	male	264	2.4	2.3	0.53	0.52
	19–2	male	370	1.8	1.6	0.13	0.12
	19–4	female	—	5.5	1.2	1.54	0.35
	19–4	female	360	2.7	2.4	0.95	0.86
			Mean	3.0±0.6	2.6±0.8	0.73±0.2	0.59±0.2
VI	21–0	male	380	2.8	2.5	0.82	0.71
	21–0	female	320	4.0	6.3	0.90	1.4
	21–1	—	340	3.5	2.2	1.3	0.83
	21–6	male	385	1.7	1.1	0.86	0.57
	21–6	male	400	3.0	2.2	1.3	0.96
	22–0	male	355	2.8	0.78	1.5	0.43
	22–2	male	540	3.8	0.78	2.5	0.52
	22–4	—	—	6.0	1.1	7.4	1.32
	23–0	male	640	3.5	1.2	4.2	1.49
	23–0	male	850	3.5	1.3	2.5	0.93
	23–6	—	750	4.0	1.1	3.3	0.93
	—	female	720	4.0	3.3	2.5	2.0
			Mean	3.6±0.3	2.0±0.5	2.4±0.6	1.0±0.1
VII	24–3	male	810	3.3	2.6	0.95	0.75
	25–1	female	600	1.2	0.43	4.3	1.6
	25–5	female	890	6.8	1.9	5.6	1.5
	26–5	male	920	1.7	0.73	5.2	2.2
			Mean	3.3±1.3	1.4±0.5	4.0±1.0	1.5±0.3

increased significantly as fetal age advanced. No significant difference of pituitary LH and HGH contents was observed in male and female fetuses.

Effects of crude hypothalamic extract on LH and HGH secretion of human fetal pituitaries in vitro

Figure 2 shows LH secretion by the fetal pituitaries *in vitro* during 7 hr of incubation. Human fetal pituitaries secreted a considerable amount of LH, but it decreased gradually as the incubation time advanced. Pituitary release of LH into the media was significantly increased by the addition of rat hypothalamic extract at 3 or 4 hr

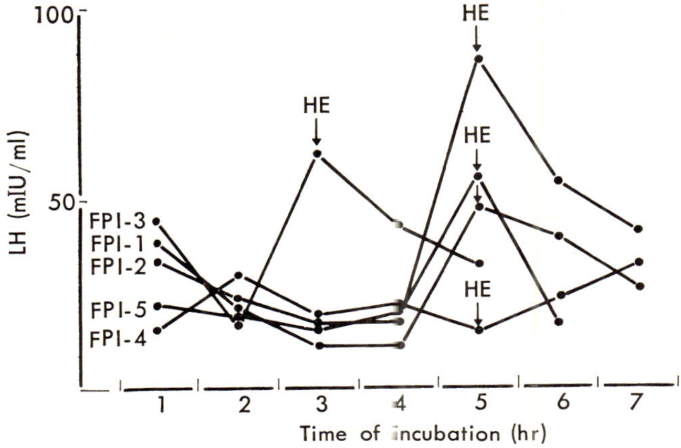

Fig. 2. Effect of crude hypothalamic extract on LH secretion of human fetal pituitaries *in vitro.*

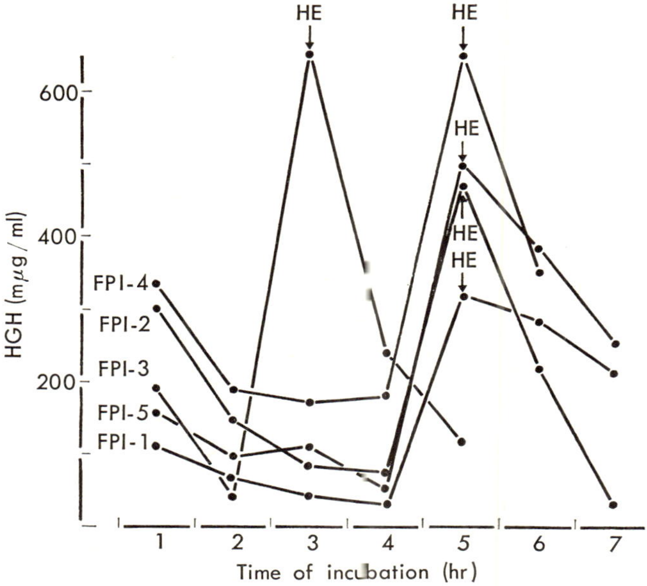

Fig. 3. Effect of crude hypothalamic extract on HGH secretion of human fetal pituitaries *in vitro.*

after the initiation of incubation, with the exception of experiment No. FPI-4.

Figure 3 shows HGH secretion by the fetal pituitaries *in vitro* during the incubation. The amount of HGH released into the media gra-

TABLE 2. Effects of crude hypothalamic extract on LH and HGH secretion of human fetal pituitaries *in vitro*

Number	Fetal age (weeks)	Fetal sex	Body weight (g)	Ratio of augmentation (HE/CE)	
				LH	HGH
FPI–1	19–6	male	360	3.2	7.2
FPI–2	22–0	female	730	4.3	6.4
FPI–3	19–0	male	370	2.8	11.2
FPI–4	22–4	—	720	0.8	4.5
FPI–5	21–4	female	500	2.2	8.2

dually decreased as the incubation time advanced. The addition of rat crude hypothalamic extract into the media at 3 or 5 hr after the initiation of incubation markedly increased the release of HGH by the fetal pituitaries.

Table 2 indicates the ratio of augmentation in pituitary LH and HGH release in response to the addition of hypothalamic extract. Pituitary release of LH and HGH after the addition of hypothalamic extract was compared with that of control cerebral cortical extract. The mean values of augmentation ratios by the addition of hypothalamic extract were 2.7 times for LH and 7.5 times for HGH, compared with those of controls.

Effects of synthetic LH-RH on LH and FSH secretion of human fetal pituitaries in vitro

Figures 4 and 5 indicate LH secretion *in vitro* by the pituitaries of fetuses in the 26th week of gestation. When 10 and 0.1 μg of synthetic LH-RH were added to the media 5 hr after the initiation of incubation in experiments No. FPI-6 and FPI-7, the pituitaries released more LH into the media in both incubations. The ratios of augmentation in pituitary LH release by the addition of synthetic LH-RH were 3.5 times and 1.6 times, respectively, when compared with that by the addition of saline solution (Table 3).

Figures 6 and 7 indicate FSH secretion during these incubations. Synthetic LH-RH also enhanced FSH release by the fetal pituitary *in vitro*. The ratios of augmentation in pituitary FSH release by the addition of 10 and 0.1 μg of synthetic LH-RH were 9.5 and 5.2 times, respectively, when compared with those by the addition of saline solution (Table 3).

When the pituitary of a human fetus in the 16th week of gestation

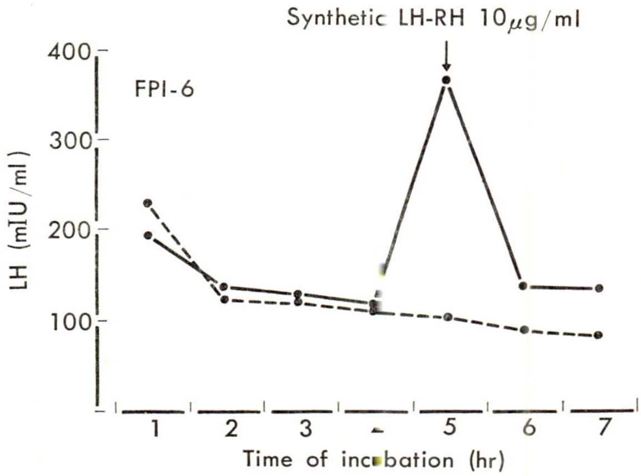

Fig. 4. Effect of synthetic LH-RH on LH secretion of human fetal pituitaries *in vitro*. ——, Exp.; - - - -, Cont.

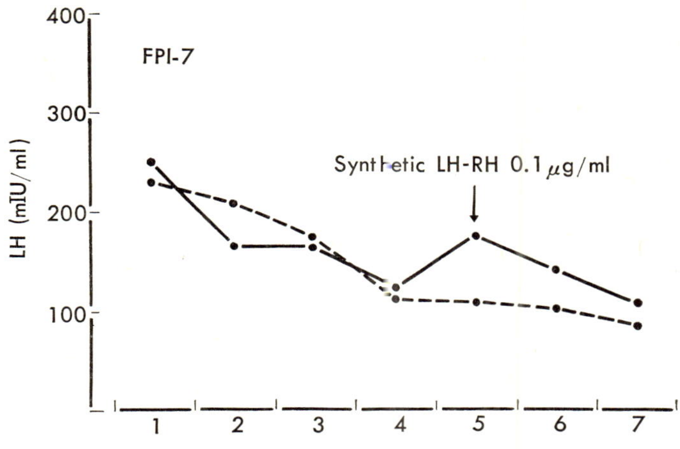

Fig. 5. Effect of synthetic LH-RH on LH secretion of human fetal pituitaries *in vitro*. ——, Exp.; - - - - -, Cont.

TABLE 3. Effects of synthetic LH-RH on LH and FSH secretion of human fetal pituitaries *in vitro*

Number	Fetal age (weeks)	Fetal sex	Body weight (g)	Synthetic LH-RH	Ratio of augmentation (Exp./Cont.)	
					LH	FSH
FPI–6	26–1	female	750	10 μg/ml	3.5	9.5
FPI–7	26–1	female	730	0.1 μg/ml	1.6	5.2

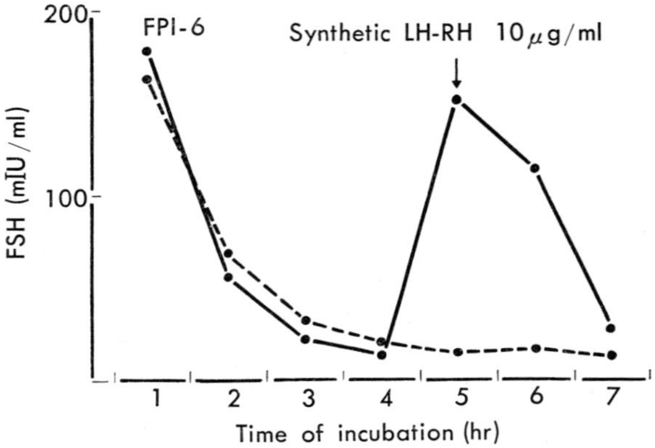

Fig. 6. Effect of synthetic LH-RH on FSH secretion of human fetal pituitaries *in vitro*. ——, Exp.; - - - -, Cont.

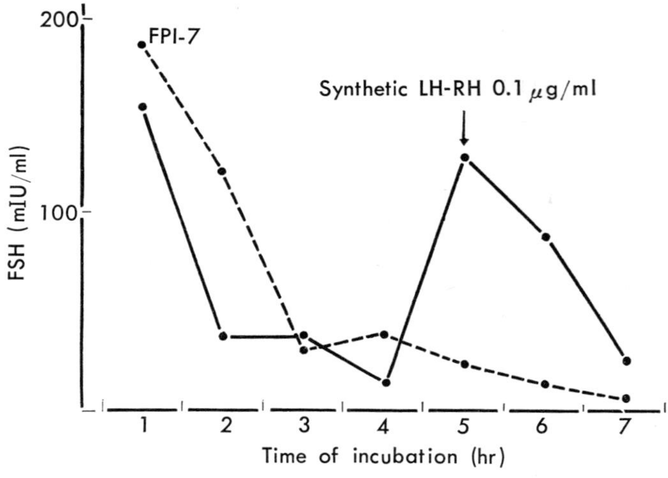

Fig. 7. Effect of synthetic LH-RH on FSH secretion of human fetal pituitaries *in vitro*. ——, Exp.; - - - - -, Cont.

was incubated, it secreted only a small amount of LH and FSH into the medium. The addition of 5 μg of synthetic LH-RH into the medium did not increase either LH or FSH release by the pituitary (Figs. 8 and 9).

DISCUSSION

The present study demonstrates that the pituitary of a human fetus

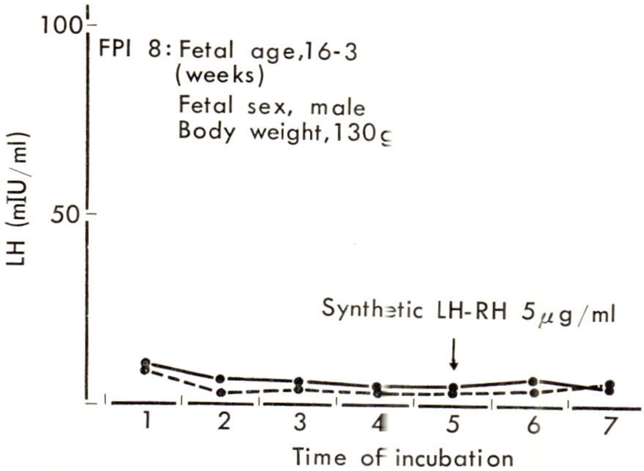

Fig. 8. Effect of synthetic LH-RH on LH secretion of human fetal pituitaries *in vitro*. ——, Exp.; - - - - -, Cont.

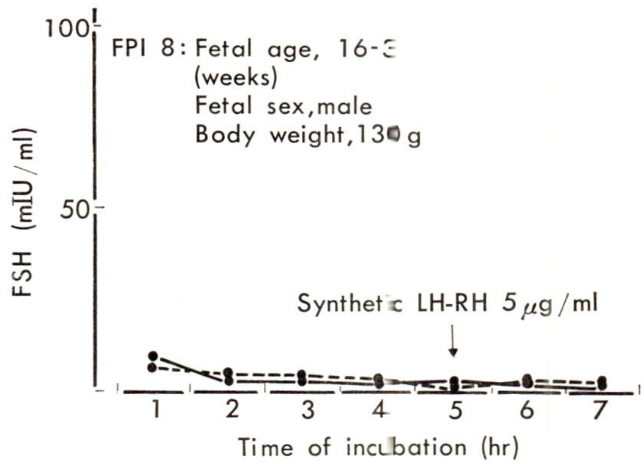

Fig. 9. Effect of synthetic LH-RH on FSH secretion of human fetal pituitaries *in vitro*. ——, Exp.; - - - - -, Cont.

in the second trimester of pregnancy contains a considerable amount of LH and HGH. The average LH and HGH contents of these fetal pituitaries were 3.4 IU/gland and 2.3 μg/gland, respectively. The pituitary content and concentration of HGH increased significantly as the fetal age advanced, although there were no remarkable changes in the pituitary content of LH. Furthermore, pituitaries incubated *in vitro* secreted LH, FSH, and HGH into the medium.

The addition of rat hypothalamic extract to the medium increased the release of LH and HGH by the pituitary. Synthetic LH-RH also enhanced the pituitary release of LH and FSH *in vitro*. However, in the pituitary of a fetus in the 16th week of gestation the secretion of LH and FSH *in vitro* was not stimulated by the addition of synthetic LH-RH. It has been reported that hypothalamic releasing factors are not species-specific in their action (Guillemin, 1964). In addition, synthetic LH-RH is known to stimulate the secretion of both LH and FSH by the pituitary (Arimura, Debeljuk & Schally, 1972; Redding, Schally, Arimura & Matsuo, 1972). Thus, the present study shows that the pituitary of the human fetus is responsive to hypothalamic releasing factors *in vitro* at least by the 19th week of fetal age.

Our data on the presence of pituitary hormones in human fetuses are in agreement with previous work reported on this subject. Immunoreactive growth hormone has been detected in the human fetal pituitary as early as 9 weeks of gestation (Parvlova *et al.* 1968). Levina (1968) also detected FSH and LH in the pituitaries of human fetuses in the 13th to 14th weeks of embryogenesis. Furthermore, incorporation of ^{14}C-labeled amino acid into immunoreactive HGH, TSH, and FSH has been demonstrated by the *in vitro* culture of pituitaries from human fetuses at 9 to 14 weeks' gestation, which may indicate synthesis of these hormones (Gitlin & Biasucci, 1969). Sex differences in the pituitary contents of LH and FSH in the human fetus have been pointed out and are attributed to the hypothalamic sex differentiation by testicular androgenic action (Levina, 1968). However, no significant sex difference in the pituitary content of LH was observed in the present study, although our method of assay was not the same as that of Levina.

The physiological significance of pituitary HGH and gonadotropin secretion in the human fetus has not been well elucidated. A considerable amount of serum HGH has been detected in a fetus in the 10th week of gestation (Kaplan & Grumbach, 1967). Whether the growth hormone secreted by the fetal pituitary is related to body size is not known, as somatic development is normal in anencephalic monsters (Nanangas, 1925). There is good evidence that fetal or neonatal ovaries are able to respond to gonadotropin stimulation (Spivak, 1934; Ober & Bernstein, 1955; Pinkerton, 1959). Also, differentiation and development of the accessory reproductive tract in the male fetus is known to be controlled by functional testes secreting androgenic hormones (Jost, 1953). However, whether gonadotropin, which stimulates fetal gonads, is produced by the fetal pituitary or by the placenta is not clear. It has been reported that gonadal development in anencephalic monsters, who lack the pituitary gland, is observed, although their testicular weight is lower than normal (Zondek, 1965). The present study suggests the possible existence of a hypothalamo-pituitary relationship during fetal life and the physiological significance of fetal pituitary HGH and gonadotropin secretion. Also, it appears that further work is required on the hypothalamic releasing factors of human fetuses.

ACKNOWLEDGEMENTS

We would like to thank the National Institutes of Arthritis and Metabolic Diseases' Endocrinology Study Section for the gift of NIAMD RIA kits for HLH and HFSH and human pituitary reference preparation. Synthetic LH-RH was synthetized and kindly provided by Dr. Yanaibara, Shizuoka Pharmacological College. This work was partly presented at the IV International Congress of Endocrinology at Washington, June 1972.

This work was supported by grant No. M72–078C from the Population Council.

REFERENCES

Antwell, W. J. (1926). The development of the hypophysis cerebri in man, with special reference to the pars tuberalis. *Am. J. Anat.* **37**, 159.

Arimura, A., Debeljuk, L. & Schally, A. V. (1972). Stimulation of FSH release *in vivo* by prolonged infusion of synthetic LH-RH. *Endocrinology* **91**, 529.

Gitlin, D. & Biasucci, A. (1969). Ontogenesis of immunoreactive growth hormone, follicle stimulating hormone, thyroid stimulating hormone, luteinizing hormone, chorionic prolactin and chorionic gonadotropin in the human conceptus. *J. Clin. Endocrinol.* **29**, 926.

Guillemin, R. (1964). Control of pituitary hormone secretion-Hypothalamic factor releasing pituitary hormones. *Recent Prog. Horm. Res.* **20**, 89.

Jost, A. (1953). Problems of fetal endocrinology: The gonadal and hypophyseal hormones. *Recent Prog. Horm. Res.* **8**, 379.

Kaplan, S. L. & Grumbach, M. M. (1967). Excerpta Medica, International Congress Series No. 142. Presented at the Int. Symp. Growth hormone, Milan, Italy, #98, Sept. 1967.

Levina, S. E. (1968). Endocrine features in development of human hypothalamus, hypophysis, and placenta. *Gen. Comp. Endocrinol.* **11**, 151.

Nanangas, J. C. (1925). A comparison of the growth of the body dimensions of anencephalic human fetuses with normal fetal growth as determined by graphic analysis and empirical formulae. *Am. J. Anat.* **35**, 455.

Ober, W. B. & Bernstein, J. (1955). Observations on the endometrium and ovary in the newborn. *Pediatrics* **16**, 445.

Palvlova, E. B., Pronina, T. S. & Skebelskaya, Y. B. (1968). Histostructure of adenohypophysis of human fetuses and contents of somatotropic and adrenocorticotropic hormones. *Gen. Comp. Endocrinol.* **10**, 269.

Pinkerton, J. H. M. (1959). Oestrogen production in the immature human ovary (Observation based upon histochemical studies). *J. Obst. Gyn. Brit. Emp.* **66**, 820.

Redding, T. W., Schally, A. V., Arimura, A. & Matsuo, H. (1972). Stimulation of release and synthesis of luteinizing hormone (LH) and follicle stimulating hormone (FSH) in tissue cultures of rat pituitaries in response to natural and synthetic LH and FSH releasing hormone. *Endocrinology* **90**, 764.

Spivak, M. (1934). Polycystic ovaries in the newborn and early infancy and their relation to the structure of the endometrium. *Am. J. Obst. Gynec.* **27**, 157.

Zondek, L. H. & Zondek, T. (1965). Observations on the testis in anencephaly with special reference to the Leydig cells. *Biol. Neonatorum* **8**, 329.

INTERRELATIONSHIPS BETWEEN THE LEVELS OF OVARIAN STEROIDS AND LUTEINIZING HORMONE

Taro Tamada, Yoshiji Tsukui, Shigeo Araki, and Seiichi Matsumoto

Interrelationships between the levels of ovarian steroids and luteinizing hormone are reviewed along with a description of studies published or in progress in our laboratory, with special emphasis on steroid-LH interaction resulting in ovulation.

Detailed studies on the time sequence of the peaks of estrogen and LH at midcycle revealed not only that the peak of estrogen precedes the LH pear but the level of estrogen at the time of the LH peak is descending or at its nadir. These findings seem to suggest strongly that dynamic change, *i.e.*, descending levels of blood estrogen from a peak may play an important role in the release of LH. The action mechanism of clomiphene is interpreted in this connection.

The stimulatory effects of progestogens is seen in LH levels of subjects receiving a single injection of progesterone or some minidose contraceptives. However, the significance of this stimulatory effect is not clear in regulation of normal menstrual cycles.

In addition, data and discussions are presented as to what causes the preovulatory estrogen peak. And it is concluded that the ' basic ' secretion of LH during the follicular phase is essential for the increase of estrogen and for subsequent hormonal events to proceed normally.

INTRODUCTION

During the past three decades, numerous investigations have been reported on the secretion of ovarian steroids and gonadotropins in humans and experimental animals throughout the normal menstrual cycle and in various artifical conditions. These investigations include studies of the urinary, plasma and pituitary levels of gonadotropins following injection and oral administration of estrogens and/or progestogens.

Few of these studies, however, have provided any concrete data on the dynamic interrelationships between the levels of sex steroids and gonadotropins. For instance, parallel measurements of the two groups of hormones in the same subjects did not appear until 1968, when Burger, Catt & Brown (1968) reported on daily measurement of plasma luteinizing hormone and urinary estrogen excretion during the menstrual cycle. Although the assay methods for estrogens were sufficiently sensitive to permit daily measurements of their excretion, the biological assays for gonadotropins required 2 to 3-day pools of urine, and therefore study of the exact time sequence of the changes occurring, particularly at ovulation, was impossible.

More recently, sensitive and specific radioimmunoassay for gonadotropins, estrogens and progestogens using small volumes of unextracted plasma have been described. (Vande Wiele, Bougmil, Dyrenfurth, Ferin, Jewelewicz, Warren, Rizkallah & Mikhail, 1970; Mishell, Nakamura, Crosignani, Stone, Kharma, Nagata & Thorneycroft, 1971; Abraham, Odell, Swerdloff & Hopper, 1972).

We also have reported on the relationships between urinary excretion of LH and estrogens serially measured in the normal (Tamada, 1970) and clomiphene treated (Tamada, Tsukui & Matsumoto, 1970; Tamada & Tsukui, 1971*b*) subjects and suggested LH excretion is preceded by estrogen excretion, and in fact, the urinary level of estrogens at the time of LH peak is not their highest but is in a declining concentration or rather at the nadir. Recently radioimmunoassay of serum gonadotropins, estrogens, and progesterone became available in our laboratory and has been used for daily and hourly measurement of these hormones around the time of ovulation in normal menstrual cycle, which results will also be described.

It has long been believed that estrogen has a stimulatory effect on LH secretion, since Hohlweg (1934) reported a single subcutaneous injection of estrogen caused ovulation in immature rats. On the other hand, it has been well established that administration of estrogen in large dosage in a rather chronic mode supresses the production and release of LH in humans (Stevens & Vorys, 1967; Odell & Swerdloff, 1968; Tsukui, 1970) and in rats (Bogdanove, 1964). These data suggest the presence of both stimulatory and inhibitory feedback effects of estrogens on LH secretion. However, the mechanism whereby estrogen exerts two different effects on LH secretion is not clear so far and neither is its role in regulation of normal ovulatory cycles. We (Ebihara, Mayuzumi, Tamada & Matsumoto,

1971) and Mayuzumi, Ebihara & Matsumoto (1971) in our laboratory reported respectively in humans and in rats injected with estrogen intravenously that the levels of LH in plasma are reversely related to presumed levels of exogenously administered estrogen.

Regarding the relationships between the levels of progestogens and gonadotropins, it is generally accepted that progestogens exert prominent inhibitory effects on LH secretion, because bioassay and radioimmunoassay of LH during the luteal phase consistently revealed a constant low level of this hormone in plasma and urine. In addition, orally administered or long-acting progestogens abolished the midcycle peak of LH (Taymor, 1964; Stevens *et al.* 1967; Jaffe & Midgley, 1969). Ovulation following progesterone injection was, however, reported in persistent estrus rats (Everett, 1948) and a plasma LH increase after a single injection of progesterone was also recognized (Nallar, Antunes-Rodorigues & McCann, 1966; Odell *et al.* 1968). We confirmed that progesterone injection and minidose oral contraceptives have stimulatory effect on LH secretion.

This report is primarily concerned with investigations in our laboratory on immunological assay data in humans and rats, both normal and exposed to exogenous steroids or anti-gonadotropic sera administration. It is essentially a record of work in progress and as such contains no final answers to the basic questions posed. There have been surprisingly rapid increases in knowledge on the level of steroids and gonadotropins in blood in the past few months. These data are also included in discussion on the regulatory mechanism of steroids on LH secretion.

INTERRELATIONSHIPS BETWEEN ESTROGENS AND LH

Levels of estrogens and LH in urine and serum during normal ovulatory cycles

Five cycles from five healthy women, aged 22–38, were studied in regard to basal body temperature, total estrogen, and LH excretion with 24 hr pool urine in one case and first morning urine in the remainder.

The total urinary estrogen excretion was measured by the rapid method of Brown, modified by Igarashi, Ishikawa & Nakamura (1968). The coefficient of variation of the method is 19.8%, and its sensitivity is 0.4 ng by the modification.

Urinary LH was assayed by a solid phase radioimmunoassay

reported by us previously (Tamada & Tsukui, 1971). This method has been scrupulously investigated in regard to specificity and reproducibility with satisfactory results. The coefficient of variation is around 10%, and sensitivity is 1.6–3.2 mIU/ml in terms of 2nd IRP-HMG. Urinary FSH was assayed by a double-antibody radioimmunoassay using purified FSH from menopausal urine as a labeling antigen and the antibody to it as an antibody. These were purchased from Serono Institute, Rome.

Urinary LH excretion

A definite elevation of urinary LH was observed in all cycles. In these five subjects, the elevation of LH could be detected for 2 to 3 days, and the peak values occurred between the 12th and 15th days of cycle or 13 to 17 days before the onset of menstruation (Table 1).

TABLE 1. Day of cycle of urinary estrogen and LH peaks and BBT rise

Subject	Urinary estrogen peak	Interval	Urinary LH peak	Interval	BBT rise*	Length of cycle (days)
A.U.	12	0	12	1	13	26
H.T.	14	1	15	1	16	28
F.M.**	11	3	14	2	16	31
H.O.	11	1	12	2	14	26
Y.A.	11	1	12	0	12	28

* The last day of low temperature; ** 24-hour pool urine.

Urinary estrogen excretion

In all five cycles studied, the estrogen levels showed a peak at midcycle with a rise during the follicular phase (Fig. 1A) or without it (Fig. 1B), a subsequent fall, and a small second rise and fall during the luteal phase. The day on which the maximum excretion occurred at midcycle varied from the 11th to the 14th day of cycle, or 14 to 17 days before the onset of the next menstrual period.

The time sequence of estrogen and LH at midcycle

In those subjects, the maximum LH value was found to occur during the same 24 hr period as this maximum estrogen value in 1 cycle, during the next 24 hr period in 3 cycles, and 3 days later in 1. Thus in all cases studied the peak in urinary LH never appeared before the peak in estrogen could be detected, the time interval be-

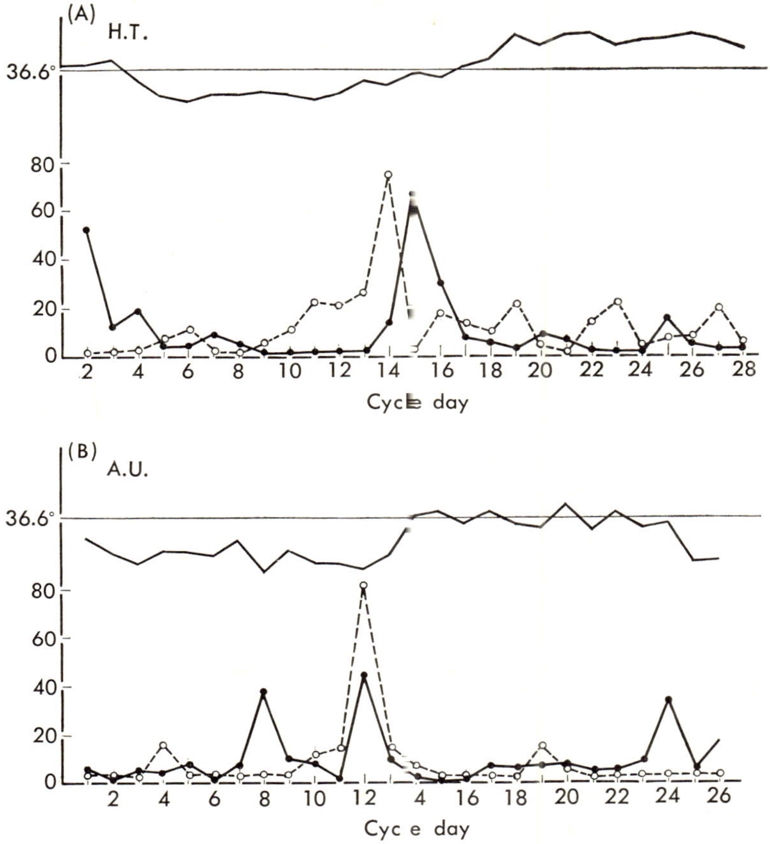

Fig. 1. Daily urinary excretion of total estrogen and LH through the menstrual cycle. - - - - -, Estrogen (μg/liter); ——, LH (IU/liter).

tween the two events being 0 to 3 days (mean 1.2) as shown in Table 1.

Burger *et al.* (1968) reported on the levels of plasma LH and urinary estrogen throughout normal menstrual cycles with special emphasis on relating the time sequence of the peaks at midcycle. They reported that the initial increase in urinary estrogens preceded that of LH, and in those cycles in which the day of maximum LH concentration could be defined precisely, this occurred either on the same day as the midcycle peak estrogen value or 1–3 days later. We compared the levels of the two hormones in a rather direct manner, by measuring the two hormones in the same urine simultaneously. Our results, however, are in good accordance with theirs. Thus it is

TABLE 2. Time sequence of peaks of estrogen and LH at midcycle

	Interval from estrogen peak to LH peak in days			
	0*	1	2	3
Authors	Number of cycles			
Burger, Catt & Brown (1968)**	2	2	1	0
Dufau *et al.* (1970)	1	1	2	1
Corker *et al.* (1971)	0	2	1	0
Mishell *et al.* (1971)	4	5	1	0
Vande Wiele *et al.* (1970)	1	1	0	0
Abraham *et al.* (1972)	2	6	1	0
Tamada (1970)***	1	3	0	1
Total	11	20	6	2

* Peaks of estrogen and LH on the same days; ** estrogen in urine and LH in plasma; *** estrogen and LH in urine. No mark, estrogen and LH in plasma.

confirmed that maximum concentration of LH never preceded the estrogen peak during the midcycle phase of the menstrual cycle. More recently, some authors have reported on simultaneous radio-immunoassay of plasma LH and estradiol-17β during the menstrual cycle (Corker, Exley, Naftolin & Macourt, 1971; Dufau, Dulmanis, Catt & Hudson, 1970; Abraham, Odell, Swerdloff & Hopper, 1972; Mishell *et al.* 1971). They observed that plasma estradiol reached a peak level prior to or on the day of the LH peak. Table 2 summarizes the data reported concerning the time sequence of the peaks of estrogen and LH at midcycle. Although the estrogen peak appears in 71.8% of all the cycles listed (39 cycles) prior to the day of LH peak, in 28.2% of the cycles both peaks are seen on the same day.

The meaning of the two groups could be entirely different in relation with interpretation of the regulatory effects of estrogens on LH secretion, because, while a descending concentration or a fall of estrogen level could be expected in the former group, an increasing concentration or higher level of estrogen will be present in the latter group, when a peak of LH occurs.

In order to investigate whether ascending or descending concentration or a certain critical amount of estrogen, regardless of dynamic changes, may play a role in triggering LH release, a more frequent monitoring than once a day of the hormones in blood will be instructive. Estradiol-17β and immunoreactive LH in serum were measured every 8 to 12 hr during the suspected period of estrogen and LH peak at the middle of several cycles. Serum estradiol and

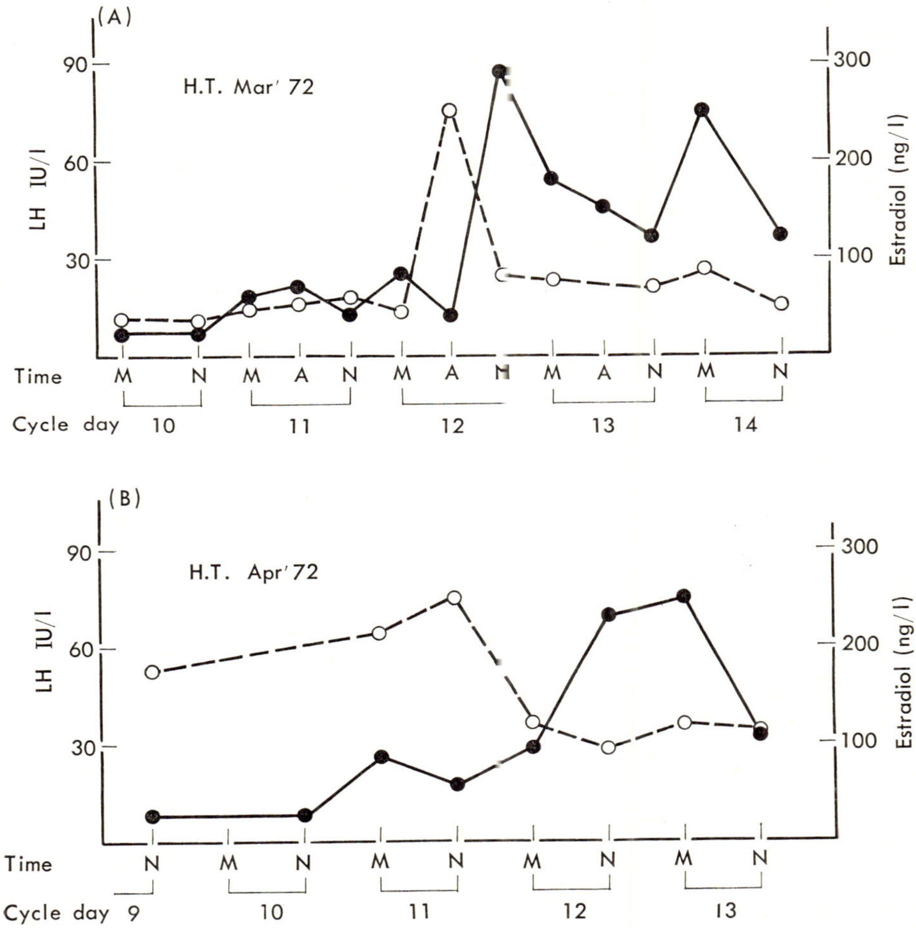

Fig. 2. Serum levels of LH and estradiol-17 β monitored every 8–12 hr at midcycles. M, N, and D indicate that the sample was taken in the morning (7.00–8.00), in the afternoon (14.00–16.00) and in the night (22.00–24.00), respectively. - - - - -, Estradiol; ——, LH.

LH were measured by specific radioimmunoassays. (Antiserum to estradiol is a gift from Dr. Caldwel , and antiserum and antigen of LH is a gift from NIH of the U.S.A.). In the 2 cycles of one case, the peaks of the two hormones were clearly delineated (Fig. 2). In the first cycle of this case (Fig. 2A), the peaks of estrogen and LH occurred on the same day, but the estrogen peak preceded the LH peak by 8 hr. In the other cycle, there were about 36 hr between the peaks of estrogen and LH, and again the former preceded the

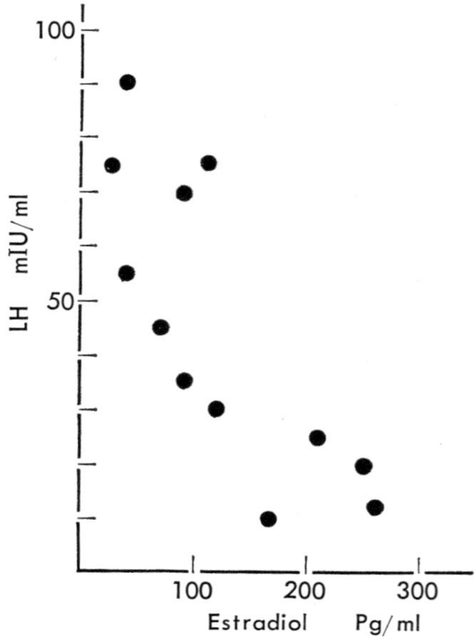

Fig. 3. Interrelationships between the values of serum LH and estradiol-17 β at the same moment during midcycles.

latter (Fig. 2B). Regardless of the interval between the peaks of estrogen and LH, the level of estrogen at the time of the LH peak was not at its highest but descending or at the basal level. In addition Fig. 3 illustrates the correlation between the serum levels of estradiol and LH at the same moment during the midcycle increase of LH (more than 30 mIU/ml) and/or of estradiol (more than 100 pg/ml) of the 2 cycles in this case. It is clearly shown that they are correlated inversely, the correlation coefficient being —0.85 (p 0.01). These results indicate that the peak of estrogen does precede the peak of LH at least in some of the cases with both peaks on the same day as revealed by daily measurements. This relationship has been clearly shown in rats by frequent simultaneous measurement of both hormones (Brown-Grant, Exley & Naftolin, 1970).

It is, therefore, probably fair to infer that a peak and the following descending level of estrogen might be a trigger of LH release at mid-cycle in the human female. This implies that a higher concentration of estrogen *per se* does not have a stimulatory effect on LH release, but a decrease of estrogen following a peak will free a suppressive effect of estrogen on LH release, thus permitting LH release from the pituitary. This possibility was investigated by measuring the plasma levels of LH frequently following exogenous estrogen injection as described in the following section.

Levels of serum LH in females injected with estrogens

Inhibitory effect of estrogens on LH secretion

The occurrence of LH surge or ovulation was reported following intravenous injection of Premarin, equine conjugated estrogen (Kupperman, Epstine, Blatt & Stone, 1958). According to Kurachi (1965), LH has become detectable 36 to 48 hr after intravenous injection of 20 mg Premarin and reached maximum in 3–4 days in urine specimens. Vande Wiele *et al.* (1970) also administered 20 mg of Premarin to 3 anovulatory patients, and found in some of the cases, a significant rise in plasma LH of the magnitude seen during the spontaneous ovulatory surge 36 hr after Premarin injection. These phenomena were taken as a stimulatory effect of estrogens on LH release by these authors. Vande Wiele *et al.* (1970) stated that the rising titer of estrogens is the trigger of the preovulatory LH release, and this relationship between estrogens and LH was incorporated into their model of the menstrual cycle.

This interpretation is, however, not easy to comprehend from the view point of homeostasis, in which a negative feedback mechanism or push and pull relationship will play a cardinal role.

We have carried out an experiment to provide more frequent measurement of plasma LH, in order to elucidate changes of LH level before the peak following Premarin administration (Ebihara *et al.* 1971). For this purpose, castrated volunteers were used because it is expected suppression of LH, if present, might be more clear in them when compared with the higher concentration of LH present prior to the treatment. In 2 cases studied, suppression of plasma LH levels is apparently shown from 2 and 4 hr after the injection, and 24 hr after injection LH started to rise, demonstrating a biphasic effect of estrogen on LH secretion (Fig. 4). This limited preliminary experiment showed that the first change following Premarin administration is a marked decrease in serum LH level.

It was suggested that exogenously administered estrogens of high concentration suppressed plasma LH first and the rise of plasma LH may start as the blood concentration of estrogen decreases. The presumptive inversed relationship between the levels of LH and estrogens seemed to indicate the presence of negative feedback mechanism involved in the regulation of LH levels following estrogen

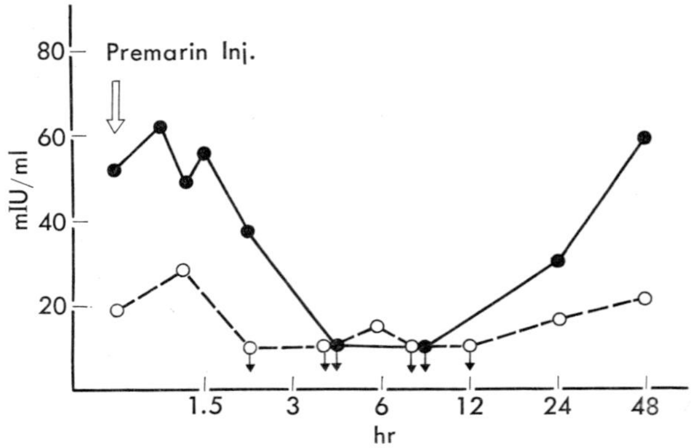

Fig. 4. Concentration of LH in serum following 20 mg Premarin i.v. injection.

administration. When reported on a similar experiment in three anovulatory women, Vande Wiele *et al.* (1970) did not find as obvious a suppression of plasma LH as in our study, because the initial levels of LH in their patients were not high and blood was taken only once or twice a day following Premarin injection.

Recently Tsai & Yen (1971) have reported a study where plasma gonadotropins were measured in pre- and post-menopausal women exposed to intravenous infusion of 17β-estradiol, and found that a uniformly suppressive effect of estradiol on the serum LH concentration appeared immediately after the start of infusion and maximal decline occurred at the end of infusion. Furthermore, Yen & Tsai (1972) performed a simultaneous measurements of LH and estradiol during and 3 days after 36 hr infusion of 17β-estradiol in a subject. According to the results, a declining pattern of LH and the inversed relationship between serum LH and estradiol was clearly shown during the infusion.

A constant suppression of serum LH concentration was also shown following the administration of ethynyl estradiol in postmenopausal subjects (Odell *et al.* 1968) and in normal menstrual cycles (Monroe, Jaffe & Midgley, 1972), estradiol (Nillius & Wide, 1970) and hexesterol or estradiol dipropionate (Tsukui, 1971).

In contrast to these is a study reported by Yen *et al.* (1971) where serum LH concentrations are shown firstly to decline, but to elevate to above the values found prior to treatment during ethynyl estradiol administration. In these subjects, the mechanism by which estrogen exerts inhibitory action initially but stimulatory action subsequently cannot be explained in a similar manner to our and other experiments mentioned above, because estrogen level in blood at the time when the serum LH level increased will be still higher in the experiment of Yen & Tsai (1971), in contrast to the experiment where a single intravenous administration of estrogen was used and estrogen level rapidly decreased in the blood. The cause of the discrepancy between the two studies is not apparent.

Stimulatory effects of estrogens on LH release

Emphasis was placed on the suppressive effects of estrogens on LH secretion in the preceding section. The suppression of LH secretion is, however, only a part of the response following the administration of estrogens, since a biphasic changes of LH, suppression first followed by augmentation, has been well documented as in the studies by Tsai *et al.* (1971) and in Fig. 4. The level of LH in the phase of augmentation sometimes surpasses the pretreatment level (Tsai *et al.* 1971; our unpublished observations in human and rat).

Meaningful interpretation involved in the stimulatory effect of estrogens on LH release is still uncertain, but the biphasic effects of estrogens on LH release observed recently may allow the following hypothesis. Estrogens work as a bar on LH releasing mechanisms, and if it is freed, LH release may abruptly start. This will fit well the most of the results mentioned above and is based on the negative feedback principle. The augmentation secondary to suppresion of LH release following the administration of estrogens need not be called as the positive feedback, because it could be considered as the results of the negative feedback freed as seen in castrated women. On the other hand, there has been a concept that a high concentration of estrogen may in some direct ways triggers the release of LH from the pituitary. If this is true, the effect of estrogens on LH release may be called a positive feedback, but it seems hardly to explain the presence of suppressive phase in the response of LH to estrogens administration.

The debate on these theories, however, is only superficial or semantic until the various physiological events involved in the regulatory mechanism between estrogens and LH are elucidated. For instance, it is not clear yet whether estrogens suppress LH release and, therefore, increase the storage of LH in the pituitary, which may produce a higher blood level when released, or estrogens not only inhibit LH release but also induce LH production.

As to whether the LH triggering effect of estrogens is a threshold phenomenon, *i.e.*, the concentration of estrogens over a critical level is necessary or the dynamic changes, increasing or decreasing, in the concentration of estradiol is essential to exert its effect could be partially answered from the results mentioned above. Since the LH surge never occurred as long as the infusion of estradiol continues (Yen *et al.* 1972; our unpublished observations), it is not the case of threshold phenomenon, but the fall from peak levels of estrogens may be responsible for the LH surge. Even if it is true, whether the duration and the level of a higher concentration of estrogen is more conducive to the size of LH surge is remained to be investigated in more details.

In addition, the interval between the administration of estrogens and the rise in LH will give some hints in investigating the physiological processes involved. It was shown that the interval was 36–48 hr, which is surprisingly long and remarkably constant (Vande Wiele *et al.* 1970; Yen *et al.* 1971). It is, however, doubtful if this is also true in normal ovulatory cycles, because a case studied by us showed only 8 hr between the peaks of estradiol and LH and many cases monitored once a day showed the peaks of both hormones on the same day.

URINARY EXCRETION OF ESTROGENS AND GONADOTROPINS IN SUBJECTS TREATED WITH CLOMIPHENE

Urinary excretion of estrogens and gonadotropins was studied following clomiphene treatment with special emphasis on relating the time sequence of the increase of estrogens and gonadotropins. (Tamada *et al.* 1970; Tamada *et al.* 1971)

Among the many previous studies of the action mechanism of clomiphene citrate (Clomid), agreement has been reached that LH and estrogen excretion increases following the medication. These reports, however, differ as to whether the increase of LH is preceded by that of estrogen or vice versa. In addition, changes of FSH levels due to medication reported so far are not consistent.

In an attempt to clarify the action mechanism of Clomid on human ovulation, we have studied the sequence of changes of LH, FSH, and estrogen excretion serially measured in 10 subjects. The patients consist of 7 cases with occasional anovulatory cycles, 2 with persistent anovulatory cycles, and 1 with infertility due to insufficient luteal phase. All the cases ovulated, and 3 of them got pregnant in the cycle where gonadotropins and estrogen were measured.

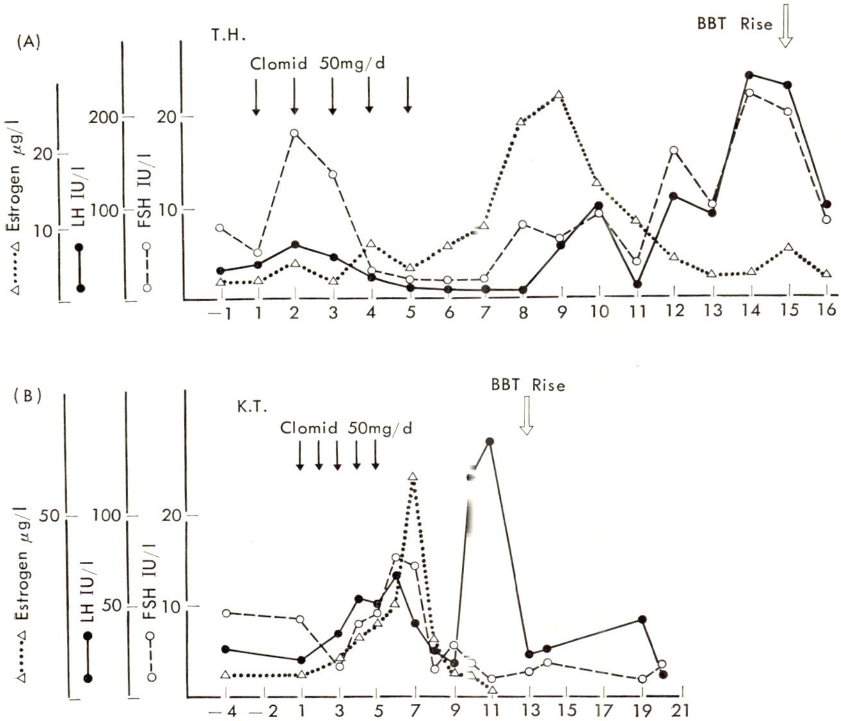

Fig. 5. Urinary FSH, LH, and estrogen levels following Clomid treatment (day 1 is the day Clomid started).

Figure 5A shows the levels of FSH, LH, and estrogen serially for the subject T.H., who has suffered from occasional anovulatory cycles. The first change due to medication was a moderate increase of FSH and LH, the peaks of which appeared on the second day of medication. Then an increase of total estrogen excretion occurred from the 6th day, reached a maximum on the 9th day, and returned to the initial low level by the 13th day. The second increase of FSH and LH was apparent from about the 9th day and reached a peak on the 14–15th day.

A similar and more rapid response was seen in another case, K.T. (Fig. 5B). Following a moderate increase of FSH and LH by day 6 of medication, a peak of estrogen was observed on day 7, and an ovulatory LH peak appeared on day 11, 2 days after that a rise in BBT occurred.

In Fig. 6, a variety of LH and estrogen excretion patterns are

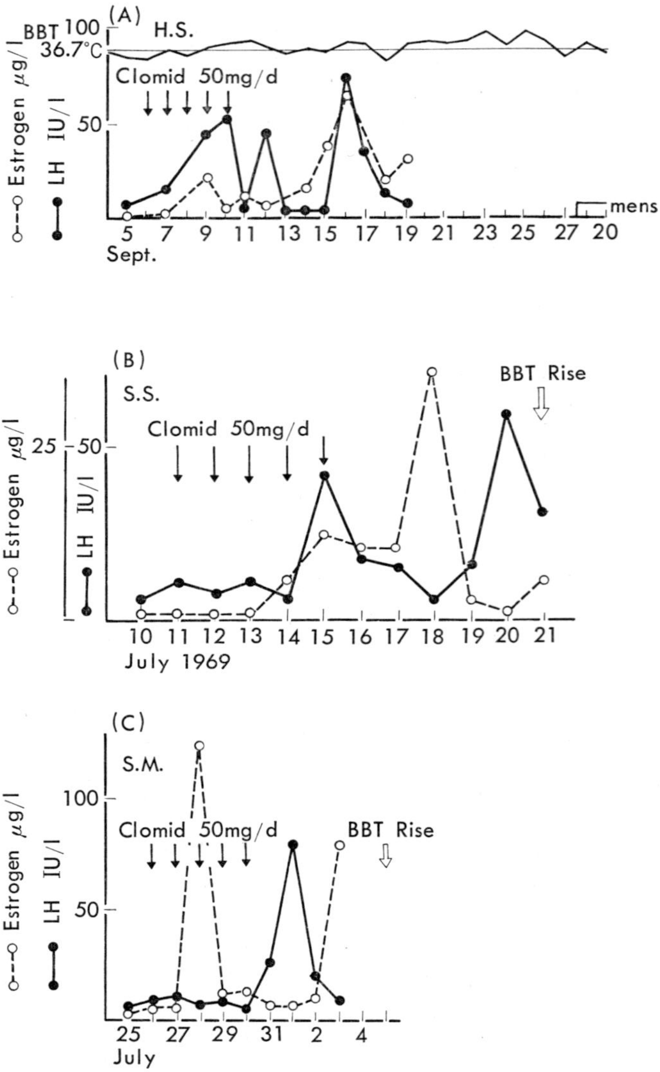

Fig. 6. Urinary LH and estrogen excretion following Clomid treatment.

shown. With subject H.S., maximum estrogen excretion appeared on the same day as an ovulatory peak of LH, which is similar to that observed at ovulation in some of normal ovulatory cycles as described above. There were some early increases of LH and estrogen observed. As for subject S.S., the ovulatory LH peak was preceded by a peak of estrogen by 2 days. There was also an early increase of

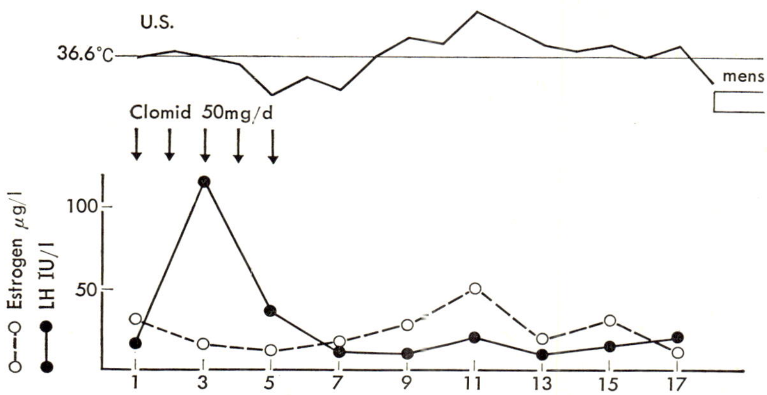

Fig. 7. Urinary LH and estrogen excretion following Clomid treatment in a case of early ovulation.

LH observed. With subject S.M., a marked increase of estrogen occurred early during the medication, but no increase of LH was observed until an ovulatory peak.

Fig. 7 shows changes of LH and estrogen in a subject (U.S.), in whom a temperature nadir occurred as early as day 7 after medication. There was only one LH peak observed on day 3, and no marked increase of estrogen appeared before ovulation. In the series reported here, ovulation occurred between days 5 and 16, the mean of which is 10.9. Therefore this is an example of a relatively rare case in which ovulation occurred within a week after the beginning of medication. In such a case, an early release of LH seems to be a trigger for ovulation as well.

A summary of the Clomid study appears in Table 3. An early increase of LH, called a " Clomid peak " by Jacobson, Marshall & Ross (1968) was observed in 7 of 10 subjects. An early FSH peak induced by Clomid appeared in 2 subjects studied. An estrogen peak was observed in all the cases except 2 subjects in whom ovulation occurred as early as days 5 and 7. The interval between the Clomid peak of gonadotropins and the estrogen peak varies from 0 to 7 days. The interval between the estrogen peak and the ovulatory peak of LH is 2.5 days on the average, ranging from 0 to 5 days. The mean interval is 2.2 days, ranging from 1 to 4 days, between the LH peak and the last day of low temperature.

According to the interval between the day of medication and ovulation, and to the excretion patterns of LH and estrogen, 2 groups of

TABLE 3. Summary of Clomid study

Subject	Early LH peak		Early FSH peak		Interval days	Estrogen peak		Interval days	Ovulatory LH peak		Interval days	BBT** day
	Day	IU	Day	IU		day	μg		day	IU		
S.K.	—	—*			—	6	125	2	8	106	2	10
S.M.	—	—			—	3	126	4	7	80	4	11
S.S.	5	42			3	8	70	2	10	59	1	11
F.S.	6	24			0	6	95	2	8	85	0	8
Y.H.	—	—			—	11	50	1	12	64	4	16
U.S.	3	115			—	1	32	—	***		4	7
K.Y.	4	40			—	1	27	—	***		1	5
H.S.	5	56										
	7	48			4	11	65	0	11	78	2	13
K.T.	6	65	6	15	1	7	58	4	11	137	2	13
T.H.	2	64	2	18	7	9	28	5	14	235	1	15

* No peak was present; ** the last day of low phase; *** early LH peak is taken as the ovulatory peak.

patients who ovulated were classified in response to Clomid administration.

In the first and smaller group, which could be designated as an early ovulatory group, ovulation occurs within 7 days after the beginning of medication and no increase of estrogen was seen before the ovulatory LH peak. In this group it seems that Clomid acts directly on the hypothalamo-pituitary system either as a releasing factor or a trigger for a releasing mechanism of LH.

In the second and larger group, possibly named as a late ovulatory one, ovulation occurs 8 days or more after the beginning of medication and an estrogen increase always precedes the ovulatory LH peak. Small to moderate increases of FSH and LH were seen during or immediately after medication in this group, hence it appears that the initial action site of Clomid is not the ovary but the hypothalamo-pituitary axis. Thus, in the latter group, the action mechanism of Clomid could be summarized as follows: Clomid will act on the hypothalamo-pituitary axis first to release gonadotropins, probably FSH as well as LH, which may cause an increase of estrogen from the ovary. Thereafter an ovulatory peak of LH and ovulation will follow as seen in the normal ovulatory cycle. The longer period between estrogen surge and LH peak in this group than in normal cycles may reflect some pathological changes of feedback mechanism in these patients. It is of interest to note in relation to the correlation of the levels of estrogen and LH, that the estrogen level is generally low at the time of LH peak in these patients, which seems to indicate more clearly an inversed correlation of the two hormones than in normal cycles.

INTERRELATIONSHIPS BETWEEN PROGESTOGENS AND LH

Systemic LH changes in patterns during antifertility treatments

In order to understand the action mechanism of the traditional and some newer regimens of oral contraceptives, gonadotropins in subjects recieving the contraceptives were serially measured before, during, and after medication. (Tsukui, 1970; Matsumoto, 1971)

The results pertaining to LH excretion in these subjects are summarized in Table 4. LH excretion was studied from two aspects, basic excretion and ovulatory peak. The former term refers to low excretion levels during proliferative (M=12.9 IU•HMG/liter) and secretory phases (M=16.8 IU•HMG/liter) of a cycle and the latter to an ovulatory peak of LH (M=82.7 IU•HMG/liter).

TABLE 4. Summary of urinary LH change in subjects on oral contraceptives

Subject	Age	Preparation	Pattern of LH excretion		BBT rise
			Basic excretion	Mid peak	
		CONVENTIONAL			
Y.F.	35	C-Quens	↑*	(−)	(−)
E.T.	32	Anovlar	↑	(−)	(−)
H.S.	35	Anovlar	→**	(−)	(−)
H.M.	33	Anovlar	→	(−)	(−)
		MINIPILLS			
N.T.	32	Ethinodiol diacetate 0.25 mg	↑	(−)	(−)
I.W.	38	Ethinodiol diacetate 0.25 mg	→	(−)	(−)
Y.F.	35	17α-ethnyl-5 estren-17β-ol 0.5 mg	↑	(−)	(+)
I.W.	38	17α-ethnyl-5 estren-17β-ol 0.5 mg	→	(−)	(−)
A.N.	31	17α-ethnyl-5 estren-17β-ol 0.5 mg	→	(+)	(+)
H.T.	35	Norethindrone 0.2 mg	↑	(+)	(+)
F.T·	28	Norethindrone 0.2 mg	→	(−)	?
H.K.	23	Norethindrone 0.4 mg	→	(−)	(−)
K.T.	30	Norethindrone 0.6 mg	↑	(+)	(+)
A.U.	26	Norethindrone 0.8 mg	↑	(−)	(−)

* Increased; ** not changed.

The ovulatory peak of LH was suppressed in all the four subjects who followed the traditional regimen. This is consistent with a lack of thermal shift in BBT. In two of those, however, moderate erratic peaks of LH were observed throughout the treatment cycles.

Ethinodiol diacetate 0.25 mg per day, when given continuously regardless of the cycles, suppressed ovulation in two females. No ovulatory LH peak was noticed in either, but there was increased basic excretion in one.

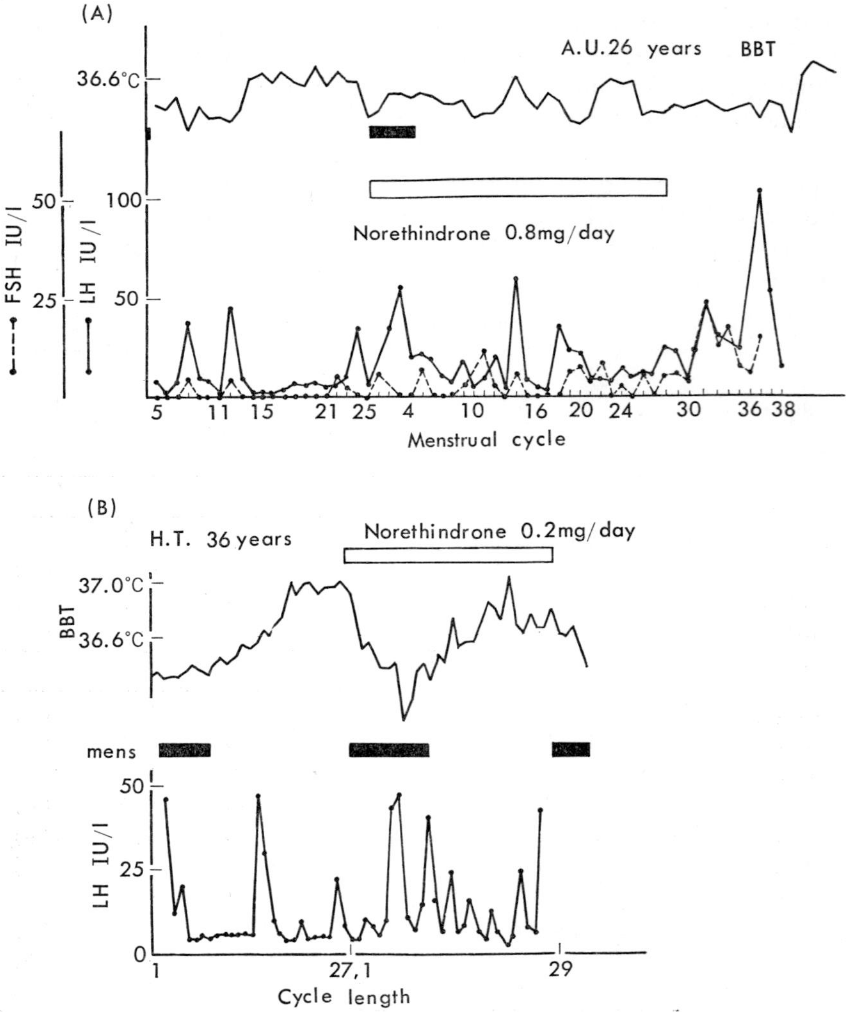

Fig. 8. Daily excretion of gonadotropins during norethindrone treatment.

A new synthetic progestin, 17α-ethynyl-5 estren-17β-ol, was given 0.5 mg per day to 3 females through a cycle. Only one of these subjects had a typical biphasic BBT, one was equivocal, and the third had no thermal shift. Abolition of the ovulatory LH peak was seen in the third subject, and increased basic excretion was observed in the first case.

The effect on LH excretion of continuous administration of various dosages of norethisterone (NET) was studied in five subjects. Two women were given 0.2 mg of NET per day from the first day of menstruation till the appearance of a menstrual flow. Basal body temperature was disturbed in both, and ovulation was apparently suppressed in one subject in whom many erratic LH peaks appeared while the ovulatory peak was suppressed. No ovulation occurred in the cases given 0.4 and 0.8 mg of NET but ovulation was observed in a woman given 0.6 mg. A suppressed LH peak and increased basic excretion were recognized in all the cases who did not ovulate. Even in the cases judged as ovulated based on the BBT shift, some erratic LH peaks were seen before and after the ovulatory peak of LH. Figure 8 demonstrates the occurrence of multiple LH peaks during the treatment cycles with norethisterone 0.2 and 0.8 mg.

There have been so many studies on gonadotropin changes during antifertility treatment that it is almost impossible to compile a list. A review by Diczfalusy (1968) summarized those data reported by 1968 in detail. While most reports emphasize a suppressive effect of contraceptives on LH secretion, Stevens *et al.* (1967) have described for the first time the stimulatory effect of estrogen on excretion of LH. Swerdloff & Odell (1969), using the radioimmunoassay technique, have reported that the changes in plasma LH levels are markedly different between combined and sequential contraceptives. According to their study, the LH peak was suppressed with a combined regimen, but many erratic peaks appeared not only during estrogen administration but also after adding progestogens to the sequential. Based on those studies, they proposed that different action mechanisms might be involved in the combined and sequential regimens.

The present study showed the stimulatory effect of minidoses of progestogens on LH excretion. Although it is still uncertain if progestogen itself or an intrinsic estrogenic property of progestogen has this effect, one of the action mechanisms of minidose pills is considered to be ascribed to activation of the hypothalamo-pituitary system.

Serum LH concentration in females following a single injection of progesterone

Odell *et al.* (1968) first showed in humans that a sharp, high FSH-LH peak occurred within 24 hr following a single injection of 10 mg of progesterone. All the subjects they studied are postmenopausal or castrated women primed with estrogens for 10 and 20 days.

The same experiment was carried out in a normal cycling and three anovulatory subjects (Fig. 9). A sharp increase of LH concentration was recognized 5 hr after the injection of 10 mg of progesterone in a normal subject who received the injection on day 11 of the cycle. The increase of LH in three anovulatory subjects, compared with that in a normal subject, were delayed or less marked. The reason for such a variety of response is not clear but may be due to the pretreatment level of estrogens according to the clinical data and Preliminary assay of estrogen.

Although the increase of LH level in response to the administration of progesterone is hardly correlated to the hormonal changes normally occurring in ovulatory cycles, a possibility is suggested that the administration of progesterone can be a test of the sensitivity of the

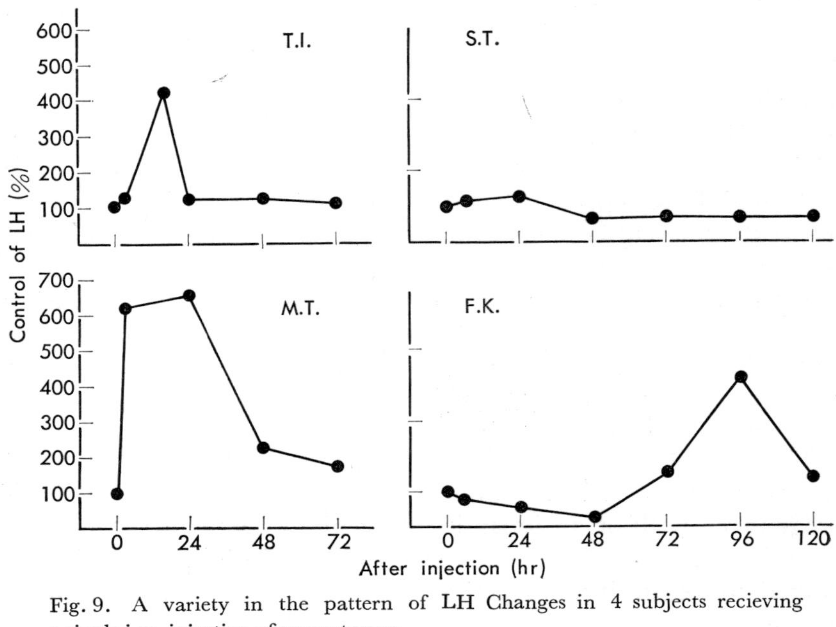

Fig. 9. A variety in the pattern of LH Changes in 4 subjects recieving a single i.m. injection of progesterone.

hypothalamo-pituitary system. Theoretically it is of interest that there seems no suppressive phase of LH concentration and augmentation occurs more rapidly than the administration of estrogens, and, therefore, a term ‘ positive feedback ’ may be applied to the response of LH to progesterone in its true sense.

THE SIGNIFICANCE OF LH ON ESTROGEN SECRETION DURING FOLLICULAR MATURATION

It has long been known that FSH alone is not capable of producing uterine growth in hypophysectomized rats (Greep, Van Dyke & Chow, 1942).

We have described, in the preceding sections, an established cause-effect relationship between the increase of estrogen in blood and LH surge at midcycle of normal cycles as well as in the subjects exposed to exogenous estrogen.

The question naturally follows, what causes such an abrupt increase of estrogen prior to LH surge. In this respect, the accumulated radioimmunoassay data on the levels of gonadotropins and estrogens either in humans or rats are not fully suggestive because there are usually no remarkable changes of FSH and LH at any time during the follicular phase as seen in the estrogen levels near midcycle. A rise in FSH during the early follicular phase was reported by some authors, but it is too far from the peak of estrogen at midcycle to correlate them directly. The levels of LH are reported to be higher in the follicular phase than in the luteal phase by some authors but this is not consistent with other reports.

The following study was undertaken to assess the role of LH on the days of metestrus (ME) and diestrus (DE) in the formation of an estrogen peak on the day of proestrus (PE) in 4 day cycling rats. Antiserum of NIH-LH S16 was produced in rabbits, 1 ml of which neutralized about 300 μg of the antigen. After absorbing 1 ml of the antiserum with 0.5 mg of NIH-FSH S8, this binds with ^{131}I-rat LH (NIAMD) but not with ^{131}I-rat FSH (NIAMD).

Less than 0.01 ml of the anti-LH serum given at ME inhibited vaginal cornification and ovulation on the expected estrus day; 0.017 ml (ED_{50}) and 0.036 ml (ED_{50}) of it inhibited an expected ovulation when given at DE and PE, respectively. The dose of anti-LH serum able to block ovulation is smaller when it is given on day DE than on day PE. These results are in good accordance with those reported by Schwartz & Ely (1970).

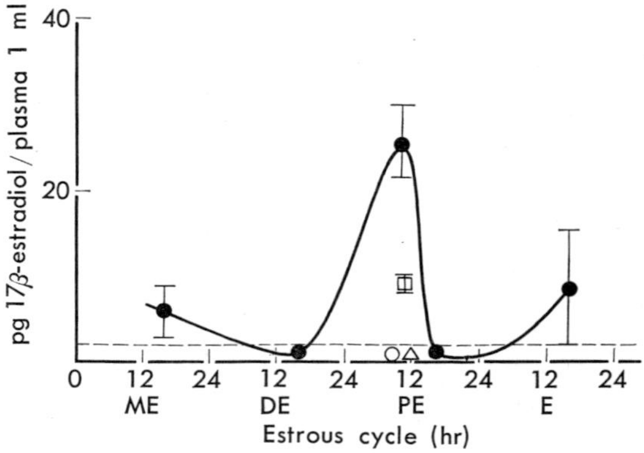

Fig. 10. The effects of anti-LH at various times of the estrous cycle on plasma 17 β-estradiol levels. Each point represents the mean, and the vertical bars represent the range. 17 β-estradiol levels under the dotted line are not detectable. ME, metestrus; DE, diestrus; PE, proestrus; E, estrus; ●——●, Control. 17 β-estradiol values in proestrus morning after the administration of anti-LH serum. □, Injection at 11.00 of ME; ○, injection at 16.00 of ME; △, injection at 11.00 of DE.

In order to study estrogen levels in the blood of rats following the anti-LH administration directly, the levels of estrogen were measured by method of vaginal triphenyltetrazolium chloride reduction in spayed mice earlier (Araki, Tamada & Matsumoto, 1972) and by a specific radioimmunoassay for estradiol-17β recently. At 11.00 and 16.00 hr on day ME and at 11.00 on day DE, 0.1 ml of the anti-LH serum absorbed with 0.5 mg NIH-FSH S8 and 0.2 ml of normal rabbit serum per ml was given. The rats were sacrificed in the morning of PE, and blood was taken from each group of rats, pooled, and measured for estrogen. Compared with the high level of estrogen usually observed in the morning of day PE with control rats, suppression of estrogen was apparent in the treated rats (Fig. 10). This clearly indicates that the basic level of LH on ME and DE is essential for blood estrogen to increase and reach a peak in the morning of PE, which stimulates the ovulatory surge of LH and thus induces ovulation.

Recently Berger & Taymor (1971) compared the effects of FSH with and without LH on the same amenorrheic patient. They found that FSH alone was ineffective in producing either an estrogen response or ovulation despite high peripheral levels of FSH. When FSH together with LH was administered in the second cycle, a marked increase in urinary estrogens and apparent ovulation occurred.

TABLE 5. The two phases of LH action pertaining to ovulation in normal ovulatory cycles

'BASIC' LH SECRETION (+FSH)
↓
PREOVULATORY ESTROGEN INCREASE
↓
FALL OF ESTROGEN LEVEL
↓
OVULATORY LH PEAK
↓
OVULATION

It was concluded that LH is necessary to induce maturation of the ovarian follicle in human beings as well.

Thus it seems that the significance of LH secretion during follicular development is well established. Since the concentration of LH in this period is low, it could be properly called as a basic LH secretion. The role of the basic LH secretion pertaining to induction of ovulation is illustrated in Table 5 where the two phases of LH action are clearly delineated, *i.e.*, the ' basic ' LH secretion is necessary for production of a preovulatory estrogen increase and the ovulatory peak of LH for ovulatory process itself in the ovary.

On the other hand, erratic LH peaks during the follicular phase may play a role in inhibition of ovulation in subjects receiving oral contraceptives, as mentioned above, and in patients with polycystic ovary syndrome (Ingersoll & McArthur, 1959; Yen, Vela & Rankin, 1970; Tamada & Matsumoto, 1971). Administration of HCG from early in the follicular phase also inhibited ovulation (Tamada & Matsumoto, 1969). It could therefore be thought that the high levels of LH (or HCG) during the follicular phase are not conducive to a normal ovulatory cycle.

Although it is apparent from the evidence mentioned above that the basic secretion of LH must be present but stay at proper, rather low concentration throughout the follicular development, the mechanism whereby the level of basic LH is controled is not known. Further studies will be needed to understand the control mechanism of basic LH secretion and to gain more insight into normal ovarian physiology.

ACKNOWLEDGEMENTS

Thanks are due to Dr. Igarashi and Miss. Ishikawa for the assay of urinary estrogens. The antiserum to estradiol is a gift from Dr. Caldwell. Human LH radioimmunoassay kit, and ovine FSH and LH is a gift from National Institute of Health, U.S.A.

Skilful technical assistance of Mrs. Ono and Miss. Ubukata is deeply appreciated.

This study was partly supported by the Ministry of Education.

REFERENCES

Abraham, G. E., Odell, W. D., Swerdloff, R. S. & Hopper, K. (1972). Simultaneous radioimmunoassay of plasma FSH, LH, progesterone, 17-hydroxyprogesterone and estradiol-17 β during the menstrual cycle. *J. clin. Endocr.* **34**, 312.

Araki, S., Tamada, T. & Matsumoto, S. (1972). Effect of LH on follicular maturation and estrogen secretion during the preovulatory phase of rats. *Folia Endocrinol. Jap.* **47**, 806 (in Japanese).

Berger, M. J. & Taymor, M. L. (1971). The role of LH in human follicular maturation and function. *Am. J. Obstet. Gynec.* **111**, 708.

Bogdanove, E. M. (1964). The role of the brain in the regulation of pituitary gonadotropin secretion. *Vitams Horm.* **22**, 205.

Brown-Grant, K., Exley, K. & Naftolin, F. (1970). Peripheral plasma estradiol and LH concentrations during the oestrous cycle of the rat. *J. Endocr.* **48**, 295.

Burger, H. G., Catt, K. J. & Brown, J. B. (1968). Relationship between plasma luteinizing hormone and urinary estrogen excretion during the menstrual cycle. *J. Clin. Endocrinol.* **28**, 1508.

Corker, C. S., Exley, D., Naftolin, F. & Macourt, D. C. (1971). The measurement of estrogens by competitive protein-binding assay. In *Hormonal Steroids,* ed. by James, V.H.T. & Martini, L., p. 248. Excerpta Med.

Diczfalusy, E. (1968). Mode of action of contraceptive drugs. *Am. J. Obstet. Gynec.* **100**, 136.

Dufau, M. L., Dulmanis, A., Catt, K. J. & Hudson, B. (1970). Measurement of plasma estradiol-17β by competitive binding assay employing pregnancy plasma. *J. Clin. Endocrinol.* **30**, 351.

Ebihara, H., Mayuzumi, R., Tamada, T. & Matsumoto, S. (1971). Studies on positive feed-back of estrogen. *Folica Endocrinol. Jap.* **46**, 49 (in Japanese).

Everett, J. W. (1948). Progesterone and estrogen in the experimental control of ovulation time and other features of the estrous cycle in the rat. *Endocrinology* **43**, 389.

Greep, R. O., van Dyke, H. B. & Chow, B. F. (1942). Gonadotropins of the swine pituitary. I. Various biological effects of purified thylakentrin (FSH) and pure metakentrin (ICSH). *Endocrinology* **30**, 635.

Hohlweg, W. (1934). Veränderungen des hypophysenvorderlappens und des Ovariums nach Behandlung mit grossen dosen von Follikelhormon. *Klin. Wochschr.* **13**, 92.

Igarashi, M., Ishikawa, J. & Nakamura, I. (1968). On a new rapid assay method for estrogen. *Igaku no Ayumi* **66**, 215 (in Japanese).

Ingersoll, F. M. & McArthur, J. W. (1959). Longitudinal studies of gonadotropin excretion in the Stein-Leventhal syndrome. *Am. J. Obstet. Gynec.* **77**, 795.

Jacobson, A., Marshall, J. R. & Ross, G. T. (1968). Plasma luteinizing hormone during clomiphene-induced ovulatory cycles. *Amer. J. Obstet. Gynec.* **101**, 1025.

Jaffe, R. B. & Midgley, A. R. (1969). Current status of human gonadotropin radioimmunoassay. *Obst. Gynec. Surv.* **24**, 200.

Kupperman, H., Epstine, J., Blatt, M. & Stone, A. (1958). Induction of ovulation in the human: Therapeutic and diagnostic importance. *Am. J. Obstet. Gynec.* **75**, 301.

Kurachi, K. (1965). On correlation between autonomous nerve and endocrine system. Abstract of a lecture presented at Jap. Obstet. and Gynecol. Society (in Japanese).

Matsumoto, S. (1971). Systemic LH changes in patterns during antifertility treatments. Excerpta Med. Int. Congr. Ser. 234b, 130.

Mayuzumi, R., Ebihara, H. & Matsumoto, S. (1971). Detailed analysis of positive estrogen feedback. *Acta Obst. Gynec. Jap.* **18**, 74.

Mishell, D. R.,Nakamura, R. M., Crosignani, P. G., Stone, S., Kharma, K., Nagata, Y. & Thorneycroft, L. H. (1971). Serum gonadotropin and steroid patterns during the normal menstrual cycle. *Am. J. Obstet. Gynec.* **111**, 60, 1971.

Monroe, S. E. Jaffe, R. B. & Midgley, A. R. (1972). Regulation of human gonadotropins. XII. Increase in serum gonadotropins in response to estradiol. *J. Clin. Endocrinol.* **34**, 342.

Nallar, R., Antunes-Rodrigues, J. & McCann, S. M. (1966). Effect of progesterone on the level of plasma LH in normal female rats. *Endocrinology* **79**, 907.

Nillius, S. J. & Wide, L. (1970). Effects of oestrogen on serum levels of LH and FSH. *Acta Endocr.* **65**, 533.

Odell, W. D. & Swerdloff, R. S. (1968). Progesterone-induced luteinizing and follicle-stimulating hormone surge in postmenopausal women: A simulated ovulatory peak. *Proc. natn. Acad. Sci. U.S.A.* **61**, 529.

Schwartz, N. B. & Ely, C. A. (1970). Comparison of hypophysectomy, antiserum to ovine LH, and ovariectomy on estrogen secretion during the rat estrous cycle. *Endocrinology* **86**, 1420.

Stevens, V. C. & Vorys, N. (1967). The regulation of pituitary function by sex steroids. *Obst. Gynec. Surv.* **22**, 781.

Swerdloff, R. S. & Odell, W. D. (1969). Serum LH and FSH levels during sequential and nonsequential contraceptive treatment of eugonadal women. *J. Clin. Endocrinol.* **29**, 157.

Tamada, T. (1970). Interrelationship of hormones. *Folia Endocrinol. Jap.* **46**, 902 (in Japanese).

Tamada, T. & Matsumoto, S. (1969). Suppression of ovulation with human chorionic hormone. *Fertil. Steril.* **12**, 840.

Tamada, T. & Matsumoto, S. (1971). Excretion of LH in the Stein-Leventhal syndrome. *Clin. Endocrinol.* **19**, 989 (in Japanese).

Tamada, T., & Tsukui, Y. (1971*a*). A new solid phase radioimmunoassay for LH in urine. *Folia Endocrinol. Jap.* **47**, 174 (in Japanese with English abstract).

Tamada, T. & Tsukui, Y. (1971*b*). Excretion of estrogens and gonadotropins following clomiphene therapy. Excerpta Med. Int. Congr. Ser. 234a, 33.

Tamada, T., Tsukui, Y. & Matsumoto, S. (1970). Action Mechanism of clomiphene. *Nippon Sanka-fujinka Gakkai Zasshi* **22**, 1032 (in Japanese).

Taymor, M. L. (1964). Effect of synthetic progestins on pituitary gonadotropin excretion. *J. Clin. Endocrinol.* **24**, 803.

Tsai, C. C. & Yen, S. S. C. (1971). Acute effects of intravenous infusion of 17-estradiol on gonadotropin release in pre- and post-menopausal women. *J. Clin. Endocrinol.* **32**, 766.

Tsukui, Y. (1970). Feedback of sex steroids on LH. *Kitakanto Igaku* **21**, 44 (in Japanese).

Vande Wiele, R. L., Bougmil, J., Dyrenfurth, I., Ferin, M., Jewelewicz, R., Warren, M., Rizkallah, T. & Mikhail, G. (1970). Mechanisms regulating the menstrual cycle in women. *Recent Prog. Horm. Res.* **26**, 63.

Yen, S. S. C. & Tsai, C. C. (1972). Acute gonadotropin release induced by exogenous estradiol during the mid-follicular phase of the menstrual cycle. *J. Clin. Endocrinol.* **34**, 298.

Yen, S. S. C. & Tsai, C. C. (1971). The biphasic pattern in the feedback action of ethynyl estradiol on the release of pituitary FSH and LH. *J. Clin. Endocrinol.* **33**, 882.

Yen, S. S. C., Vela, P. & Rankin, J. (1970). Inappropriate secretion of FSH and LH in polycystic ovarian disease. *J. Clin. Endocrinol.* **30**, 435.

FEEDBACK OF ESTROGEN IN THE HYPOTHALAMIC CONTROL OF GONADOTROPHIN SECRETION*

KINJI YAGI and YUKIKO SAWAKI

The feedback of estrogen in the control of gonadotrophin secretion in the female rat was studied. It has been demonstrated that the hypothalamus includes receptor cells for plasma estrogen. Some neurons in the suprachiasmatic area of the medial preoptic (POSC) nucleus, the anterior hypothalamic (HA), and acurate (ARC) nuclei respond specifically to plasma estrogen. These neurons are presumably of slow adaptation in nature. In cyclic female rats unit activity in the POSC was high on the second day of diestrus and the day of proestrus. Unit activity in the ARC was high in proestrous rats. Ovariectomy depressed the activity of POSC units and augmented the activity of ARC units. There are neural connections between the POSC and the ARC and between the ARC and the median eminence. Electrical stimulation of the POSC and ARC has been found to bring about facilitatory effects on gonadotrophin secretion by the adenohypophysis. These results suggest a negative feedback effect of estrogen through the ARC and a positive effect through the POSC on the activity of the adenohypophysial-ovarian system.

INTRODUCTION

It is well established that the hypothalamus controls adenohypophysial functions by means of neurohumoral factors and the adenohypophysis in turn controls estrogen secretion by the ovary through gonadotrophins (see, for example, Harris, 1955; Everett, 1964; Flerkó, 1966; Davidson, 1969). A number of studies have provided considerable evidence for the feedback mechanism of estrogen in the hypothalamic control of gonadotrophin secretion by

* This work was carried out in the 1st Department of Physiology, Yokohama City University.

the adenohypophysis. Implants of estrogen, for example, in hypothalamic areas including the arcuate nucleus have been shown to induce gonadal atrophy in rats (Lisk, 1960) which was probably due to a reduction of gonadotrophin secretion. It has been demonstrated that the anterior hypothalamus is indispensable for occurrence of compensatory ovarian hypertrophy induced by unilateral ovariectomy (Flerkó & Bárdos, 1961; Halász & Gorsky, 1967; Köves & Halász, 1969). Ovariectomy induces an increase in the plasma level of luteinizing hormone (LH) in the rat (Taleisnik & McCann, 1961; Ramirez & McCann, 1963; McCann & Ramirez, 1964; Parlow, 1964; Gay & Sheth, 1972). Estrogen injection (Ramirez *et al.* 1963; McCann *et al.* 1964; Parlow, 1964) or an implant of estrogen in the hypothalamus (McCann *et al.* 1964) suppresses the increase in the plasma LH level after ovariectomy. Ovariectomy in rats with hypothalamic lesions failed to induce the rise of plasma LH level (Taleisnik *et al.* 1961). These observations suggest a *negative feedback* mechanism of estrogen in the sense that the hypothalamo-adenohypophysial system functions to compensate a deviation of plasma estrogen from the normal level. This mechanism was sometimes termed 'inhibitory' feedback (Flerkó, 1966; Davidson, 1969).

There are some observations to suggest a *positive feedback* mechanism of estrogen within the hypothalamus, in the sense that plasma estrogen facilitates LH secretion by the adenohypophysis and the elevated plasma LH in turn stimulates estrogen secretion by the ovary (see Everett, 1964; Flerkó, 1966; Davidson, 1969; Kobayashi, Hara & Miyake, 1969 *a*, *b*). This mechanism is also called 'stimulatory' feedback (Flerkó, 1966; Davidson, 1969). For example, chronic administration of estrogen to immature rats has been demonstrated to induce luteinization of the ovaries (Hohlweg, 1934) only in rats with the adenohypophysis *in situ* (Westman & Jacobsohn, 1938). A single injection of estrogen on the second day of diestrus in rats with 5-day cycles has been shown to cause ovulation a day earlier than in controls (Everett, 1948). A single injection of estrogen facilitates LH secretion by the adenohypophysis in prepubertal (Ramirez & Sawyer, 1965), adult (Kobayashi *et al.* 1969 *a*), and pregnant or pseudopregnant (Presl, 1961) rats. Daily administration of estrogen has been found to cause LH and follicle stimulating hormone surge on the 5th day of injection in ovariectomized rats (Swerdloff, Jacobs & Odell, 1972). Ovariectomy on the second day of diestrus in rats with 4-day cycles has been reported to abolish the decrease in adenohypophysial content of LH that is expected to occur after the ovulating surge of LH (Schwartz, 1964). Ovariectomy in the morning at proestrus has been shown to reduce the amount of LH released during the ovulating surge of gonadotrophins on the same day (Lawton, 1972). These observations suggest a possible positive feedback mechanism of estrogen.

Neural mechanisms within the hypothalamus for these two types of estrogen feedback are, however, largely unclear as yet. The purpose of the present study is to describe and discuss experimental results so far obtained on these mechanisms in the female rat.

ESTROGEN RECEPTOR CELLS IN THE HYPOTHALAMUS

Receptor cells to sense the plasma level of estrogen must be present somewhere for positive or negetive feedback mechanisms to exist. Furthermore, the receptor cells must be in functional connection with the tubero-infundibular neurosecretory neurons that secrete gonadotrophin releasing factors in the median eminence. Implants of estrogen in the arcuate nucleus of the hypothalamus induce atrophy of the ovaries and uteri and resulted in a decrease in nucleolar size of arcuate nucleus (Lisk, 1960; Lisk & Newlon, 1963). Implants of estrogen in the anterior hypothalamus or the medial basal hypothalamus including the arcuate nucleus suppress an increase in LH release that is induced following ovariectomy (Chowers & McCann, 1965). Estrogen implanted in the median eminence reduces the content of luteinizing hormone releasing factor (LRF) in the median eminence (McCann *et al.* 1964).

Flerkó & Bardos (1960, 1961) observed that lesions in the suprachiasmatic area of the anterior hypothalamus resulted in pituitary hypertrophy and abolition of compensatory ovarian hypertrophy. Surgical isolation of the medial basal hypothalamus has been demonstrated to abolish the compensatory ovarian hypertrophy, but it did not prevent the increase in pituitary and serum LH content after castration (Halász *et al.* 1967). Köves *et al.* (1969) claimed in a study of surgical isolation that the ventral part of the anterior hypothalamus is the indispensable neural structure for occurrence of compensatory ovarian hypertrophy. Authors of these studies insist that the estrogen sensitive cells are probably located both in the anterior hypothalamus and the medial basal hypothalamus.

Radioautographic studies have demonstrated uptake of estrogen by neurons in the ventral part of preoptic area, the anterior hypothalamus, and the medial basal hypothalamus (Michael, 1962, 1965; Eisenfeld & Axelrod, 1965; Kato & Villee, 1967 *a*; Pfaff, 1968 *a*, *b*; Stumpf, 1968). Being inhibited by pretreatment with unlabeled estrogen, the uptake of labeled estrogen by hypothalamic neurons is considered specific to estrogen (Eisenfeld & Axelrod, 1966; Kato & Villee, 1967 *b*; Vértes & King, 1971). Estradiol-binding macromolecules were also found in the hypothalamus (Eisenfeld, 1969; Ginsburg, Marriott & Thomas, 1970).

Recently estrogen responsive neurons have been demonstrated in the medial preoptic and medial basal hypothalamic areas of castrated female rats (Yagi, 1973; see following section). This fact satisfies one of the necessary conditions for estrogen receptor cells to exist in these areas of the brain.

All observations stated above support the concept that receptor cells specifically sensitive to estrogen exist in the hypothalamus. A question, however, remains to be answered as to whether each individual receptor cell functions as a receptor in the estrogen feedback mechanism. Identification of receptor cells for estrogen feedback requires further study of whether a single neuron specifically responds to locally applied estrogen and whether the estrogen sensitive neuron is in functional connection with the tubero-infundibular neurosecretory neurons that secrete gonadotrophin releasing factors (cf. Yagi, Iwasaki, Sawaki & Satow, 1971).

THE EFFECTS OF ESTROGEN ON PREOPTIC AND HYPOTHALAMIC UNIT ACTIVITY

In female rats the medial preoptic and anterior hypothalamic areas and the medial basal hypothalamus are involved in the negative feedback effects of plasma estrogen on the gonadotrophin secretion by the adenohypophysis (Köves *et al.* 1969; McCann *et al.* 1964). The possible existence of a neural circuit for estrogen feedback in the hypothalamus leads to the expectation that some neurons should change their electrical activity following elevation or reduction of peripheral plasma estrogen level.

Estrogen responsive neurons have been suggested to exist in these areas (Lincoln, 1967; Lincoln & Cross, 1967). Recently Yagi (1970, 1973) has reported that single preoptic and hypothalamic units respond to intravenously administered estrogen in the castrated female rat. Ten of 26 units recorded from the suprachiasmatic part of the medial preoptic nucleus (POSC), 9 of 23 units recorded

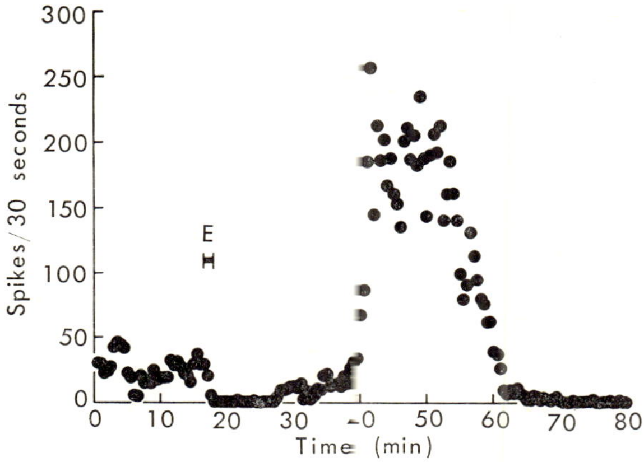

Fig. 1. An increase in firing rates following an intravenous administration of 50 μg of 17 β-estradiol in a unit recorded from the arcuate nucleus of the hypothalamus in a female rat 2 weeks after ovariectomy. The mark E indicates the injection of estrogen. Total volume of the injected solution was 0.2 ml (from Yagi, 1973).

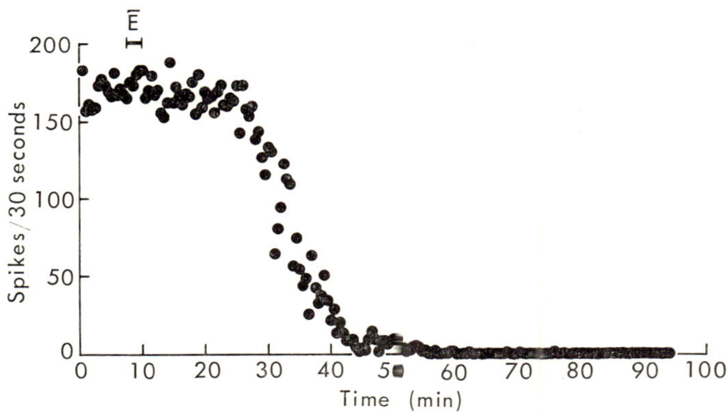

Fig. 2. A decrease in firing rates following an intravenous injection of 50 μg of 17 β-estradiol. The unit was recorded in the medial preoptic nucleus of the ovariectomized rat. No unit that was inhibited by estrogen restored its activity within the observation period of up to 150 min (from Yagi, 1973).

from the anterior hypothalamic nucleus (HA), and 4 of 21 units recorded from the arcuate nucleus of the hypothalamus (ARC) responded to the intravenously administered estrogen (5 or 50 μg) as shown in Fig. 1. Ten POSC units and 9 ARC units decreased their firing rates after the estrogen injection (Fig. 2). The depressed activity level was not restored within the observation period of up to

150 min. The remaining units (POSC, 6; HA, 14; ARC, 8) did not change their activity level after estrogen injection. These results demonstrate the presence of two types of estrogen responsive neurons in the preoptic and hypothalamic areas which show either a transitory increase or a long-lasting decrease in firing rates following an intravenous estrogen injection. It seems likely that the anterior hypothalamic nucleus does not include the neurons that are depressed by estrogen. The distribution of recorded units, however, does not necessarily reflect the actual distribution of responsive neurons in these diencephalic areas, since the probability for a spontaneously firing unit to be detected depends among other things on its firing frequency.

Latency and duration of observed responses varied very widely. However, mean latency and mean duration calculated for each nucleus were quite similar in all 3 groups (Table 1). Mean firing rates during 300 s before the estrogen injection were lower in the units that were excited by estrogen than in the units that were in-

TABLE 1. Latency and duration of responses to intravenously administered estrogen

Nucleus	Latency (min)	Duration (min)
POSC	16±14 (15)	32±21 (6)
HA	15±11 (7)	32±17 (7)
ARC	16±10 (12)	23±11 (4)

The results on both types of units, those excited and those inhibited by estrogen, are included in calculation of the mean latency. The mean duration was derived from responses with an increase in firing rates. The dose of estrogen was 50 μg.

Mean±S.D. of an observation. Number in parentheses indicates the number of units.

POSC, suprachiasmatic area of the medial preoptic nucleus; HA, anterior hypothalamic nucleus; ARC, arcuate nucleus of the hypothalamus (from Yagi, 1973).

TABLE 2. Mean firing rates before injection of 50 μg of estrogen

Nucleus	Units excited by estrogen (spikes/s)	Units inhibited by estrogen (spikes/s)	Unresponsive units (spikes/s)
POSC	1.8±1.1 (7)	3.6±2.2 (8)	5.7±5.2 (5)
HA	0.4±0.4 (7)		1.4±2.2 (11)
ARC	0.8±0.7 (4)	3.1±1.4 (8)	2.1±1.4 (6)

Each value was obtained from the number of spikes in 300 s before estrogen injection. The results are expressed as the mean±S.D. The number in parentheses is the number of units examined (from Yagi, 1973).

hibited by the hormone (Table 2). As for the latency and duration of responses to plasma estrogen comparable data have not been reported as yet. Barraclough & Cross (1963) reported that the maximum depression of the response of hypothalamic neurons to cervical probing was reached between 15 and 40 min after an intravenous injection of 400 μg of progesterone in the rat, the recovery being complete 1 h after the injection. An EEG change induced in the female rat by intravenously injected progesterone (100–400 μg) was found to have a latency between 10 and 140 s and a duration between 10 and 45 min (Ramirez, Komisaruk, Whitmoyer & Sawyer, 1967). These results reported on the effect of progesterone are comparable to the present results in regard to the time course of responses (Table 1). The results on mean firing rates (Table 2) resemble the observation reported by Lincoln *et al.* (1967) that hypothalamic units that were inhibited by the pain stimulus had higher firing rates than those excited by the stimulus.

Spontaneous firing rates recorded in 11 units (POSC, 4; ARC, 7) for up to 180 min fluctuated somewhat as shown in Fig. 3. The mean activity level, however, remained constant within narrow limits throughout the observation period. An intravenous infusion of warm Locke's solution performed at 60 min from the onset of recording did not produce any appreciable change in mean firing rates

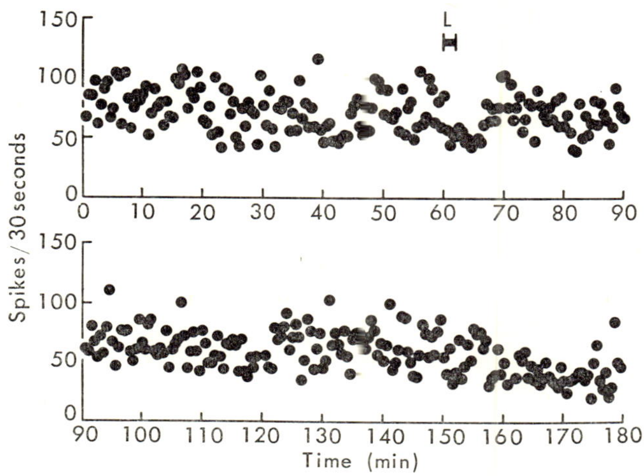

Fig. 3. Control experiment with an intravenous injection of 0.2 ml of Locke's solution. No appreciable change in the mean activity level was observed. The unit was recorded in the medial preoptic nucleus of a female rat 2 weeks after castration (from Yagi, 1973).

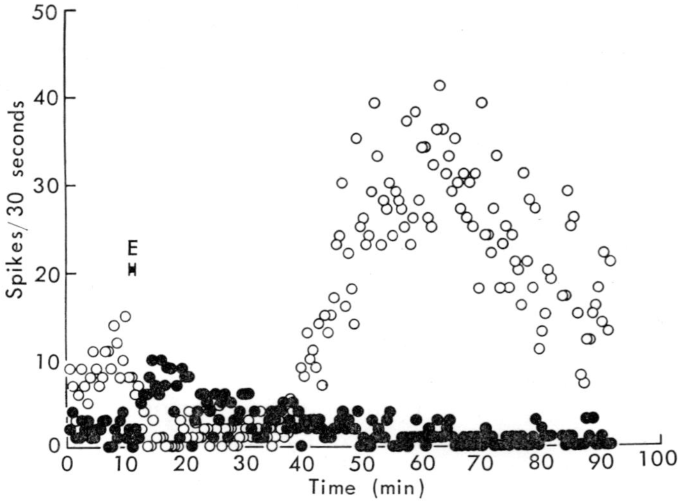

Fig. 4. Differential responses in 2 simultaneously recorded units. One unit recorded in the anterior hypothalamic nucleus (open circles) increased the firing rates after an intravenous injection of 50 μg of estrogen, and the other unit recorded in the arcuate nucleus (filled circles) did not respond to the injection (from Yagi, 1973).

(Fig. 3). In 15 out of 22 pairs of units recorded simultaneously from the same rat one of the two units responded to estrogen, while another unit of the pair did not respond to the hormone (Fig. 4). The effects of both 5 μg of cortisone acetate and 5 μg of 17 ß-estradiol on firing rates were studied in a total of 12 units (POSC, 5; HA, 4; ARC, 3). Four units (POSC, 3; HA, 1) responded with an increase in firing rates to the estrogen with no response to the glucocorticoid (Fig. 5). Three units responded with a decrease in firing rates to the estrogen and one ARC unit decreased its firing rates after the glucocorticoid. The remaining 4 units did not respond to either of these steroids. None of these units responded to both steroids. These observations suggest that the observed effects of estrogen on firing rates of certain preoptic and hypothalamic units were specific.

All of 5 units examined responded with an increase in firing rates to each of 50 and 400 μg of estrogen administered successively in this order at an interval of about 60 min or more (Fig. 6). The duration was longer and the latency was shorter in the response to 400 μg of estrogen than in the response to 50 μg of the hormone in each unit. The plasma concentration of estrogen 60 min after an injection is estimated to be 4.5×10^{-5} times lower than the concentration immediately after the injection by assuming a turnover time of 6 min for a presumed exponential decay of plasma estrogen (Pearlman, 1957). Repetitive responses of certain units to successively ad-

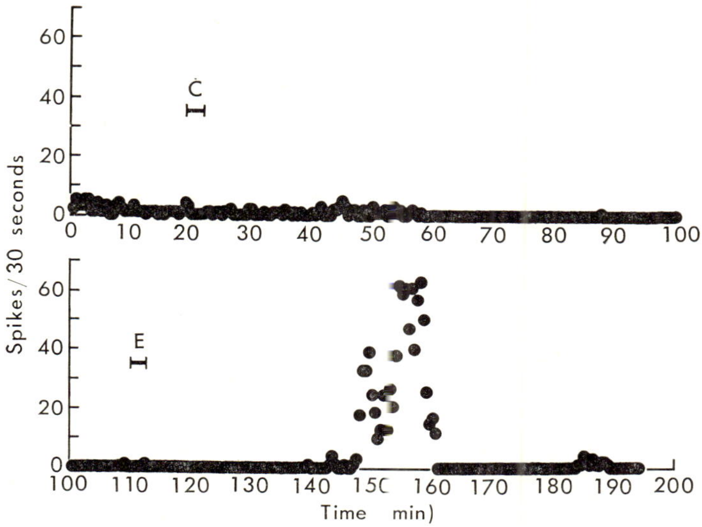

Fig. 5. Effects of intravenous injections of glucocorticoid and estrogen on firing rates of a preoptic unit in an ovariectomized rat. C and E represent the injection of 5 μg of cortisone acetate and 5 μg of 17 β-estradiol, respectively. Note an increase in firing rates after estrogen injection (from Yagi, 1973).

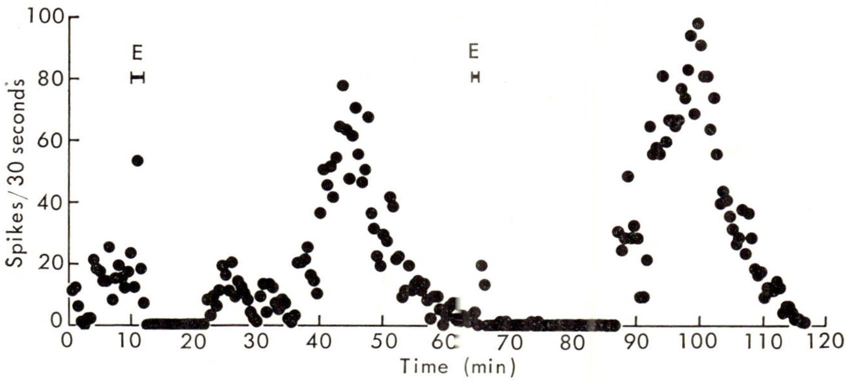

Fig. 6. Increases in firing rates in response to successively administered estrogen. The unit was recorded in the anterior hypothalamic nucleus of the castrated female rat. The first and second doses were 50 and 400 μg, respectively (from Yagi, 1973).

ministered estrogen with an increase in firing rates, therefore, probably suggest that the neurons that are excited by estrogen adapt slowly.

The dose of estrogen used in the present experiments (5 or 50 μg) is considerably larger than the 'physiological' dose for chronic administration of estrogen estimated as 0.015 to 0.175 μg/day in adolescent female rats (Byrnes & Meyer, 1951) or as 0.12 to 0.18 μg/100 g·day in adult spayed rats (Ramirez *et al.* 1963) on the basis of negative feedback effects of estrogen on the gonadotrophin secretion. However, the result that mean firing frequency under the condition of no ovarian estrogen in the blood was lower in the units which were excited by estrogen than in the units which were inhibited (Table 2) seems to indicate that the observed responses were produced as the result of physiological effects of estrogen. In further experiments (Yagi & Sawaki, 1972), successive injections of 5, 50, and 400 μg of estrogen were examined for the same rat. An increase in firing rates was not dose dependent, but dose-duration relationship was obtained (Fig. 7). Although reproducibility of the response was not examined, extrapolation of this relationship to 0 min

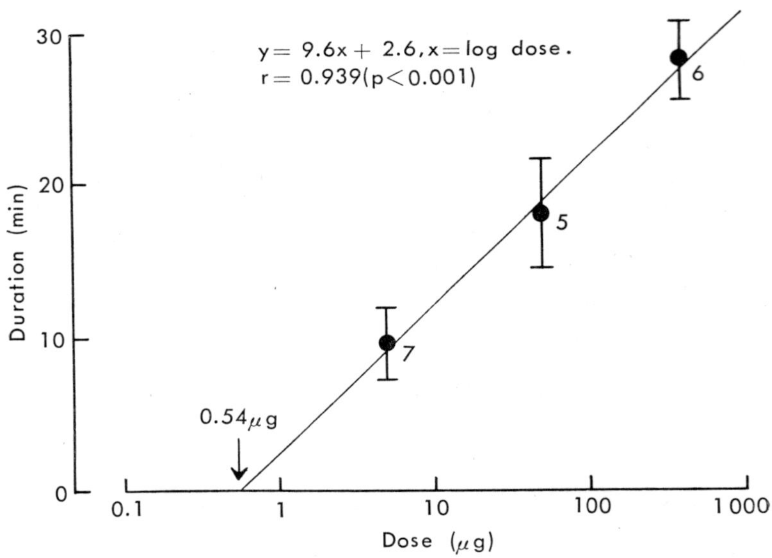

Fig. 7. Relation of duration of response with an increase in firing rates induced by estrogen to the dose. Five, 50, and 400 μg of estrogen were injected in this order in the same rat. Units were recorded in the anterior hypothalamic and medial preoptic nuclei. Numbers beside circles indicate the number of examinations.

duration gives a value of 0.54 μg. This amount is comparable to the ‘ physiological ’ dose referred above.

A number of experiments with lesions (Flerkó & Bárdos, 1959; Taleisnik *et al.* 1961), electrical stimulation (Critchlow, 1958; Everett, 1965; Harris & Ruf, 1970; Clemens, Shaar, Kleber & Tandy, 1971), and surgical isolation (Köves & Halász, 1970; Butler & Donovan, 1971; Kaasjager, Woodbury, van Dieten & van Rees, 1971) of localized areas within the diencephalon have demonstrated that POSC is the neural structure triggering ovulation. The arcuate nucleus of the rat hypothalamus have been shown to include the tubero-infundibular neurons by the experiments with lesions (McCann *et al.* 1964) and surgical isolation (Halász *et al.* 1967), and by the method of electrophysiological identification (Yagi & Sawaki, 1970 *b*; Harris, Makara & Spyer, 1971; Sawaki & Yagi, 1973). A neural connection between POSC and ARC has been demonstrated (Tejasen & Everett, 1967; Yagi & Sawaki, 1970 *a*; Dyer & Cross, 1972). It has been reported that both POSC and ARC areas are involved in the feedback effects of plasma estrogen on LH secretion (Lisk, 1960; Flerkó *et al.* 1961; McCann *et al.* 1964; Köves *et al.* 1969). It is, therefore, highly probable that the estrogen responsive neurons demonstrated in the preoptic and hypothalamic areas of female rats by the present experiments are involved in the hypothalamic control of functions of adenohypophysial-ovarian system.

Plasma estrogen has been shown to have the negative feedback effects on LH secretion (McCann *et al.* 1964; Chowers *et al.* 1965). It has been demonstrated that an increase in plasma estrogen in the morning of proestrus is necessary for an ovulatory surge of LH in the afternoon of the same day (Kobayashi *et al.* 1969 *b*). It has been also demonstrated that exogenous estrogens facilitate LH surge in castrated female rats (Swerdoloff *et al.* 1972) as well as in cyclic rats (Kobayashi *et al.* 1969 *a*). Since electrical stimulation of the medial preoptic and medial basal hypothalamic areas has been found to induce the secretion of luteinizing hormone releasing factor (Harris *et al.* 1970) and LH (Clemens *et al.* 1971) as well as ovulation, it seems likely that an increase in electrical activity of these areas will bring about an increase in LH secretion. Therefore, it seems reasonable to think that the slowly adapting unit which increased their activity level following an estrogen injection is involved in the facilitatory effects of estrogen on LH secretion, and that the units showed a long-lasting depression after the injection are involved in the negative feedback effects of estrogen.

NEURAL CONNECTIONS BETWEEN HYPOTHALAMIC NUCLEI AND THE MEDIAN EMINENCE

The neurons responsive to plasma estrogen exist in the medial preoptic and arcuate nuclei, as stated above. Presence of neural connections between these neurons and axon terminals in the median eminence is a necessary prerequisite for these neurons in the POSC and ARC to constitute a neural circuit in the feedback control of gonadotrophin secretion. A number of studies, mostly based on lesions and stimulations of the preoptic and medial basal hypothalamic areas, have suggested the presence of neural connections between these diencephalic structures and the median eminence (Assenmacher, 1963; Everett, 1964; Flerkó, 1966; Barraclough, 1967; Davidson, 1969).

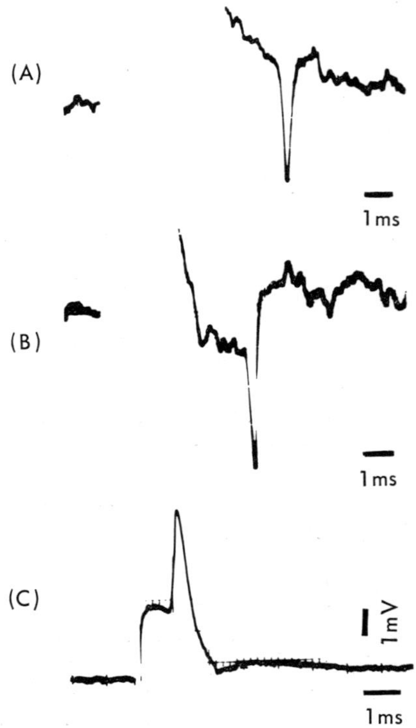

Fig. 8. Unit responses to electrical stimulation of the median eminence in female rats. Units were recorded in the suprachiasmatic (A), ventromedial (B), and arcuate (C) nuclei of the hypothalamus in the female rat. Vertical calibration is applicable to C. Amplitude of spikes of A and B is about 0.2 mV (from Sawaki & Yagi, 1973).

Histological evidence suggests that the tubero-infundibular axons with their terminals in the external layer of the median eminence

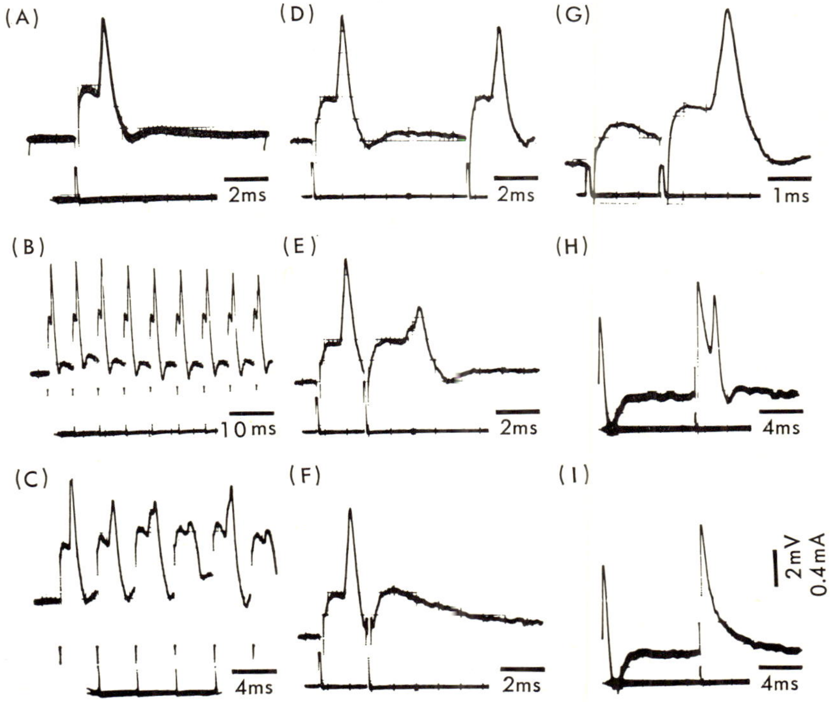

Fig. 9. Responses of arcuate units to stimulation of the median eminence in the female rat. A, Superimposed record of 10 successive responses to repetitive stimuli at 1 Hz showing constancy in latency. B, Responses to repetitive stimuli at 160 Hz. C, Responses to repetitive stimuli at 286 Hz showing responsiveness to high-frequency stimulation. D, E and F, Responses to paired stimuli with threshold intensity. Note that threshold and latency of response to the test stimulus increased as the interval shortened in D and E. F shows an absolute refractory period. G, Showing that a subliminal conditioning stimulus does not affect the threshold and latency of responses to the test stimulation. H and I, Cancellation of antidromically conducted impulses by collision with spontaneously occurring spike. Stimulation of the median eminence induced responses in an arcuate unit (H) at an interval just longer than the critical length of time following a spontaneously occurring spike and failed to produce a response (I) at an interval just shorter than the critical value. The oscilloscope sweep was triggered by the vertical signal of spontaneous spike (spike in the left end of the trace), and the interval between the spontaneous spike and the stimulation was controlled. Lower trace in each pair of traces indicates current intensity of stimulus. A to G were recorded in the same arcuate unit (modified from Sawaki & Yagi, 1973).

originate from cell bodies in the arcuate and periventricular nuclei of the hypothalamus (Flerkó, 1966). Electron-dense granules within the cytoplasm of cell bodies in the arcuate nucleus are similar in size and shape to those observed in the median eminence (Kobayashi, Kobayashi, Yamamoto & Kaibara, 1967).

Recently electrophysiological techniques have been applied to investigate neural connections between the POSC, ARC, and the median eminence (Yagi *et al.* 1970 *a*, *b*; Yagi *et al.* 1971; Harris *et al.* 1971; Dyer *et al.* 1972; Sawaki *et al.* 1973). In our study (Sawaki *et al.* 1973) electrical stimulation of the median eminence induced unit spikes in the suprachiasmatic (2 units), ventromedial (7 units), and arcuate (35 units) nuclei (Fig. 8). Spontaneously firing units of the supraoptic (17 units), medial preoptic (17 units), and anterior hypothalamic (20 units) nuclei did not respond to the same stimulation. The observed responses, however, do not necessarily represent antidromically conducted impulses, since impulses conducted along axon collaterals with excitatory synapses, if any, of the tubero-infundibular tract may have activated neurons in these nuclei. Further examination regarding this possibility was made for the arcuate units.

Some arcuate units responded to 0.1 ms pulses applied to the median eminence at 1 Hz with constant latency and invariable threshold (Fig. 9A). Units with constant latency responded in one to one fashion to repetitive stimuli of more than 100 Hz (Fig. 9B) and up to 286 Hz (Fig. 9C). Failure of antidromic invasion of the soma and dendrites by an impulse is the most likely explanation for the small and slow spikes in the fourth and sixth responses in Fig. 9C (Sawaki *et al.* 1973, and see below). Constancy in latency of unit responses to stimulation and the following of repetitive, high frequency stimulation by the unit have been used as the main criteria to identify cell bodies of neurosecretory neurons in the hypothalamo-neurohypophysial system (Ishikawa, Koizumi & Brooks, 1966; Yagi, Azuma & Matsuda, 1966; Dyball & Koizumi, 1969; Kelly & Dreifuss, 1970; Novin, Sundsten & Cross, 1970; Yamashita, Koizumi & Brooks, 1970; Barker, Crayton & Nicoll, 1971 *a*, *b*, *c*; Dyball, 1971; Dyball & Dyer, 1971; Dreifuss & Kelly, 1972 *a*, *b*; Koizumi & Yamashita, 1972; Lincoln & Wakerly, 1972). These criteria, however, are only two of the necessary conditions for the observed response to be an ' antidromically ' conducted impulse.

Temporal facilitation of transmission at excitatory synapses has been reported (Eccles, 1964). Even a single impulse produces potentiation of transmitter release by a subsequent impulse in the neuromuscular junction (Hubbard, 1963). These phenomena with regard to excitatory synapses leads to the expectation that if the observed response were a consequence of synaptic activation, threshold would be lower and latency would be shorter (Coombs, Curtis & Eccles, 1957) for the response to a test stimulus following a conditioning stimulus in experiments with paired stimuli than for the control response. Both threshold and latency of the response to test stimulus increased as the interval between conditioning and test stimuli was shortened to less than 6 ms in all of 11 arcuate units examined (Fig. 9D and E, and Fig. 10). A reduction of excitability and an increase in latency after the conditioning stimulus just above threshold is not attributable to recurrent inhibition which has been proved to last for several tens of milliseconds in the hypothalamo-neurohypophysial system (Kandel, 1964; Dreifuss *et al.* 1972 *a*; Koizumi *et al.* 1972), since the difference in time course between the observed reduction of excitability and recurrent inhibition is so large. Further

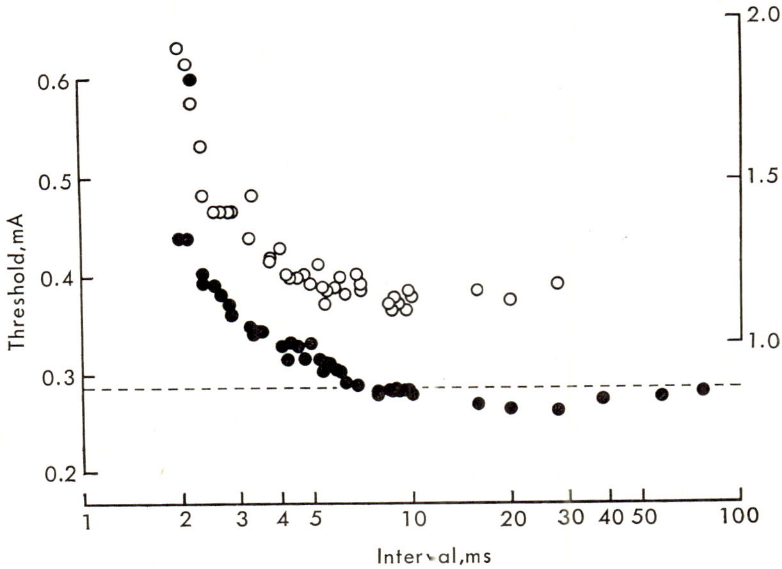

Fig. 10. Threshold and latency to the test stimulus given at various intervals from the preceding conditioning stimulus of suprathreshold intensity. Note that both threshold and latency increased as the interval was shortened to less than 6 ms. The test stimuli with an interval less than 2 ms did not produce a response in this unit. The unit was same as that of Fig. 9A (from Sawaki & Yagi, 1973).

results supporting this view showed that a conditioning stimulus with an intensity just below threshold did not alter the threshold and latency of responses in 6 units produced by the test stimulus at any interval, even shorter than the length of the absolute refractory period (Fig. 9G).

It is highly probable that an inflection in the rising phase of the spike induced by the test stimulus at a rather short interval (Fig. 9E) indicates dissociation of the action potential of the soma and dendrites from that of the initial segment of the axon due to a difference in excitability between these parts of the neuron (Eccles, 1964; Koizumi *et al.* 1972). Therefore, slow and small spikes observed among the responses to repetitive stimuli (Fig. 9C) may be attributable to the failure of invasion of the soma and dendrites by antidromically conducted impulses. The results of experiments with paired stimuli, therefore, satisfy another necessary condition for the response produced by stimulation of the median eminence to be an antidromically conducted impulse in that neither temporal facilitation nor potentiation by the test stimulus was observed.

Stimulating pulses of threshold intensity were applied to the median eminence at various intervals following spontaneous spike discharge. Stimulating pulses given at longer intervals than a critical length of time always induced action potentials in the arcuate nucleus (Fig. 9H). However, within the range of intervals shorter than this critical value, the stimulation failed to produce responses (Fig. 9I). The critical length of intervals was constant in each unit and ranged between 8 and 12.9 ms in 5 arcuate units examined. The abolition of evoked impulses indicates cancellation of antidromic impulses by collision with spontaneously occurring orthodromic impulses. It is, therefore, suggested that observed action potentials induced by the stimulation of the median eminence represent an ‘antidromic’ invasion of the arcuate neurons by the impulses primarily evoked in the median eminence. Cancellation of induced spikes by collision has been used as one of the criteria for an induced action potential to be an antidromically conducted impulse (Kelly *et al.* 1970; Harris *et al.* 1971; Dreifuss *et al.* 1972 *a*; Koizumi *et al.* 1972; Dyer *et al.* 1972). These observations, however, do not completely exclude the possibility of an alternative explanation for the abolition of induced action potentials, namely, that the collision might have occurred in axon collaterals of the tubero-infundibular tract, which may have excitatory synapses on arcuate neurons.

Unit impulses produced in the arcuate nucleus by stimulation of the median eminence satisfy the criteria of constancy in latency, one to one responsiveness to high-frequency repetitive stimuli, absence of facilitation following a conditioning stimulus and cancellation of evoked impulse by collision with a spontaneously occurring spike. It is, therefore, concluded that cell bodies of the tubero-infundibular axons are located in the arcuate nucleus of the rat hypothalamus.

The median eminence includes axon terminals containing catecholamines as well as those containing releasing factors (Fuxe & Hökfelt, 1966; Björklund, Flack, Hromek, Owman & West, 1970; Clementi, Ceccarelli, Cerati, Demonte, Felici, Motta & Pecile, 1970). Recent investigations disclosed that the tubero-infundibular dopamine neurones modify the control by releasing factors of adenohypophysial functions (Lichtensteiger, 1971; Keller & Lichtensteiger, 1971). Evidence presented in this study, therefore, strongly supports the concept that the arcuate nucleus of the hypothalamus includes neurons of the final common pathway of the hypothalamo-adenohypophysial system.

Electrical stimulation of the ARC through a concentric bipolar electrode induced unit responses in the POSC with a constant latency. Responses of such POSC units to repetitive stimuli with high frequency and to paired stimuli were much the same as those in the ARC units (Yagi *et al.* 1972 *a*; Dyer *et al.* 1972). It is suggested that the neurons in the POSC probably innervate the ARC.

UNIT ACTIVITY IN THE PREOPTIC AND HYPOTHALAMIC AREAS OF CYCLIC AND CASTRATED FEMALE RATS

Electrical stimulation of either the medial preoptic and anterior hypothalamic area (POM/HA) or the medial basal hypothalamus (MBH) including the arcuate nucleus has been demonstrated to induce the release of LRF and LH in female rats (Harris *et al.* 1970; Clemens *et al.* 1971; Keller *et al.* 1971). In female rats LH secretion has been shown to increase on the day of proestrus during the estrous cycle (Kobayashi *et al.* 1968; Brown-Grant, Exley & Naftolin, 1970).

It has been reported that the POM/HA and MBH areas in female rats are involved in the feedback action of plasma estrogen on the hypothalamo-adenohypophysial control of ovarian functions (McCann *et al.* 1964; Köves *et al.* 1969). Estrogen responsive neurons have been demonstrated to exist in these areas of the diencephalon in female rats (Yagi, 1973). It has been shown that the rates of estrogen secretion by the ovary intends to increase on the second day of diestrus and reaches to a peak on the day of proestrus in female rats with 4-day estrous cycles (Hori, Ide & Miyake, 1968; Brown-Grant *et al.* 1970).

Recently Yagi & Sawaki (1971) have reported the unit activity in the POSC and ARC of cyclic and castrated female rats anesthetized with urethane. The distribution of firing rates of the units recorded from the POSC at each stage of the estrous cycle and after castration is shown in Fig. 11. The decreased percentage of units with firing

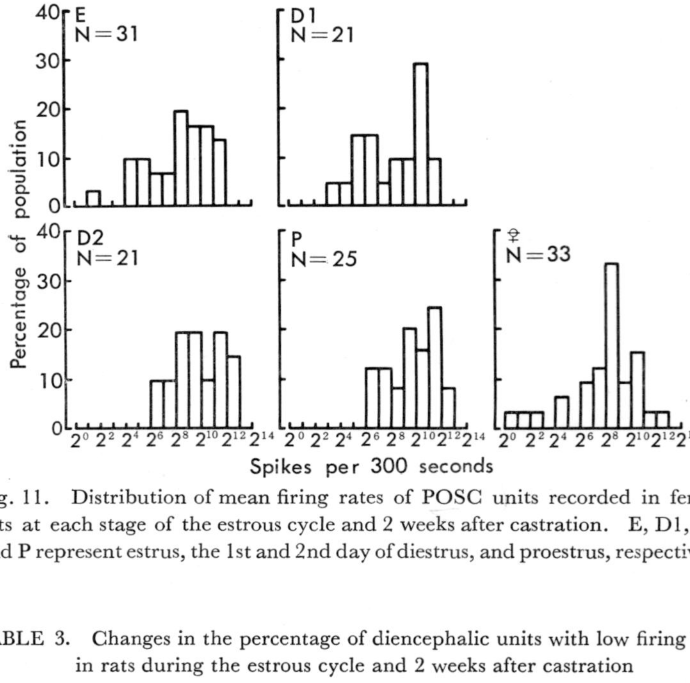

Fig. 11. Distribution of mean firing rates of POSC units recorded in female rats at each stage of the estrous cycle and 2 weeks after castration. E, D1, D2, and P represent estrus, the 1st and 2nd day of diestrus, and proestrus, respectively.

TABLE 3. Changes in the percentage of diencephalic units with low firing rates in rats during the estrous cycle and 2 weeks after castration

Nucleus	E	D1	D2	P	Castrated rats
POSC	22.6	23.9	0	0	15.1
ARC	17.6	25.9	15	6.1	5.1

Figures in the Table represent percentages of units with firing rates lower than 64 spikes/300 s for POSC units and 16 spikes/300 s for ARC units. E, estrus; D1 and D2, the 1st and 2nd day of diestrus; P, proestrus; POSC, the suprachiasmatic part of the medial preoptic nucleus; ARC, the arcuate nucleus of the hypothalamus.

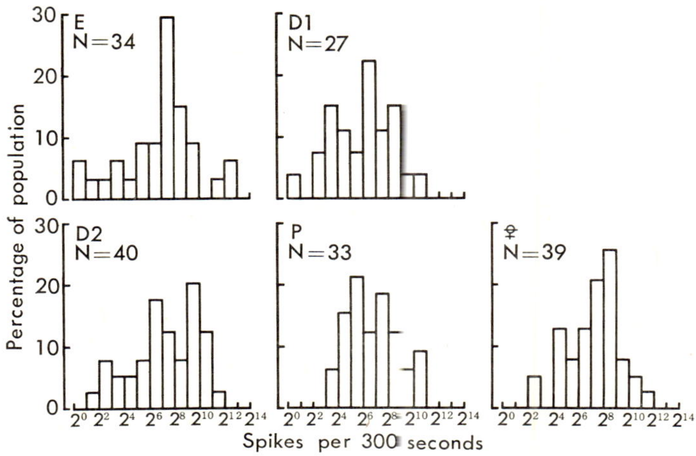

Fig. 12. Distribution of mean firing rates of ARC units recorded in female rats at each stage of the estrous cycle and 2 weeks after castration.

rates lower than 64 spikes per 300 s was observed on the second day of diestrus and the day of proestrus (Table 3). Histograms of firing rates of ARC units at each stage of the estrous cycle are shown in Fig. 12. As for the ARC units with firing rates lower than 16 spikes per 300 s, the percentage was low in rats at proestrus and after castration (Table 3).

Distributions of number of firings per 1 s fitted to the Poisson distribution in 7 of 11 POSC units and 6 of 13 ARC units examined by χ^2-test. It has been reported that firing patterns of hypothalamic units are not statistically homogeneous (Oomura, Kimura, Ooyama, Maeno, Iki & Kuniyoshi, 1964; Kawakami & Saito, 1969). It is, therefore, difficult to estimate by a simple method the probability for a unit to fire once during the time the moving electrode tip is within a space surrounding a firing neuron. Assuming that a microelectrode tip is moving at a speed ($\bar{v}$) through a spherical space with a radius (r) in which the electrode can pick up unit spikes, the mean number ($\bar{N}_i$) of firings of a neuron i with a mean firing rate ($\bar{f}_i$) is represented as $\bar{N}_i=\bar{f}_i \cdot 4r/3\bar{v}$. In our experiments r and $\bar{v}$ were about 20 μm and 0.79 μm/s, respectively. As the frequency of occurrence of a unit with $\bar{f}_i$ was normalized by $1/\bar{N}$, mean activity levels are approximately estimated for both the POSC and ARC (Fig. 13).

Mean levels of unit activity in the POSC were high on the second day of diestrus and the day of proestrus when plasma estrogen level

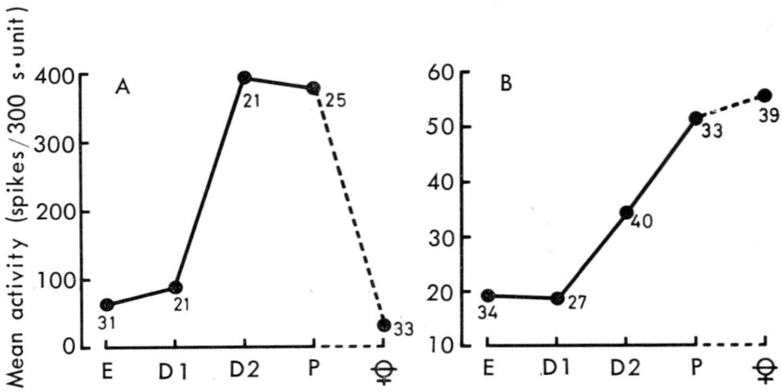

Fig. 13. Mean activity levels of spontaneously firing units during the estrous cycle and after castration. A, Preoptic (POSC) units; B, arcuate (ARC) units. The mean values were calculated based on the normalized data (see text). It should be noted that ovariectomy depressed the activity level of the POSC, and it augmented the activity level of the ARC.

is high (Fig. 13A). Mean levels of unit activity in the ARC, however, were high at proestrus and after castration when the rate of LH secretion is high (Ramirez *et al.* 1963; Hori *et al.* 1968; Brown-Grant *et al.* 1970) irrespectively of plasma estrogen level. These results agree with the results on the multiple unit activity reported by Kawakami, Terasawa & Ibuki (1970) and on the single unit activity reported by Dyer *et al.* (1972) in that the electrical activity in these areas is highest at proestrus during the estrous cycle.

Lesions in the POSC block ovulation and produce persistent estrus with the polycystic ovaries (Flerkó *et al.* 1959, 1960; Taleisnik *et al.* 1961). Results of surgical isolation of localized areas in the diencephalon have been found completely consistent with the consequences of lesioning (Halász *et al.* 1967; Tejasen *et al.* 1967; Köves *et al.* 1970; Kaasjager *et al.* 1970; Butler *et al.* 1971; Rodgers & Schwartz, 1972). Electrical stimulation of the POSC in anesthetized rats has been shown to induce an increase in LRF activity in the hypophysial portal plasma (Harris *et al.* 1970), an elevation of plasma LH level (Clemens *et al.* 1971; Keller *et al.* 1971), and ovulation (Critchlow, 1958; Barraclough & Gorsky, 1961; Everett, 1965; Barraclough, 1966; Harris *et al.* 1970; McDonald & Gilmore, 1971). These observations suggest that an increase in electrical activity of the POSC probably enhances gonadotrophin secretion by the adenohypophysis. It is, therefore, concluded that the POSC is involved in the positive feedback of plasma estrogen.

Anesthetic dose of urethane has been demonstrated to block ovulation (Haller & Barraclough, 1968; Fink & Harris, 1970; Dyer *et al.* 1972; Lincoln & Kelly, 1972). It is, therefore, likely that the elevated unit activity of the POSC observed in rats anesthetized with urethane does not represent the neural stimulus, which causes an ovulatory surge of LH in the proestrous afternoon, as pointed out by Dyer *et al.* (1972). An explanation for the increased unit activity of the POSC is that the elevated plasma estrogen may open wider the functional ' gate ' in the POSC through which the unidentified neural stimulus is transmitted to tubero-infundibular neurons. In other words both the unit activity elevated by plasma estrogen and another neural stimulus which is probably timed by a biological clock and readily depressed by anesthetics may provide input signals to a hypothetical mechanism in the POSC which is analogous to an 'AND ' circuit of logical circuits and produces the neural stimulus to cause LH surge.

The elevated unit activity of ARC observed in rats under the condition of enhanced LH secretion is completely consistent with the conclusion drawn in the previous section of this article that the ARC includes neurons of the final common pathway of the hypothalamo-adenohypophysial system. Mean activity level of the ARC was higher in castrated female rats than in proestrous rats. It is well known that the ARC in female rats is involved in the negative feedback of estrogen (Lisk, 1960; McCann *et al.* 1964). It is, therefore, concluded that the hypothalamo-adenohypophysial system functions in its control of LH secretion as a ' servo control system ' (Grodins, 1963) under the influence of negative feedback effects of plasma estrogen.

CONCLUSION

Observations so far reported and described above can be summarized as follows. (1) The hypothalamus contains receptor cells to concentrate plasma estrogen. (2) Slow-adapting neurons specifically responsive to estrogen are included in the POSC, HA, and ARC. (3) Unit activity in the POSC was high on the second day of diestrus and the day of proestrus, when the plasma estrogen level is high during the estrous cycle. Ovariectomy lowered the unit activity. (4) Unit activity in the ARC was high in proestrous and castrated rats, both of which have the high rate of LH secretion. (5) There are neural connections between the POSC and the ARC and between the ARC and the median eminence. (6) Electrical stimulations of the POSC and ARC produce facilitatory effects on gonadotrophin secretion by the adenohypophysis.

Consideration of these observations *in toto* gives us a hypothesis that plasma estrogen plays the role of the feedback parameter in giving a positive feedback effect through the POSC and a negative feedback effect through the ARC on the activity of the adenohypophysial-ovarian system. In other words plasma estrogen may open wider in the POSC and may close tighter in the ARC the functional ‘ gate ’ through which neural information controlling the release of gonadotrophin releasing factors is mediated to the median eminence.

Feedback effects of progesterone were not discussed in this article, mainly becasuse the preovulatory increase in progesterone secretion is considered to be a consequence of elevated LH secretion at proestrus (Uchida, Kadowaki & Miyake, 1969 *a*, *b*; Barraclough, Collu, Massa & Martini, 1971).

ACKNOWLEDGEMENTS

This work was supported by Grants No. 87013, 0803, and 60803 from the Ministry of Education of Japan.

REFERENCES

Assenmacher, I. (1963). Les regulations hypothalamiques de la fonction gonadotrope. *Acta neuroveg.* **25**, 339–382.

Barker, J. L., Crayton, J. W. & Nicoll, R. A. (1971 *a*). Supraoptic neurosecretory cells: antidromic modulation. *Science, N. Y.* **171**, 206–207.

Barker, J. L., Crayton, J. W. & Nicoll, R. A. (1971 *b*). Supraoptic neurosecretory cells: adrenergic and cholinergic sensitivity. *Science, N. Y.* **171**, 208–210.

Barker, J. L., Crayton, J. W. & Nicoll, R. A. (1971 *c*). Noradrenaline and acetylcholine responses of supraoptic neurosecretory cells. *J. Physiol.* **218**, 19–32.

Barraclough, C. A. (1966). Modification in the CNS regulation of reproduction after exposure of prepubertal rats to steroid hormones. *Recent Prog. Horm. Res.* **22**, 503–539.

Barraclough, C. A. (1967). Modifications in reproductive function after exposure to hormones during the prenatal and early postnatal period. In *Neuroendocrinology*, vol. 2, ed. Martini, L. & Ganong, W. F., pp. 61–99. New York: Academic Press.

Barraclough, C. A., Collu, R., Massa, R. & Martini, L. (1971). Temporal interrelationship between plasma LH, ovarian secretion rates and peripheral plasma progestin concentrations in the rat: effects of Nembutal and exogenous gonadotropins. *Endocrinology* **88**, 1437–1447.

Barraclough, C. A. & Cross, B. A. (1963). Unit activity in the hypothalamus of the cyclic female rat: effect of genital stimuli and progesterone. *J. Endocr.* **26**, 339–359.

Barraclough, C. A. & Gorsky, R. A. (1961). Evidence that the hypothalamus is responsible for androgen-induced sterility in the female rat. *Endocrinology* **68**, 68–79.

Björklund, A., Flack, B., Hromek, F., Owman, C. & West, K. A. (1970). Identification and terminal distribution of the tubero-hypophyseal monoamine fibre system in the rat by means of stereotaxic and microspectrofluorometric techniques. *Brain Res.* **17**, 1–23.

Brown-Grant, K., Exley, D. & Naftolin, F. (1970). Peripheral plasma oestradiol and luteinizing hormone concentrations during the oestrous cycle of the rat. *J. Endocr.* **48**, 295–296.

Butler, J. E. M. & Donovan, B. T. (1971). The effect of surgical isolation of the hypothalamus on reproductive function in the female rat. *J. Endocr.* **49**, 293–304.

Byrnes, W. W. & Meyer, R. K. (1951). Effect of physiological amounts of estrogen on the secretion of follicle stimulating and luteinizing hormones. *Endocrinology* **49**, 449–460.

Chowers, I. & McCann, S. M. (1965). Content of luteinizing hormone-releasing factor and luteinizing hormone during the estrous cycle and after changes in gonadal steroid titer. *Endocrinology* **76**, 700–708.

Clementi, F., Ceccarelli, B., Cerati, E., Demonte, M. L., Felici, M., Motta, M. & Pecile, A. (1970). Subcellular localization of neurotransmitters and releasing factors in the rat median eminence. *J. Endocr.* **48**, 205–213.

Clemens, J. A., Shaar, C. J., Kleber, J. W. & Tandy, W. A. (1971). Areas of the brain stimulatory to LH and FSH secretion. *Endocrinology* **88**, 180–184.

Coombs, J. S., Curtis, D. R. & Eccles, J. C. (1957). The generation of impulses in motoneurones. *J. Physiol.* **139**, 232–249.

Critchlow, V. (1958). Ovulation induced by hypothalamic stimulation in the anesthetized rat. *Am. J. Physiol.* **195**, 171–174.

Davidson, J. M. (1969). Feedback control of gonadotropin secretion. In *Frontiers in Neuroendocrinology*, ed. Ganong, W. F. & Martini, L., pp. 343–388. London: Oxford Univ. Press.

Dreifuss, J. J. & Kelly, J. S. (1972 *a*). Recurrent inhibition of antidromically identified rat supraoptic neurones. *J. Physiol.* **220**, 87–103.

Dreifuss, J. J. & Kelly, J. S. (1972 *b*). The activity of identified supraoptic neurones and their response to acetylcholoine applied by iontophoresis. *J. Physiol.* **220**, 105–118.

Dyball, R. E. J. (1971). Oxytocin and ADH secretion in relation to electrical activity in antidromically identified supraoptic paraventricular units. *J. Physiol.* **214**, 245–256.

Dyball, R. E. J. & Dyer, R. G. (1971). Plasma oxytocin concentration and paraventricular neurone activity on rats with diencephalic islands and intact brains. *J. Physiol.* **216**, 227–235.

Dyball, R. E. J. & Koizumi, K. (1969). Electrical activity in the supraoptic and paraventricular nuclei associated with neurophypohysial hormone release. *J. Physiol.* **201**, 711–722.

Dyer, R. G. & Cross, B. A. (1972). Antidromic identification of units in the preoptic and anterior hypothalamic areas projecting directly to the ventromedial and arcuate nuclei. *Brain Res.* **43**, 254–258.

Dyer, R. G., Pritchett, C. J. & Cross, B. A. (1972). Unit activity in the diencephalon of female rats during the oestrous cycle. *J. Endocr.* **53**, 151–160.

Eccles, J. C. (1964). *The Physiology of Synapses*, pp. 101–113. Berlin: Springer-Verlag.

Eisenfeld, A. J. (1969). Hypothalamic oestradiol-binding macromolecules. *Nature* **224**, 1202–1203.

Eisenfeld, A. J. & Axerlod, J. (1965). Selectivity of estrogen distribution in tissues. *J. Pharmac. exp. Ther.* **150**, 469–475.

Eisenfeld, A. J. & Axerlod, J. (1966). Effect of steroid hormones, ovariectomy, estrogen pretreatment, sex and immaturity on the distribution of ^{3}H-estradiol. *Endocrinology* **79**, 38–42.

Everett, J. W. (1948). Progesterone and estrogen in the experimental control of ovulation time and other features of the estrous cycle in the rat. *Endocrinology* **43**, 389–405.

Everett, J. W. (1964). Central neural control of reproductive functions of the adenohypophysis. *Physiol. Rev.* **44**, 373–431.

Everett, J. W. (1965). Ovulation in rats from preoptic stimulation through platinum electrodes. Importance of duration and spread of stimulus. *Endocrinology* **76**, 1195–1201.

Fink, G. & Harris, G. W. (1970). The luteinizing hormone releasing activity of extracts of blood from the hypophysial portal vessels of rats. *J. Physiol.* **208**, 221–241.

Flerkó, B. (1966). Control of gonadotropin secretion in the female. In *Neuroendocrinology*, vol. 1, ed. Martini, L. & Ganong, W. F., pp. 613–668. New York: Academic Press.

Flerkó, B. & Bárdos, V. (1959). Zwei verschiedene Effekte experimenteller Läsion des Hypothalamus auf die Gonaden. *Acta neuroveg.* **20**, 248–262.

Flerkó, B. & Bárdos, V. (1960). Pituitary hypertrophy after anterior hypothalamic lesion. *Acta endocr., Copenh.* **35**, 375–380.

Flerkó, B. & Bárdos, V. (1961). Absence of compensatory ovarian hypertrophy in rats with anterior hypothalamic lesions. *Acta endocr., Copenh.* **36**, 180–184.

Fuxe, K. & Hökfelt, T. (1966). Further evidence for the existence of tubero-infundibular dopamine neurons. *Acta physiol. scand.* **66**, 245–246.

Gay, V. L. & Sheth, N. A. (1972). Evidence for release of LH in castrated male and female rats. *Endocrinology* **90**, 158–162.

Ginsburg, M., Mariott, J. & Thomas, P. J. (1970). Oestradiol-binding macromolecules in the hypothalamus. *J. Physiol.* **209**, 39–41 *P.*

Grodins, F. S. (1963). *Control Theory and Biological Systems.* New York: Columbia Univ. Press.

Halász, B. & Gorsky, R. A. (1967). Gonadotrophin hormone secretion in female rats after partial or total interruption of neural afferents to the medial basal hypothalamus. *Endocrinology* **80**, 608–622.

Haller, E. W. & Barraclough, C. A. (1968). Hypothalamic regulation of ovulation: effects of urethane and progesterone. *Proc. Soc. exp. Biol. Med.* **129**, 291–295.

Harris, G. W. (1955). *Neural Control of the Pituitary Gland.* London: Arnold.

Harris, G. W. & Ruf, K. B. (1970). Luteinizing hormone releasing factor in the rat hypophysial portal blood collected during electrical stimulation of the hypothalamus. *J. Physiol.* **208**, 243–250.

Harris, M. C., Makara, G. B. & Spyer, K. M. (1971). Electrophysiological identification of neurones of the tubero-infundibular system. *J. Physiol.* **218**, 86–87 P.

Hohlweg, W. (1934). Veränderungen des Hypophysenvorderlappens und des Ovariums nach Behandlungen mit grossen Dosen von Follikelhormon. *Klin. Wschr.* **13**, 92–95.

Hori, T., Ide, M. & Miyake, T. (1968). Ovarian estrogen secretion during the estrous cycle and under the influence of exogenous gonadotropins in rats. *Endocr. jap.* **15**, 215–222.

Hubbard, J. I. (1963). Repetitive stimulation at the mammalian neuromuscular junction, and the mobilization of transmitter. *J. Physiol.* **169**, 641–662.

Ishikawa, T., Koizumi, K. & Brooks, C. McC. (1966). Electrical activity recorded from the pituitary stalk of the cat. *Am. J. Physiol.* **201**, 427–431.

Kaasjager, W. A., Woodbury, D. M., van Dieten, J. A. M. J. and van Rees, G. P. (1971). The role played by the preoptic region and the hypothalamus in spontaneous ovulation and ovulation induced by progesterone. *Neuroendocrinology* **7**, 54–64.

Kandel, E. R. (1964). Electrical properties of hypothalamic neuroendocrine cells. *J. gen. Physiol.* **47**, 691–718.

Kato, J. & Villee, C. A. (1967 *a*). Preferential uptake of estradiol by the anterior hypothalamus of the rat. *Endocrinology* **80**, 567–575.

Kato, J. & Villee, C. A. (1967 *b*). Factors affecting uptake of estradiol-6, 7-^{3}H by the hypophysis and hypothalamus. *Endocrinology* **80**, 1133–1138.

Kawakami, M. & Saito, H. (1969). The analysis of inter-spike interval fluctuation of hypothalamic unit activity in response to luteinizing hormone and oxytocin. *Jap. J. Physiol.* **19**, 243–259.

Kawakami, M., Terasawa, E. & Ibuki, I. (1970). Changes in multiple unit activity of the brain during the estrous cycle. *Neuroendocrinology* **6**, 30–48.

Keller, P. J. & Lichtensteiger, W. (1971). Stimulation of tubero-infundibular neurones and gonadotrophin secretion. *J. Physiol.* **219**, 385–401.

Kelly, J. S. & Dreifuss, J. J. (1970). Antidromic inhibition of identified rat supraoptic neurones. *Brain Res.* **22**, 406–409.

Kobayashi, F., Hara, K. & Miyake, T. (1968). Luteinizing hormone concentrations in pituitary and in blood plasma during the estrous cycle of the rat. *Endocr. jap.* **15**, 313–319.

Kobayashi, F., Hara, K. & Miyake, T. (1969 *a*). Effects of steroids on the release of luteinizing hormone in the rat. *Endocr. jap.* **16**, 251–260.

Kobayashi, F., Hara, K. & Miyake, T. (1969 *b*). Causal relationship between luteinizing hormone release and estrogen secretion in the rat. *Endocr. jap.* **16**, 261–267.

Kobayashi, T., Kobayashi, T., Yamamoto, K. & Kaibara, M. (1967). Electron microscopic observation on the hypothalamo-hypophysial system in the rat. II. Ultrafine structure of the median eminence and of the nerve cells of the arcuate nucleus. *Endocr. jap.* **14**, 158–177.

Koizumi, K. & Yamashita, H. (1972). Studies of antidromically identified neurosecretory cells of the hypothalamus by intracellular and extracellular recordings. *J. Physiol.* **221**, 683–705.

Köves, K. & Halász, B. (1969). Data on the location of the neural structures indispensable for the occurrence of ovarian compensatory hypertrophy. *Neuroendocrinology* **4**, 1–11.

Köves, K. & Halász, B. (1970). Location of the neural structures triggering ovulation in the rat. *Neuroendocrinology* **6**, 180–193.

Lawton, I. E. (1972). Facilitatory feedback effects of adrenal and ovarian hormones on LH secretion. *Endocrinology* **90**, 575–579.

Lichtensteiger, W. (1971). Effect of electrical stimulation of the fluorescence intensity of catecholamine-containing cells. *J. Physiol.* **218**, 63–84.

Lincoln, D. W. (1967). Unit activity in the hypothalamus, septum and preoptic area of the rat: characteristics of spontaneous activity and the effect of oestrogen. *J. Endocr.* **37**, 177–189.

Lincoln, D. W. & Cross, B. A. (1967). Effect of oestrogen on the responsiveness of neurones in the hypothalamus, septum and preoptic area of rats with light-induced persistent oestrus. *J. Endocr.* **37**, 191–203.

Lincoln, D. W. & Kelly, W. A. (1972). The influence of urethane on ovulation in the rat. *Endocrinology* **90**, 1594–1599.

Lincoln, D. W. & Wakerly, J. B. (1972). Accelerated discharge of paraventricular neurosecretory cells correlated with reflex release of oxytocin during suckling. *J. Physiol.* **222**, 23 *P*.

Lisk, R. D. (1960). Estrogen-sensitive centers in the hypothalamus of the rat. *J. exp. Zool.* **145**, 197–207.

Lisk, R. D. & Newlon, M. (1963). Estradiol: evidence for its direct effect on hypothalamic neurons. *Science, N. Y.* **139**, 223.

McCann, S. M. & Ramirez, V. D. (1964). The neuroendocrine regulation of hypophyseal luteinizing hormone secretion. *Recent Prog. Horm. Res.* **20**, 131–170.

McDonald, P. G. & Gilmore, D. P. (1971). The effect of ovarian steroids on hypothalamic thresholds for ovulation in the female rat. *J. Endocr.* **49**, 421–429.

Michael, R. P. (1962). Estrogen-sensitive neurons and sexual behavior in the female cats. *Science, N.Y.* **136**, 322–323.

Michael, R. P. (1965). Oestrogens in the central nervous system. *Brit. med. Bull.* **21**, 87–90.

Novin, D., Sundsten, J. W. & Cross, B. A. (1970). Some properties of antidromically activated units in the paraventricular nucleus of the hypothalamus. *Expl Neurol.* **26**, 330–341.

Oomura, Y., Kimura, K., Ooyama, H., Maeno, T., Iki, M. & Kuniyoshi, M. (1964). Reciprocal activities of the ventromedial and lateral hypothalamic areas of cats. *Science, N.Y.* **143**, 484–485.

Parlow, A. F. (1964). Differential action of small doses of estradiol on gonadotrophins in the rat. *Endocrinology* **75**, 1–8.

Pearlman, W. H. (1957). Circulating steroid hormone levels in relation to steroid hormone production. In *Hormones in the Blood, CIBA Foundation Colloquia on Endocrinology*, vol. 11, ed. Wolstenholme, G. E. W. & Millar, E. C. P., pp. 233–251. London: Churchill.

Pfaff, D. W. (1968 *a*). Autoradiographic localization in the rat brain after injection of tritiated sex hormone. *Science, N.Y.* **161**, 1355–1356.

Pfaff, D. W. (1968 *b*). Uptake of ^{3}H-estradiol by the female rat brain. An autoradiographic study. *Endocrinology* **82**, 1149–1152.

Presl, J. (1961). The time relationship between oestrogen administration and the secretion of luteinizing hormone in rats. *Acta endocr., Copenh.* **36**, 443–454.

Ramirez, V. D., Komisaruk, B. R., Whitmoyer, D. I. & Sawyer, C. H. (1967). Effects of hormones and vaginal stimulation on the EEG and hypothalamic units in rats. *Am. J. Physiol.* **212**, 1376–1384.

Ramirez, V. D. & McCann, S. M. (1963). Comparison of the regulation of luteinizing hormone (LH) secretion in immature and adult rats. *Endocrinology* **72**, 452–464.

Ramirez, V. D. & Sawyer, C. H. (1965). Advancement of puberty in the female rat by estrogen. *Endocrinology* **76**, 1158–1168.

Rodgers, C. H. & Schwartz, N. B. (1972). Diencephalic regulation of plasma LH, ovulation, and sexual behavior in the rat. *Endocrinology* **90**, 461–465.

Sawaki, Y. & Yagi, K. (1973). Electrophysiological identification of cell bodies of tubero-infundibular neurones in the rat. *J. Physiol.* **230**, 1–11.

Schwartz, N. B. (1964). Acute effects of ovariectomy on pituitary LH, uterine weight, and vaginal cornification. *Am. J. Physiol.* **207**, 1251–1259.

Stumpf, W. E. (1968). Estradiol-concentrating neurons: topography in the hypothalamus by dry-mount autoradiography. *Science, N.Y.* **162**, 1001–1003.

Swerdoloff, R. S., Jacobs, H. S. & Odell, W. D. (1972). Synergistic role of progestogens in estrogen induction of LH and FSH surge. *Endocrinology* **90**, 1529–1536.

Taleisnik, S. & McCann, S. M. (1961). Effects of hypothalamic lesions on the secretion and storage of hypophyseal luteinizing hormone. *Endocrinology* **68**, 263–272.

Tejasen, T. & Everett, J. W. (1967). Surgical analysis of the preoptico-tuberal pathway controlling ovulatory release of gonadotropins in the rat. *Endocrinology* **81**, 1387–1396.

Uchida, K., Kadowaki, M. & Miyake, T. (1969 *a*). Ovarian secretion of progesterone and 20α-hydroxypregn-4-en-3-one during rat estrous cycle in chronological relation to pituitary release of luteinizing hormone. *Endocr. jap.* **16**, 227–237.

Uchida, K., Kadowaki, M. & Miyake, T. (1969 *b*). Acute effects of various gonadotropins and other pituitary hormones on preovulatory ovarian progestin secretion in hypophysectomized rats. *Endocr. jap.* **16**, 239–249.

Vértes, M. & King, R. J. B. (1971). The mechanism of oestradiol binding in rat hypothalamus: effect of androgenization. *J. Endocr.* **51**, 271–282.

Westman, A. & Jacobsohn, D. (1938). Endoklinologische Untersuchungen an Ratten mit durchtrennten Hypophysenstiel. III. Über die luteinizierende Wirkung des Follikelhormons. *Acta Obstet. scand.* **18**, 115–123.

Yagi, K. (1970). Effects of estrogen on the unit activity of the rat hypothalamus. *J. physiol. Soc. Jap.* **32**, 692–693.

Yagi, K. (1973). Changes in firing rates of single preoptic and hypothalamic units following an intravenous administration of estrogen in the castrated female rat. *Brain Res.* **53**, in press.

Yagi, K., Azuma, T. & Matsuda, K. (1966). Neurosecretory cell: capable of conducting impulse in rats. *Science, N.Y.* **142**, 778–779.

Yagi, K., Iwasaki, S., Sawaki, Y. & Satow, Y. (1971). Electrophysiological studies on the neurosecretory neuron. *Med. J. Osaka Univ.* **21**, 75–90.

Yagi, K. & Sawaki, Y. (1970 *a*). Neural mechanism in the rat hypothalamus. *J. physiol. Soc. Jap.* **32**, 496.

Yagi, K. & Sawaki, Y. (1970 *b*). On the localization of neurosecretory cells controlling adenohypophysial function. *J. physiol. Soc. Jap.* **32**, 621–622.

Yagi, K. & Sawaki, Y. (1971). Changes in the electrical activity of the hypothalamus during sexual cycle and the effect of castration on it in the female rat. *J. physiol. Soc. Jap.* **33**, 546–547.

Yagi, K. & Sawaki, Y. (1972). Duration of responses of hypothalamic units to the intravenous estrogen in relation to its dose in the ovariectomized rat. *J. physiol. Soc. Jap.* **34**, 528–529.

Yamashita, H., Koizumi, K. & Brooks, C. McC. (1970). Electrophysiological studies of neurosecretory cells in the cat hypothalamus. *Brain Res.* **20**, 462–466.

Yoshinaga, K., Hawkins, R. A. & Stocker, J. F. (1969). Estrogen secretion by rat ovary *in vivo* during the estrous cycle and pregnancy. *Endocrinology* **85**, 103–112.

CONTROL MECHANISM OF NEURO-ENDOCRINE SYSTEM

BIOLOGICAL FACTORS CONTROLLING THE SYNTHESIS AND RELEASE OF ADENOHYPOPHYSIAL HORMONES

Kiyoshi Yamamoto

Two separate pituitary functions—synthesizing and releasing growth hormone and prolactin—were determined in the rat by a recently developed *in vitro* method. The parameters utilized in estimating these functions were $(p+m)/100$ g body weight for the synthesis function and $100 \times m/(p+m)$ for the release function, where p is the radioactivity of ^{14}C-leucine incorporated into the hormone newly synthesized and retained in the pituitary, and m the radioactivity of the hormone newly synthesized and released into the incubation medium. The alterations in these functions due to age, sex, estrous cycle, starvation, cold exposure, thyroidectomy, and thyroid hormone supplement are described. The experimental results explained, in most cases, the accumulated data on the changes in the pituitary content and plasma level of growth hormone or prolactin under various experimental conditions. In discussing the physiological significance and mechanisms of the experimental findings, it is emphasized that the synthesis function and the release function frequently change in inverse directions, indicating that these functions are regulated by more than one controlling factor in the hypothalamus. The role of growth hormone and also probably of prolactin in metabolic readjustment against altered external environment is also stressed. Growth hormone and prolactin should be regarded not merely as transient, stress-reactive factors but also as long-lasting, metabolically significant factors.

INTRODUCTION

Since the pioneer studies by Sokoloff's and Tata's schools around 1961, it has been recognized that thyroid hormones exert their biological effects, *e.g.*, stimulation of oxygen consumption, by stimulating protein synthesis in many target tissues (Yamamoto, in press). In contrast, in the anterior pituitary it has been known that thyroidectomy causes hypertrophy and stimulates oxygen consumption, suggesting depressed protein synthesis by thyroid hormones. As the mechanisms of thyroid hormone actions have been one of the major subjects in my laboratory for many years (Yamamoto, 1971), this peculiarity of the anterior pituitary led us to studies which indicated the fact that thyroidectomy stimulates and thyroid hormone injection depresses the synthesis of the whole protein in this tissue (Tonoue & Yamamoto, 1967 *a*, *b*). Since this hormone effect developed within 3 hr *in vivo* and was demonstrable even *in vitro*, it was assumed that thyroid hormones probably act directly on the pituitary at the translation step of protein synthetic processes, unlike other tissues in which the hormones mainly act at the transcription step. Our findings were soon confirmed by other workers (Lee, Bowers & Miller, 1968).

Our next interest was to isolate each of the protein hormones of the anterior pituitary and study the pituitary functions of synthesizing and releasing these hormones after thyroidectomy and thyroid hormone supplement, and further under various other biological conditions. In many studies carried out so far, incorporation of labeled amino acids into the whole protein extracted only from the pituitary was taken as a satisfactory index of the protein synthesis function, and the released portion of the labeled protein was not taken into consideration.

On the other hand, a large number of studies have appeared on the alterations in the pituitary content and the plasma level of anterior pituitary hormones, including growth hormone (GH) and prolactin, under various experimental conditions. The hormones were determined by bioassay or radioimmunoassay. However, since the pituitary hormone content is a function of the rates of hormone synthesis and release, and the plasma level a function of the rates of release and peripheral utilization of the hormone, these data could only indirectly reveal the pituitary functions of hormone synthesis and release. Therefore, the interpretations of the data have been arbitrary. For example, the decreased pituitary hormone content has been interpreted as showing either decreased synthesis or increased release, depending on the purpose of the workers. The elevated plasma hormone level has been generally accepted to indicate increased hormone release from the pituitary, though peripheral utilization is also an important determinant factor (Kwa, Feltkamp, van der Gugten & Verhofstad, 1970).

Tissue culture studies have also been utilized for investigating pituitary functions *in vitro*. When the amounts of the hormone retained in the pituitary and released into the culture medium were both increased, the data were interpreted as indicating stimulated hormone synthesis and release functions. However, the question of whether the experimental conditions favored the general activities of the pituitary cells or selectively activated the specific function of hormone synthesis or release always remains in this type of experiment.

Under these circumstances, direct and discriminative determination of the two distinct pituitary functions of hormone synthesis and release was considered worthwhile as well as interesting. Fortunately, VanderLaan's school had reported that minute amounts of GH and prolactin can easily be isolated from individual rat anterior pituitary by employing polyacrylamide gel disc electrophoresis (Lewis, Cheever & VanderLaan, 1965). This method, when combined with the isotope incorporation method, was expected to afford a useful means for determining the pituitary functions of synthesis and release of GH and prolactin. The first trial was reported by MacLeod & Abad (1968), but they did not determine the hormones released from the individual anterior pituitary. Later, a satisfactory method was developed after several preliminary studies on the necessary improvements (Yamamoto, Taylor & Cole, 1970). A series of experiments has since been carried out and is now being continued with rats under various internal or external conditions. Some of the data obtained thus far are assembled and briefly discussed below.

PRINCIPLES, PROCEDURES, AND LIMITATIONS OF THE METHOD

General principles

Individual anterior pituitaries, isolated from rats that had received various treatments or been kept under different biological conditions, were incubated *in vitro* with ^{14}C-leucine. The anterior pituitary obtained from a conditioned animal was expected to continue to function *in vitro* for several hours as it does *in vivo*. In our experiments, usually 3 hr incubation was employed.

The purpose of our study was to determine separately the two distinct functions of hormone synthesis and release. Therefore, GH and prolactin were isolated from the anterior pituitary and from the corresponding incubation medium after sufficient incorporation of labeled leucine.

In order to evaluate the pituitary functions, the following parameters were chosen and employed:

Hormone synthesis function: $(p+m)/100$ g body weight,

Hormone release function: $100 \times m/(p+m)$,

where p is the radioactivity (cpm or dpm) of the hormone newly synthesized and retained in the pituitary, and m is that of the hormone newly synthesized and released into the medium during incubation. Therefore, the term $(p+m)$ represents the sum total of the hormone newly synthesized during the incubation period. The term $100 \times m/(p+m)$ is the percent release of the newly synthesized hormone. When m and $(p+m)$ change in parallel, the percent release does not differ. This is interpreted as showing no change in the release function, whether the amount of released hormone (m) is increased or decreased.

Though this paper deals mainly with these two functional parameters, other measures such as p/rat, m/rat, $(p+m)$/rat, and $m/100$ g body weight were determined in each experiment and are reproduced in some of the following tables and figures. The following expressions are also used for convenience in the following figures: " Synthesis " for $(p+m)$ per rat or per 100 g body weight, " release " for m per rat or per 100 g body weight, and " percent release " for $100 \times m/(p+m)$.

The expressions of $(p+m)$ or $m/100$ g body weight are helpful in comparing groups of animals with different body weights. Further, these expressions are useful measures in discussing the physiological significance of the data, since GH or prolactin is synthesized and released to meet the need of peripheral tissues, and the peripheral action of the hormone is closely related to the plasma hormone level and, consequently, to the plasma volume, which parallels the body weight in most cases.

The rate of *de novo* synthesis of protein in many tissues has usually been expressed in terms of cpm or dpm of incorporated labeled amino acid per unit amount of tissue protein. This is, however, not suitable for pituitary protein hormones and may lead sometimes to erroneous conclusions, since the hormonal proteins are released more or less actively during the incubation period. The expression cpm or dpm/mg tissue has also been used frequently. With rat anterior pituitary, however, this is also unsuitable, since slight variations in the amount of fluid adhering to the small pituitary tissue cause considerable error in weighing, and attempts to completely remove the fluid might cause tissue damage, which should be avoided especially when the hormone release function is determined.

Routine procedures

Rats of known age were obtained from commercial sources and kept in our animal room usually for 1 week or more in order to accustom the animals to the new regulated environment. Animals were sacrificed by instantaneous decapitation, since this is the best sacrifice method with the least influence on the anterior pituitary (Neill, 1970). Each isolated anterior pituitary was incubated with ^{14}C-leucine as described elsewhere (Yamamoto *et al.* 1970; Ieiri, 1971). After the incubation, the anterior pituitary was taken out, washed to remove adhering isotopes, and homogenized with 0.2 ml of distilled water. To the incubation medium, a piece of rat anterior pituitary tissue (approximately 2 mg in wet weight) was added as " carrier ", and a homogenate was prepared.

Growth hormone and prolactin in each homogenate were fractionated by polyacrylamide gel disc electrophoresis (Lewis *et al.* 1965; Yamamoto *et al.* 1970). In order to locate the bands of the hormones, the gel column was placed in a 7% acetic acid solution containing 0.002% amido black, instead of the usual 1% dye. As soon as GH and prolactin bands became clearly visible (usually within 10 min), the column was taken out, and the stained hormone bands were dissected. Before liquid scintillation counting of the radioactivity of each isolated hormone, the dissected gel block was washed first with glacial acetic acid containing 0.25% phosphotungstic acid and then with a glacial acetic acid Cellosolve mixture (1: 1, v/v) after the method of Ieiri (1971).

This routine method had the following advantages: Addition of " carrier " pituitary to the incubation medium enabled fractionation of a minute amount of the labeled hormone released from a single anterior pituitary. Homogenization of the pituitary with distilled water and leaving the homogenate overnight under ice-cold conditions permitted high recovery (presumably over 80%) of pituitary hormones stored in the secretory granules, without causing any additional incorporation of labeled leucine into the hormones. Staining with a highly diluted dye solution made the hormone locations distinct without causing any notable quenching at radioactivity counting. By this simplified staining method, the time and trouble frequently met during destaining were eliminated. Thorough washing of the gel block eliminated contamination of free labeled leucine almost completely without notable loss of hormonal proteins, and, by removing strong quenchers (phosphotungstic acid and water), gave high counting efficiencies (approximately 90%).

It was essential to employ groups of animals of the same age (same days after birth) in one experiment, since the pituitary functions under study change rapidly day by day in a certain age range (Fig. 1). In many of our experiments, animals were sacrificed on a certain day between 70 and 80 days after birth.

In some experiments, the pituitary functions of synthesis and release of the whole protein were effectively and easily determined by employing acrylamide after the method of Ieiri (1971).

The results are expressed in means $\pm$ standard errors. Asterisks attached to the values indicate statistically significant differences, *, **, and *** showing differences at 5, 1 and 0.1% levels, respectively.

Limitations

The application of our method is limited to GH and prolactin at present. Different procedures are to be developed to fractionate the other pituitary hormones, the content of which is much less than these two.

By this method, the pituitary function of hormone release is determined only on labeled, newly synthesized hormones. With other protein hormones, such as insulin, it has been known that newly synthesized hormones are released more readily than stored hormones. Since newly synthesized GH and prolactin are released at similar rates to those of the stored hormones in normal rats (Ieiri *et al.* unpublished data), the following discussion assumes this parallelism. However, the parallelism is to be checked under various experimental conditions.

The present method gives direct data on the pituitary functions of hormone synthesis and release, and it is clear that these two distinct functions are the most important factors in deciding pituitary contents and also plasma levels of the hormones. As shown below, the changes in the estimated functions coincide and explain well the changes in the pituitary content and plasma level in most cases, but these two groups of parameters should not be directly correlated to each other. For example, since the amounts of stored GH and prolactin are usually large, a slight stimulation of the release function may cause a considerable loss of the stored hormones, which may sometimes exceed the amount of hormones newly synthesized due to a relatively strong stimulation of the synthesis function, and this may result in lowered pituitary contents. Similar discrepancies of various types are expected, and the changes in pituitary contents or plasma levels cannot always be judged from alterations in the anterior pituitary functions.

VARIATIONS IN PITUITARY FUNCTIONS UNDER SEVERAL BIOLOGICAL CONDITIONS

Age and sex

Employing incorporation of labeled amino acids, several studies have reported the synthesis of the whole protein or some protein hormones in the rat anterior pituitary (see Yamamoto *et al.* 1970). In these experiments, however, the whole protein or each protein

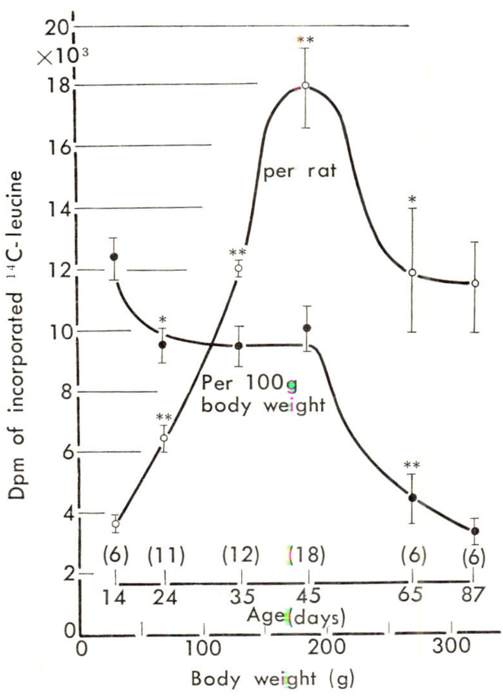

Fig. 1. GH synthesis in male rats at various ages. Dpm per 100 g body weight represents GH synthesis function. Numerals in parentheses show the number of rats used. Vertical bars indicate standard errors of the means. Asterisks indicate statistically significant differences from the foregoing value (from Yamamoto *et al.* 1970).

TABLE 1. Synthesis and release of GH in the anterior pituitaries

Group number	Days after birth	Body wt (g)	Dpm/rat	
			Retained in pituitary	Released into medium
1	30	105.3±2.8	8217±1698	398±19
2	46	228.3±6.3	14013± 643*	299±19**
3	78	303.0±3.7	9613±1082*	280±27
4	30	79.3±2.3††	5866±1861	95±31††
5	46	141.0±3.6††	6309±1068††	80±25††
6	78	223.8±1.5††	6791± 822	97±19††

Each anterior pituitary from 12 male and 12 female rats (4 per age group) was incubated tion, respectively. *, ** Statistically significant differences from the value immediately sponding age groups of different sexes at 5 and 1% levels, respectively (from Yamamoto

hormone was extracted from the pituitary usually by chemical methods, and the radioactivity was counted. As mentioned above, however, the loss of released hormonal protein into the medium should not be neglected. For instance, as much as 1/2 of labeled prolactin is lost from the pituitary during 3 hr incubation (Fig. 3). On the other hand, studies on differences due to age and sex have been lacking. Therefore, male and female rats at ages ranging from 14 to 120 days were studied, and the pituitary functions of synthesis and release of GH and prolactin were determined, taking the released parts of the hormones into consideration (Yamamoto *et al.* 1970).

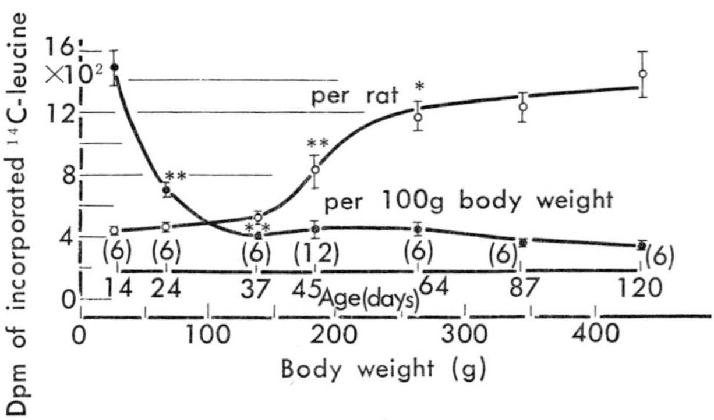

Fig. 2. Prolactin synthesis in male rats at various ages. Terms and symbols are the same as in Fig. 1 (from Yamamoto *et al.* 1970).

of male and female rats at three different ages

Total synthesized	Percent release[b]	Dpm/100 g body wt Retained in pituitary	Released into medium	Total synthesized[a]
Male				
8594±1709	4.9±0.9	7803±1613	359±16	8061±1422
14310± 647*	2.1±0.1*	6124± 282	130± 6**	6261± 136
9893±1099*	2.8±0.2*	3173± 357**	93±10*	3275± 381**
Female				
5960±1855	1.9±0.7†	7397±2347	124±43††	7521±2082
6395±1053††	1.6±0.4	4382± 757	62±16††	4444± 651†
6869± 843	1.1±0.3††	3034± 367	35±12††	3077± 394

with labeled leucine (0.1 μCi) for 2 hr. [a, b] Indicate synthesis function and release func-
above. †, †† Statistically significant differences between values obtained from corre-
et al. 1970).

Figure 1 shows that GH synthesis strikingly fluctuates in the course of aging in male rats. GH synthesis ($p+m$) per rat reaches a maximum approximately 45 days after birth. However, the GH synthesis function, *i.e.*, $(p+m)/100$ g body weight, is higher as the rat is younger. The function drops from 14 to 24 days after birth and again approximately 45 days after birth.

In contrast, as shown in Fig. 2, prolactin synthesis per rat increases continuously with age, the increase being especially rapid during the period of sexual maturation between 40 and 50 days after birth. However, the prolactin synthesis function is high 14 days after birth, decreases rapidly until about 35 days, and does not change thereafter.

Tables 1 and 2 show the differences between the males and females of 3 different age groups. Although the sex differences are not always clear with young, 30-day-old rats, GH synthesis and release functions are generally lower and prolactin synthesis and release functions markedly higher in the females. As seen in Table 1, the GH synthesis function in the females decreases earlier with age than in the males. The GH release function, *i.e.*, percent release, is markedly high in young males and decreases sharply with age. In the females, the release function is lower than in the males and decreases gradually with age.

Table 2 shows sex differences in prolactin. In contrast to the males, in which both prolactin synthesis and release functions decrease with age, in the females the prolactin synthesis function signifi-

TABLE 2. Pituitary functions of prolactin synthesis and release in male and female rats at three different ages

Group number	Days after birth	Synthesis function (dpm)	Release function (%)
		Male	
1	30	849± 45	26.0±4.0
2	46	552± 62**	16.3±0.9*
3	78	517± 85	17.8±0.5
		Female	
4	30	1243±326	23.8±0.9
5	46	2949±398*, ††	22.5±4.3
6	78	3862±305††	22.0±2.3

Experimental conditions and symbols are the same as in Table 1 (modified from Yamamoto *et al.* 1970).

cantly increases, and the prolactin release function does not decrease with age. Thus, the sex differences are greater in adolescent and adult rats.

The physiological significance of these findings would be as follows: For about 4 weeks after birth, the age period during which growth is most rapid and does not differ much between the sexes, GH synthesis function is high, and sex differences are not found, though GH release function is higher in the males corresponding to a somewhat greater body weight gain in this sex. Thereafter, the males grow at a considerably faster rate than the females. During this period, GH synthesis and release functions are higher in the males. These functional differences are obviously important in causing the larger body weight in the males. The higher GH synthesis function in the males also explains the higher pituitary GH content in the males reported previously (MacLeod, Abad & Eidson, 1969; Jones, Fisher, Lewis & VanderLaan, 1965).

In regard to prolactin, the synthesis function decreases in the males but increases drastically in the females with age. The release function also decreases with age in the males but not in the females. These facts are consistent with the clear and important role of this hormone in the females. The high functions of prolactin synthesis and release in adolescent and adult females indicate the importance of this hormone throughout the sexual life in females. The high pituitary prolactin content in the adult females (Jones *et al.* 1965) is due to the high prolactin synthesis function in this sex.

The mechanisms of the age and sex differences described here are not yet thoroughly clarified. Since estrogen injection into the rat causes a drastic increase of prolactin synthesis (MacLeod *et al.* 1969), the sex and age differences in prolactin synthesis (and probably also in prolactin release) may be caused by different levels of circulating estrogens. Whether testosterone plays a role in the males or not is yet obscure, since this hormone shows an effect similar to that of estrogens. The mechanisms of age and sex differences in GH have not yet been clarified.

Estrous cycle

The activity of the hypothalamo-hypophysio-gonadal axis changes markedly, in various aspects, during the reproductive cycles of animals and humans. Concerning prolactin in the rat, the pituitary content and plasma level of the hormone have been determined (see Ieiri, Akikusa & Yamamoto, 197). However, these data differ considerably in different reports and, further, did not provide any direct estimates of the pituitary functions of prolactin synthesis and release. Based on these considerations, the changes in synthesis and release of prolactin (and of GH) were investigated in adult female rats showing regular 4-day estrous cycles.

Each estrous stage was determined carefully employing various suitable indices. Two series of experiments were carried out under identical experimental conditions. In one series the anterior pituitary was isolated in the morning between 10: 00 and 11: 00 a.m. (Ieiri *et al.* 1971), and in the other in the afternoon between 4: 00 and 5: 00 p.m. (Ieiri, Nobunaga & Yamamoto, 1972) from rats showing typical estrous stages.

The data of the morning experiment are shown in Fig. 3 and

TABLE 3. Synthesis and release of prolactin during the estrous cycle (morning data)

Stages of estrous cycle	Synthesis (cpm/100 g body wt)[a]	Release (cpm/100 g body wt)
Proestrus (6)[b]	6918±971*	2375±502**
Estrus (5)	11721±658	5875±467
Metestrus (5)	5968±570**	1878±155**
Diestrus (7)	5691±670**	1729±222**

[a] Indicates synthesis function. [b] Number of rats. Asterisks show statistically significant differences from values at estrus (modified from Ieiri *et al.* 1971).

Table 3. As seen in Fig. 3, prolactin synthesis and release per rat are markedly high only at estrus. The prolactin synthesis function (Table 3) is also strikingly higher at estrus than at 3 other stages of

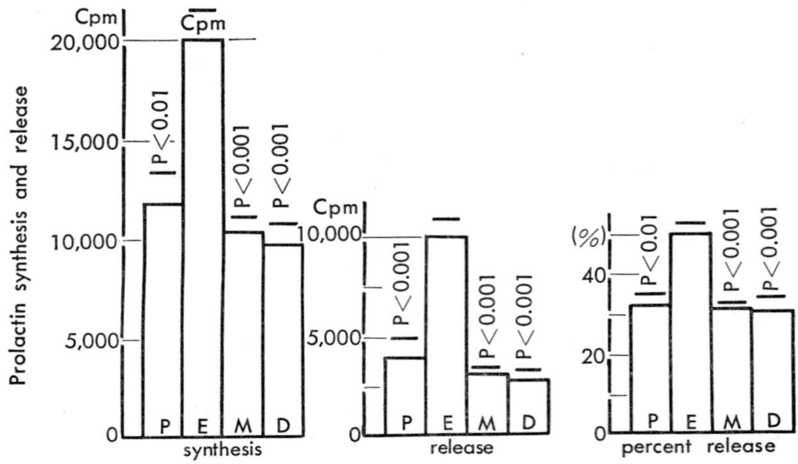

Fig. 3. Prolactin synthesis, release (cpm per rat), and percent release during the estrous cycle of the rat (morning data). The percent release represents prolactin release function. P, E, M, and D indicate proestrus, estrus, metestrus, and diestrus, respectively. The P value attached to each column shows statistical significance of the differences between the estrus stage and each of the other stages (Ieiri *et al.* 1971).

TABLE 4. Pituitary functions of prolactin synthesis and release during the estrous cycle (afternoon data)

Stages of estrous cycle	Number of rats	Body weight (g)	Synthesis function (cpm)	Release function (%)
Proestrus	6	211±6	12971±1548	44.8±0.9
Estrus	6	219±4	13932±1510	48.2±4.1
Metestrus	6	219±6	7717± 478	44.7±3.9
Diestrus	6	206±2	6428± 225	35.3±2.0
P–E	P values		NS	NS
P–M			<0.02	NS
P–D			<0.001	<0.01
E–M			<0.01	NS
E–D			<0.001	<0.02
M–D			<0.05	<0.05

P, E, M, and D are proestrus, estrus, metestrus, and diestrus, respectively. NS means nonsignificant (modified from Ieiri *et al.* 1972).

the estrous cycle. No notable differences are found among 3 groups of rats at proestrus, metestrus, and diestrus. The prolactin release function (percent release) (Fig. 3) is also remarkably high only at estrus.

The data of the afternoon experiment are shown in Table 4. The prolactin synthesis function is significantly higher both at proestrus and estrus than at metestrus and diestrus. The prolactin release function is significantly higher at proestrus, estrus, and metestrus than at diestrus.

Combining these two series of data, the changes over one estrous cycle in pituitary functions of synthesizing and releasing prolactin are illustrated in Fig. 4. These functions rise rapidly between the

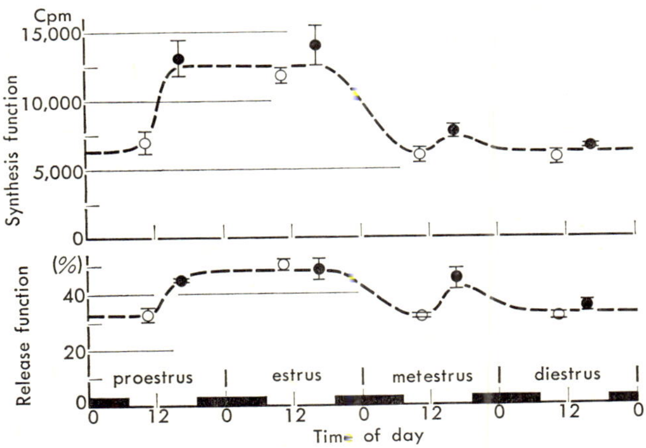

Fig. 4. Fluctuations in pituitary functions of prolactin synthesis (cpm per 100 g body weight) and release (percent release) during the estrous cycle of the rat.

morning and the afternoon of the day of proestrus, the elevated functional states continue until the evening of the day of estrus, and the functions finally decline towards the morning of the day of metestrus. Transient elevations of prolactin synthesis and release functions in the afternoon of the day of metestrus might indicate the presence of a circadian rhythm of the functions on this day.

These cyclic functional fluctuations explain, at least partly, the cyclic changes in the pituitary content and plasma level of prolactin reported before (see Ieiri *et al.* 1971). Since prolactin utilization in peripheral tissues is markedly influenced by the plasma level of estrogens (Kwa *et al.* 1970), better understanding of the changes in

TABLE 5. Pituitary functions of GH synthesis and release during the estrous cycle

Stages of estrous cycle	Synthesis function (cpm)		Release function (%)	
Proestrus				
morning	8873± 966		11.5±1.8	
afternoon		10228± 742		8.9±0.7
Estrus				
morning	10234±1559		9.3±1.3	
afternoon		8646± 486		10.3±1.0
Metestrus				
morning	6645± 620		11.6±1.0	
afternoon		8749± 369		10.9±0.8
Diestrus				
morning	9014±1291		8.0±1.6	
afternoon		8966± 577		12.2±0.8*

* Significantly different from the value at proestrus (modified from Ieiri *et al.* 1971, 1972).

TABLE 6. Pituitary functions of synthesis and release of the whole protein during the estrous cycle (afternoon data)

Stages of estrous cycle	Synthesis function (cpm)	Release function (%)
Proestrus	59716±2907	15.3±0.7*
Estrus	51511±2378	18.0±0.7
Metestrus	44145±1748*	14.5±1.1*
Diestrus	41142±1586**	11.6±0.5***

Asterisks show significant differences from the values at estrus.

the plasma prolactin level will be obtained when the cyclic changes in prolactin utilization are known.

The physiological significance of the cyclic changes in prolactin synthesis and release functions is well understood; increased synthesis and release of prolactin in the afternoon of the day of proestrus elevate the plasma level of the hormone, and the elevated hormone is utilized for maintaining the corpus luteum.

Concerning the mechanisms of stimulation of prolactin release, it is generally accepted that the elevated plasma level of estrogens in the morning of the day of proestrus acts as the trigger. In rats with 4 day cycles, plasma estradiol reaches its peak in the morning of proestrus (Brown-Grant, Exley & Naftolin, 1970). Injection or *in vitro* addition of estrogens stimulates prolactin synthesis (MacLeod *et al.* 1969). The present results indicate that estrogens stimulate both prolactin synthesis and release functions of the anterior pituitary. The action sites of estrogens would be the hypothalamus and anterior pituitary, since these tissues contain estrogen receptors (see Kato, 1970).

In contrast to prolactin, GH synthesis and release functions do not appear to be under clear cyclic fluctuation (Table 5). Only in the afternoon of diestrus is the release function significantly higher than at proestrus.

The function of the whole protein synthesis is high both at proestrus and estrus, whether the determination is carried out in the morning (Ieiri *et al.* 1971) or in the afternoon (Table 6). The high synthesis function in the morning of proestrus is ascribed to increased synthesis of FSH and LH, since the prolactin synthesis function is not elevated in this period. The whole protein release function is the highest at estrus in the afternoon experiment (Table 6).

Starvation

Retardation of animal growth due to starvation has been partly explained by the lack of GH, as was suggested by the low pituitary content and plasma level of this hormone (see Akikusa, 1971). The effects of starvation on rat pituitary prolactin have not been reported. In our laboratory, the pituitary functions of synthesis and release of GH, prolactin, and the whole protein were determined after 1, 3, and 5 days of starvation using young adult male rats of the same age (Akikusa, 1971).

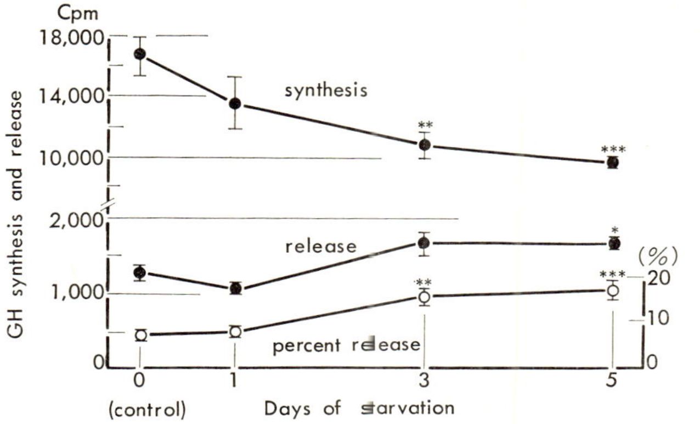

Fig. 5. Effects of starvation on GH synthesis, release (per 100 g body weight), and percent release in male rats. Synthesis and percent release represent GH synthesis and release functions, respectively (Akikusa, 1971).

Figure 5 shows the results on GH. The GH synthesis function decreases and the GH release function (percent release) increases with increasing days of starvation. Despite the decreased synthesis function, the amount of GH newly synthesized and released into the medium (release in Fig. 5) increases due to the stimulated release function.

In regard to prolactin, the synthesis function increases gradually with increasing days of starvation, but the release function (percent release) remains constant throughout the starvation period (Fig. 6). The amount of prolactin newly synthesized and released (release in Fig. 6), therefore, increases with starvation days.

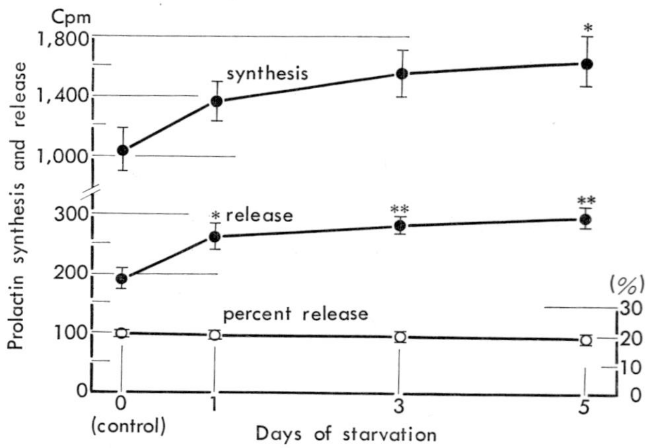

Fig. 6. Effects of starvation on prolactin synthesis, release (per 100 g body weight), and percent release in male rats. Terms in the figure are the same as in Fig. 5 (Akikusa, 1971).

The observed alterations in GH synthesis and release functions can explain the changes in pituitary GH content and plasma GH level reported before (see Akikusa, 1971). The decreased pituitary content is due to both depressed GH synthesis function and stimulated GH release function. The lowered plasma GH level would be mainly due to markedly depressed GH synthesis function, since the GH release function is stimulated. Although no data are available concerning prolactin, the pituitary content and plasma level of prolactin would increase following starvation, according to the present results.

The physiological significance of the pituitary functions altered by starvation is not well known yet. The stimulated GH release function may be of some physiological importance. This stimulated function may compensate, to some extent, for the depressed GH synthesis function which may possibly be caused by undernutrition, and may minimize the fall of the plasma GH level and consequently of the plasma glucose level. The effects of 5 day starvation on blood components are shown in Table 7.

TABLE 7. Effects of 5 day starvation on serum free fatty acids (FFA), glucose, and protein

	FFA (μEq/l)[b]	Glucose (mg/dl)	Protein (g/dl)
Control (6)[a]	328±20	113±5	6.1±0.1
Starvation (6)	848±89***	96±9**	6.0±0.2

[a] Number of animals; [b] calculated as palmitate

Concerning prolactin, the elevated serum free fatty acid (FFA) level (Table 7) may be, at least partly, ascribable to the increased prolactin synthesis, since this hormone is reported to increase FFA mobilization in the dog (Winkler, Rathgeb, Steele & Altszuler, 1971).

The mechanisms of alterations in pituitary functions by starvation are also not well understood. Although a reduced hypothalamic content of GH-releasing factor (GH-RF) has been found in starved rats (Meites & Fiel, 1965; Dickerman, Negro-Vilar & Meites, 1969), and this reduction was considered to be due to less availability of amino acids and energy needed for GH-RF synthesis, such a reduction may reflect either reduced synthesis or increased release of GH-RF. The present finding of a stimulated GH release function favors the latter assumption.

The effects of undernutrition on the anterior pituitary are different from cell type to cell type. For example, the hormone synthesis function of GH cells is depressed, but this function is stimulated in prolactin cells by starvation. The hormone release function is activated in GH cells, but such an activation does not occur in prolactin cells. Provided such is the case with the hypothalamus, it is possible that GH-RF synthesis is not influenced, but that its release is stimulated by starvation and consequently the GH release function is stimulated. The effect of GH-RF on GH synthesis has not been conclusively elucidated. The depressed GH synthesis function caused by starvation may not be the result of decreased GH-RF synthesis and/or release. The elevated serum level of FFA (Table 7) may be one of the factors of this GH synthesis depression, since *in vivo* release of GH was found to be decreased by lipid administration in the monkey (Blackard, Boylen, Hinson & Nelson, 1969).

The mechanisms of stimulation of prolactin synthesis by starvation cannot be correlated directly with decreased PIF release and/or synthesis, since the lack of change in the prolactin release function does not indicate the possibility.

The pituitary functions of whole protein synthesis and release are shown in Table 8. The synthesis function decreases only slowly,

TABLE 8. Effects of stavation on pituitary functions of synthesis and release of the whole protein[a]

Days of starvation	Number of rats	Body weight (g)		Synthesis function (cpm)	Release function (%)
		Initial	Final		
0 (control)	6	233±9	262±10	40100±2020	6.3±0.4
1	6	233±8	231± 8	37960±1950	6.0±0.5
3	6	230±3	191± 3	35190±2380	7.7±0.8
5	6	230±3	167± 3	33920±1090*	6.2±0.2

[a] Chemically extracted sample (modified from Akikusa, 1971).

and the release function does not change. The decreased synthesis may be mainly due to the decreased GH synthesis. As the release function of GH increases and that of prolactin does not change following starvation, the lack of change in the release function of the whole pituitary protein indicates that the function of releasing other pituitary hormones is decreased by starvation.

Cold exposure

The effects of various stressful stimuli, including cold exposure, on the pituitary content and the plasma level of GH and prolactin have been reported (see Yamamoto & Ieiri, 1972). In many of these studies, the effects have apparently been regarded as transient, physiologically unimportant reactions of the anterior pituitary, and therefore, have been studied only in short-term experiments of a few hours. However, based on the well-known, profound metabolic effects of GH on blood glucose and FFA, GH is expected to play a role in metabolic readjustment when animals are exposed to a cold environment. Although a similar role of prolactin is not widely known, Winkler *et al.* (1971) recently demonstrated that prolactin increases the rates of mobilization and utilization of FFA. It was, therefore, interesting to expose rats to cold of various degrees and durations and study the changes in the pituitary functions of synthesis and release of these hormones. In our study (Yamamoto & Ieiri, 1972), male adult rats were exposed to mild (6°C) or severe cold (1°C) for 2, 4, and 12 hr and 7 days.

Effects of acute cold exposure (2, 4 hr) on GH were not evident (Yamamoto & Ieiri, 1972). However, when rats are exposed to cold for up to 12 hr, the GH synthesis function is significantly decreased by exposure at both 6 and 1°C, and the GH release function (percent release) is increased at 1°C (Fig. 7). The results obtained by 7 day exposure at 6 and 1°C, shown in Fig. 8, are substantially the same as those of 12 hr exposure shown in Fig. 7.

The prolactin synthesis function is more sensitive to cold than the GH synthesis function. The function is increased markedly by only 2 hr exposure at 1°C and by 4 hr exposure at 6°C (Fig. 9).

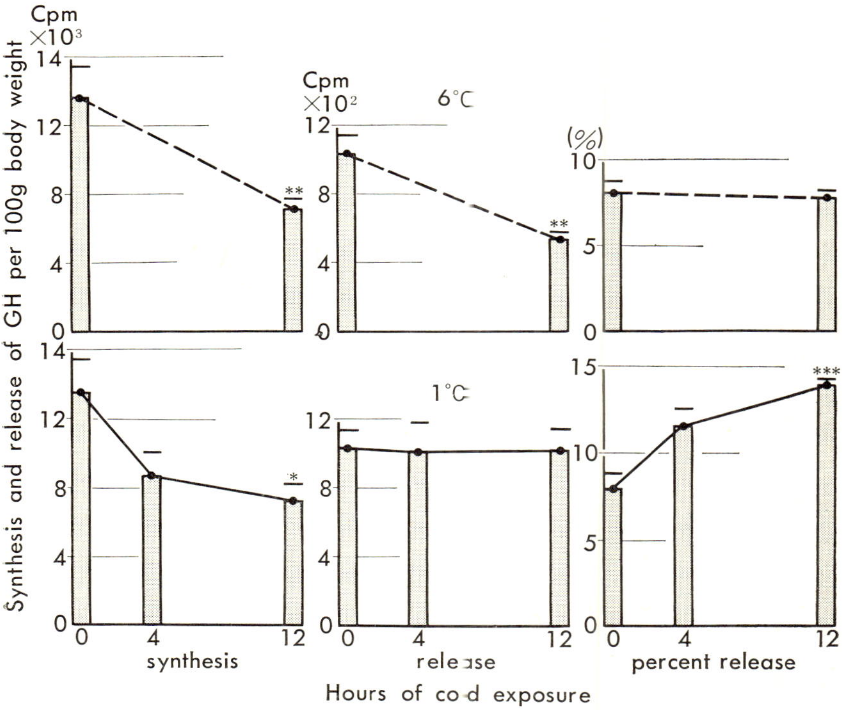

Fig. 7. GH synthesis, release (cpm per 100 g body weight) and percent release in male rats exposed to subacute cold. Synthesis and percent release represent GH synthesis and release functions, respectively (Yamamoto & Ieiri, 1972).

By subacute exposure for 12 hr, prolactin synthesis function remains high, and prolactin release function (percent release) is increased at 1°C (Fig. 10). Chronic exposure for 7 days showed a similar tendency of increased functions of prolactin synthesis and release as seen in Fig. 8.

The findings described above are consistent with the reports that rat pituitary GH content decreases by cold exposure for a few hours (see Yamamoto & Ieiri, 1972). The decrease has been ascribed to the increased release of the hormone, but thepre sent study showed that the main cause is decreased GH synthesis, which is more sensitive to cold than GH release.

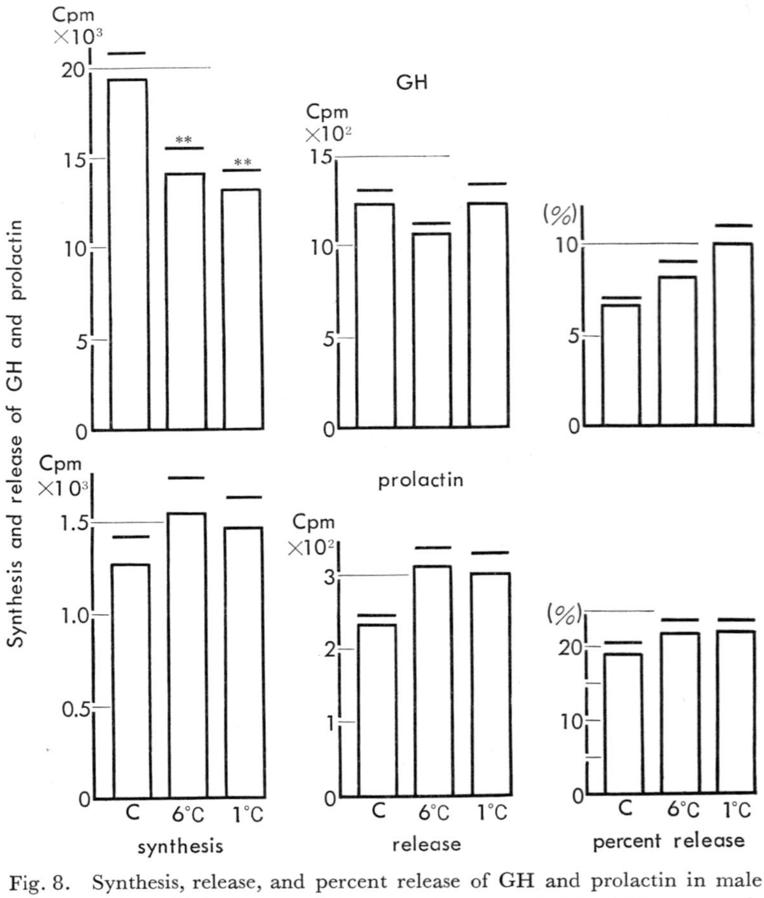

Fig. 8. Synthesis, release, and percent release of GH and prolactin in male rats exposed to cold for 7 days. Terms are the same as in Fig. 7 (Yamamoto & Ieiri, 1972).

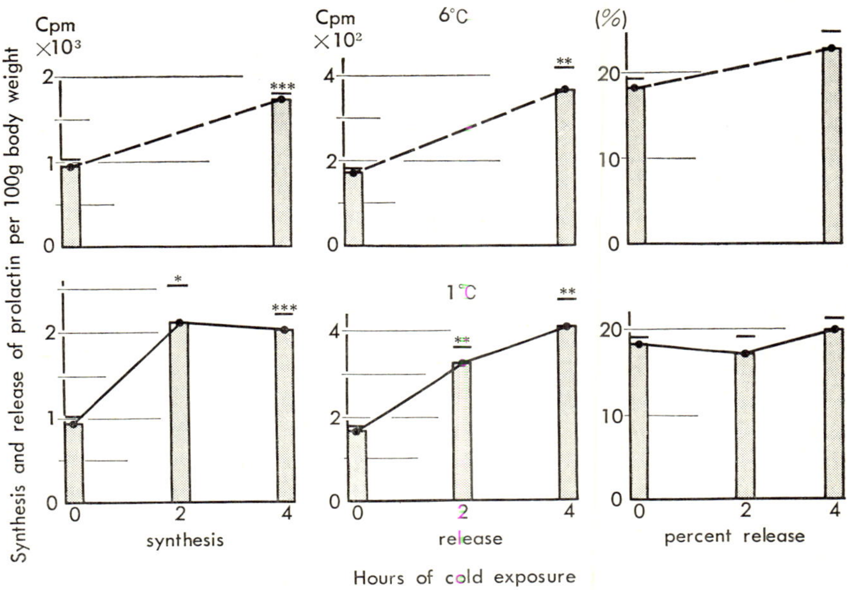

Fig. 9. Prolactin synthesis, release, and percent release in male rats exposed to acute cold. Terms are the same as in Fig. 7 (Yamamoto & Ieiri, 1972).

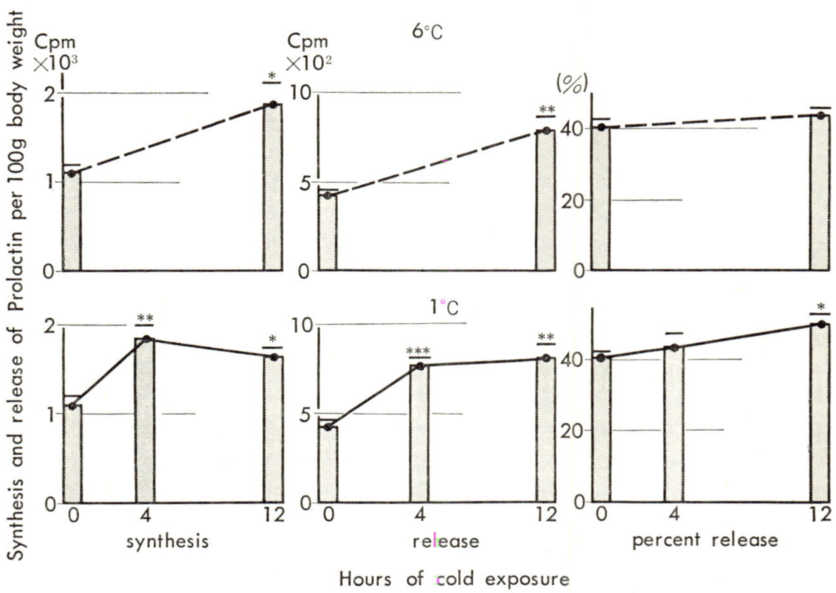

Fig. 10. Prolactin synthesis, release, and percent release in male rats exposed to subacute cold. Terms are the same as in Fig. 7 (Yamamoto & Ieiri, 1972).

Prolactin has also been reported to be influenced by some stressful stimuli (see Yamamoto & Ieiri, 1972), but the effect of cold exposure has not been reported. Based on the present results, it is presumed that the pituitary prolactin content and the plasma level would be elevated by cold exposure, since prolactin synthesis and release functions are both stimulated.

The physiological significance of the present findings would be as follows: The effects of cold exposure on the pituitary functions of synthesis and release of GH and prolactin continue as long as 7 days. This suggests strongly that the altered pituitary functions play a metabolically significant role during prolonged cold exposure. Although the GH synthesis function is depressed by cold exposure, this depression is compensated for by the stimulated release function, and probably results in no marked drop of the plasma GH level. Increased functions of prolactin synthesis and release may maintain or rather elevate the plasma prolactin level, and the prolactin may also play an important role. Table 9 shows that the levels of serum

TABLE 9. Effects of acute and subacute cold exposure on serum FFA, glucose, and protein

Experimental conditions	FFA (μEq/liter)[a]	Glucose (mg/dl)	Protein (g/dl)
Controls at 25°C (6)[b]	600±41	119±5	6.1±0.2
4 hr at 1°C (6)	900±49***	129±4	6.7±0.1
12 hr at 1°C (6)	797±29**	132±9	6.6±0.3
12 hr at 6°C (6)	603±32	123±5	6.4±0.1

[a] Calculated as palmitate; [b] number of animals (Yamamoto & Ieiri, 1972).

FFA, glucose, and protein are well maintained or elevated by acute or subacute cold exposure. The levels were found within the normal range even after 7-day exposure to severe cold (Yamamoto & Ieiri, 1972). These facts suggest strongly the roles of GH and prolactin in metabolic readjustment during cold exposure. Further, it is interesting to consider that polyphagia observed in cold-exposed rats (Yamamoto & Ieiri, 1972) might be correlated with stimulated prolactin synthesis and release, since, in migrating birds, prolactin markedly stimulates food intake (Farner, King & Stetson, 1969). Although further detailed experiments are needed to substantiate these assumptions, GH and prolactin should be regarded not merely as transient, stress-reactive factors but also as long-term factors acting on plasma components during cold exposure.

In order to discuss the detailed mechanisms involved in the present findings, the data available are still insufficient. GH-RF and PIF would mediate the action of cold stimulus on the pituitary functions. There are two ways by which cold stimulus could affect these hypothalamic factors. One is a direct neural influence and the other an indirect influence from altered plasma levels of blood glucose and FFA, well-known regulators of GH release from the anterior pituitary. Since blood glucose remains at the normal level throughout the cold exposure period, it is unlikely that the GH release function is altered by a change in glucose level. The elevated level of FFA was observed in the course of cold exposure, but the changes in pituitary functions of GH synthesis and release did not parallel the changes in the plasma FFA level. Therefore, it is tentatively inferred that cold stimulus may act on the hypothalamus through neural mechanisms to alter GH-RF and PIF and, consequently, GH (and prolactin) synthesis and release.

On the other hand, however, the present results showed that GH synthesis and release functions do not change in parallel: Exposure to mild cold for 12 hr depresses GH synthesis but does not alter its release; severe cold exposure for 12 hr depresses GH synthesis but stimulates its release; cold exposure for 7 days depresses GH synthesis but tends to stimulate its release. Such nonparallel or inverse changes make it improbable that the pituitary functions of GH synthesis and release are regulated only by one factor, the GH-RF. Since GH-RF surely stimulates GH release, GH synthesis may be independent of GH-RF or be more strongly regulated by a factor(s) other than GH-RF. TSH-releasing hormone (TRH) is interesting

TABLE 10. Effects of cold exposure on pituitary functions of synthesis and release of the whole protein (modified from Yamamoto & Ieiri, 1972)

Experimental conditions	Synthesis function (cpm)	Release function (%)
Controls at 25°C	35486±2042	11.7±0.6
4 hr at 1°C	31730±2210	11.4±0.5
12 hr at 1°C	30721±1131*	11.3±0.3
12 hr at 6°C	29349±1264*	11.1±0.5
Controls at 25°C	28686±1998	9.2±0.8
7 days at 1°C	22065±1189*	11.3±1.2
7 days at 6°C	23715±1365	9.5±0.8

in this connection, since it was reported that TRH depresses GH synthesis and stimulates prolactin synthesis (Tashjian, Barowsky & Jensen, 1971). Cold exposure may act on the hypothalamus, stimulate TRH synthesis and/or release, and consequently depress GH synthesis and stimulate prolactin synthesis.

The pituitary functions of synthesis and release of the whole protein are listed in Table 10. The synthesis function decreases significantly after 12-hr cold exposure at both 6 and 1°C. The release function does not change. Chronic cold exposure for 7 days showed similar effects. The decreased synthesis function may be a reflection of the decreased GH synthesis function.

Thyroidectomy and thyroid hormone supplement

Thyroidectomy increases incorporation of labeled amino acids into the whole pituitary protein, and thyroid hormone suppresses the increase (Tonoue & Yamamoto, 1967*a*, *b*, 1968*a*; Lee *et al.* 1968; Matsuzaki, 1970). Concerning GH, the pituitary content drastically decreases in hypothyroid states, and the decrease is normalized by thyroid hormone administration. The changes in pituitary prolactin content so far reported have been inconsistent (see Ieiri, 1971). Recent studies in our laboratory strongly suggested that GH synthesis function is markedly decreased by thyroidectomy and returns to normal after thyroxine injection (Suzuki & Shibasaki, 1970;

TABLE 11. Effects of thyroidectomy on pituitary functions of synthesis and release of GH and prolactin

Experiment number		Body weight (g) Initial	Final	Synthesis function (cpm)	Release function (%)
				GH	
1	Control	152±6	207±6	11581±1908	9.2±0.9
	Thyrex	151±3	180±5	7728± 673	13.0±1.2*
2	Control	222±3	272±5	11079± 719	13.6±1.2
	Thyrex	226±9	248±8	8740± 513*	14.3±0.9
				Prolactin	
1	Control	152±6	207±6	673± 132	19.0±3.0
	Thyrex	151±3	180±5	1204± 95***	11.1±0.5*
2	Control	222±3	272±5	1025± 108	17.0±0.8
	Thyrex	272±9	248±8	1420± 106*	11.2±0.8***

Six rats comprised each experimental group. Thyrex means thyroidectomy (modified from Ieiri, 1971).

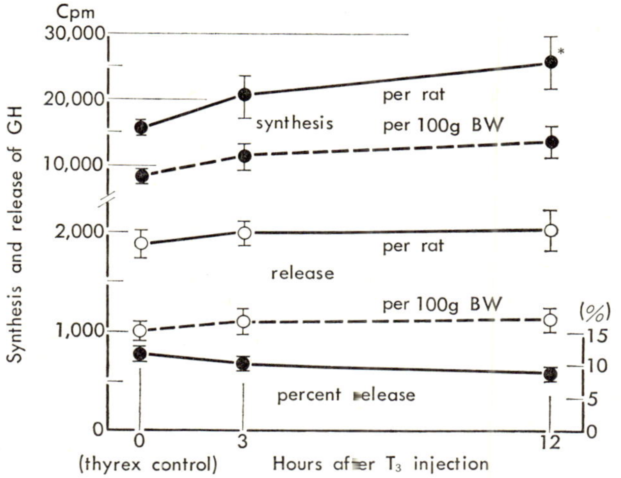

Fig. 11. Effects of a single injection of T_3 (20 μg per 100 g body weight) on GH synthesis, release, and percent release in thyroidectomized male rats. Synthesis per 100 g body weight and percent release represent synthesis and release functions, respectively. Values in thyroidectomized rats (0 time control) were obtained 12 days after the operation (Ieiri, 1971).

Matsuzaki, 1970). In the present study, the effects of thyroidectomy and triiodothyronine (T_3) administration on the pituitary functions of synthesis and release of GH, prolactin and the whole protein were investigated, using young adult male rats of the same age (Ieiri, 1971).

Table 11 shows that the GH synthesis function is decreased, and the GH release function increased by thyroidectomy. Figure 11 illustrates the effects of T_3 injection on thyroidectomized rats. The GH synthesis function, decreased by thyroidectomy, increases gradually after T_3 injection. The GH release function (percent release), increased by thyroidectomy, decreases slowly after T_3 injection.

The effects of thyroidectomy on the pituitary functions of prolactin synthesis and release are also shown in Table 11. In contrast to GH, the prolactin synthesis function is increased, and the prolactin release function decreased by thyroidectomy. The increased prolactin synthesis function decreases within 3 hr after T_3 injection, but the T_3 effect on the prolactin release function is not consistent in repeated experiments (Ieiri, 1971).

The effects of thyroidectomy, *i.e.*, depressed GH synthesis function and stimulated GH release function, readily explain the marked decrease of the pituitary GH content after thyroidectomy previously reported (see Ieiri, 1971). The increased GH synthesis function and the decreased release function by T_3 injection explain well the increasing pituitary GH content after thyroid hormone replacement. The changes in the pituitary prolactin content caused by thyroidectomy or thyroid hormone replacement have been different from report to report, many reasons being considered (Ieiri, 1971). The present results suggest that the content would be increased by thyroidectomy.

The physiological significance of the present findings is, to some extent, clear with GH, but not with prolactin. Retardation of growth caused by thyroidectomy can be ascribed to the lack of thyroid hormones and the markedly reduced GH synthesis, although the stimulated GH release function may more or less compensate for the decreased GH supply due to depressed synthesis.

In regard to the mechanisms of thyroid hormone actions, the following is of some interest: The effects of injected thyroid hormones on GH and prolactin cells develop early, in some cases within 3 hr. In contrast, a stimulation of protein synthesis in tissues other than the anterior pituitary is evident usually after 36 hr, and this delayed effect was considered to indicate that the hormones mainly accelerate the transcription step of the protein synthesis in these tissues (Yamamoto, in press). In the anterior pituitary, thyroid hormones appear to accelerate the translation step. However, since *in vitro* effects of T_3 are not observable on GH and prolactin synthesis (Ieiri, 1971), it is likely that thyroid hormones do not act directly on GH or prolactin cells but act on the hypothalamus, and hypothalamic factors (GH-RF, PIF) may influence pituitary protein synthesis at the translation step. Other than GH-RF or PIF, TRH may be an important factor in regulation of GH or prolactin synthesis, since this hypothalamic hormone has been reported to depress GH synthesis and stimulate prolactin synthesis *in vitro* (Tashjian *et al.* 1971).

Figure 12 summarizes the effects of thyroidectomy and T_3 injection on the pituitary functions of whole protein synthesis and release. The whole protein synthesis increases after thyroidectomy and returns to normal after T_3 injection. The release function (percent release) does not change notably after thyroidectomy, but decreases after T_3 injection. The stimulated synthesis function after thyroidectomy may be mainly due to increased TSH synthesis (Tonoue & Yamamoto, 1968*b*). Nonhormonal proteins may also contribute, since pituitary hypertrophy occurs. The lack of change in the protein release function after thyroidectomy may be the result of combined

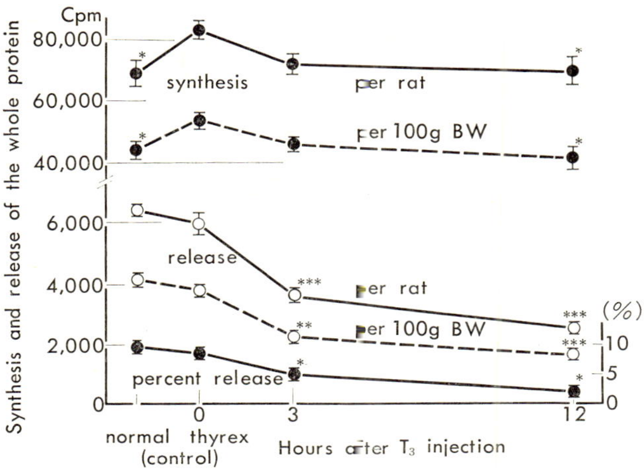

Fig. 12. Effects of thyroidectomy and T_3 injection on synthesis, release, and percent release of the whole pituitary protein in male rats. Experimental conditions and terms are the same as in Fig. 11 (Ieiri, 1971).

TABLE 12. Effects of T_3 *in vitro* on pituitary functions of synthesis and release of the whole protein

Experimental animal	Days after birth	Body weight (g)	T_3 concentration (M)	Synthesis function (cpm)	Release function (%)
Normal males	75	285±6	0	23682± 782	12.7±1.1
		284±6	10^{-7}	25070±1587	10.1±0.6
		281±6	10^{-6}	20070±1881	8.4±1.2*
		280±7	10^{-5}	23266±2376	6.6±1.0*
Thyroidectomized males	75	230±10	0	26832±2377	8.1±0.6
		230±9	10^{-7}	32421±2664	5.3±0.5**
		231±9	10^{-6}	29014±1382	4.9±0.3***
		231±9	10^{-5}	29605±2334	4.3±0.2***
Normal males	33	82±2	0	73203±5020	11.1±0.8
		83±2	10^{-7}	68115±5815	9.0±0.4*
		82±2	10^{-6}	73132±3889	6.4±0.3***
		84±1	10^{-5}	64919±4082	5.7±0.4***
Normal females	33	68±1	0	72146±4165	13.6±0.9
		67±1	10^{-7}	70163±3483	9.0±1.0**
		68±2	10^{-6}	67896±3606	8.0±0.4***
		68±2	10^{-5}	70540±5637	7.4±0.5***

Six rats comprised each experimental group (modified from Ieiri, 1971).

effects on release of various protein hormones. The decreased release function caused by T_3 injection may be due mainly to the decreased TSH release function.

In vitro addition of T_3 produces no effects on the whole protein synthesis function but a significant depression of the protein release function (Table 12). These facts support, on the one hand, the above-mentioned assumption that thyroid hormones act only indirectly on pituitary protein synthesis, and strongly suggest, on the other hand, that thyroid hormones affect directly the anterior pituitary to depress TSH release, though such direct effects are not observable on GH or prolactin release (Ieiri, 1971).

SUMMARY AND COMMENTS

Effects of some biological factors on the anterior pituitary functions of GH and prolactin synthesis and release are summarized in Table 13. Surveying this table, it is clear that under many biological conditions the hormone synthesis and release functions do not change in parallel. Such nonparallel or inverse changes are particularly evident with GH. The GH synthesis function is decreased, and the release function increased by starvation, cold exposure, and thyroidectomy. Thyroid hormone causes the synthesis function to in-

TABLE 13. Modifications of anterior pituitary functions under various biological conditions

Biological conditions	Anterior pituitary function			
	GH synthesis	GH release	Prolactin synthesis	Prolactin release
Sex				
male	Active	Low	Less active	High
female	Less active	Lower	Active	Higher
Aging				
male	Decrease sharply		Decrease	
female	Decrease gradually		Increase	No change
Estrous cycle (female)	No change		High in proestrus afternoon and on estrus day	
Starvation (male)	Decrease	Increase	Increase	No change
Cold exposure (male)	Decrease	Increase	Increase	Increase
Thyroidectomy (male)	Decrease	Increase	Increase	Decrease
Thyroid hormone (male)	Increase	Decrease	Decrease	?

crease and the release function to decrease. As stated before, these findings indicate that the pituitary functions of GH synthesis and release are separately regulated by more than one hypothalamic factor of GH-RF. These biological conditions may influence GH-RF and other factor(s) at the same time and alter the two functions in inverse directions. The interesting role of TRH was mentioned in regulating GH and prolactin synthesis during cold exposure, after thyroidectomy and thyroid hormone replacement. The role of other possible hypothalamic factors, such as the GH-inhibiting factor (Krulich, Dhariwal & McCann, 1968) and prolactin releasing factor (Mishkinsky, Khazen & Sulman, 1968), is also interesting in this respect, but further studies under various biological conditions are necessary.

It is also clear from Table 13 that the GH and prolactin synthesis functions move in inverse directions: Under conditions in which the GH synthesis function is active or increased, the prolactin synthesis function is less active or decreased, and vice versa. Further, as mentioned above, a similar relationship probably exists between the pituitary functions of GH and TSH synthesis. These facts are apparently in accord with the classic shift hypothesis. It is likely that under conditions such as starvation, cold exposure, and thyroidectomy, or in female animals, the GH synthesis function is depressed and pituitary amino acids become more readily available for the synthesis of TSH, prolactin, or other protein hormones needed in such circumstances. Since GH is the most abundant and its synthesis is the most active among other pituitary hormones, the depression of GH synthesis would particularly favor the conversion of amino acids to the synthesis of other pituitary hormones.

The metabolic role of prolactin, the synthesis of which is increased during starvation, cold exposure, or thyroidectomy, has not been demonstrated so well as that of GH, which has been extensively studied. If prolactin plays a role in metabolic readjustment against these unfavorable conditions, the increased prolactin synthesis would be a change that meets the physiological need. Our experimental results strongly suggest this. Many important metabolic actions of prolactin are already widely known in lower vertebrates and birds. In mammals, studies are still few and further investigations are needed.

The release rate (percent release) of prolactin is very high, sometimes reaching 50% during 3 hr incubation and usually is approximately 10 times that of GH. However, the prolactin release is not a spontaneous process, since under low temperature or anaerobic conditions the release ceases (Matsuzaki & Yamamoto, unpublished data). Therefore, the prolactin release *per se* is an active biological process. It is interesting that the prolactin release mechanisms that are highly active *in vitro* are inhibited by PIF *in vivo*, and GH release mechanisms that are rather inactive *in vitro* are stimulated by GH-RF *in vivo*.

ACKNOWLEDGEMENTS

The author wishes to express his hearty appreciation to his coworkers, and to Mrs. Ieiri, and Mr. and Mrs. Kakegawa for their technical assistance. This work was supported in part by the Ministry of Education, Japan, and the Cancer Association of Greater New Orleans, U.S.A.

REFERENCES

Akikusa, Y. (1971). Effect of starvation on synthesis and release of growth hormone and prolactin in the rat anterior pituitary. *Endocr. Jap.* **18**, 411–416.

Blackard, W. G., Boylen, C. T. ,Hinson, T. C. & Nelson, N. C. (1969). Effect of lipid and ketone infusions on insulin-induced growth hormone elevation in rhesus monkeys. *Endocrinology* **85**, 1180–1185.

Brown-Grant, K., Exley, D. & Naftolin, F. (1970). Peripheral plasma oestradiol and luteinizing hormone concentrations during the oestrous cycle of the rat. *J. Endocr.* **48**, 295–296.

Dickerman, E., Negro-Vilar, A. & Meites, J. (1969). Effects of starvation on plasma GH activity, pituitary GH and GH–RF levels in the rat. *Endocrinology* **84**, 814–819.

Farner, D. S., King, J. R. & Stetson, M. H. (1969). The control of fat metabolism in migrating birds. In *Progress in Endocrinology*, ed. Gual, C., pp. 152–157. Amsterdam: Excerpta Medica Found.

Ieiri, T. (1971). Effects of thyroidectomy and triiodothyronine (T_3) on the synthesis and release of growth hormone (GH) and prolactin. *Jap. J. Physiol.* **21**, 551–562.

Ieiri, T., Akikusa, Y. & Yamamoto, K. (1971). Synthesis and release of prolactin and growth hormone *in vitro* during the estrous cycle of the rat. *Endocrinology* **89**, 1533–1537.

Ieiri, T., Nobunaga, T. & Yamamoto, K. (1972). Fluctuations in the synthesis and release of prolactin by the anterior pituitary of the rat during the oestrous cycle. *J. Endocr.* **55**, 451–452.

Jones, A. E., Fisher, J. N., Lewis, U. J. & VanderLaan, W. P. (1965). Electrophoretic comparison of pituitary glands from male and female rats. *Endocrinology* **76**, 578–583.

Kato, J. (1970). Estrogen receptors in the hypothalamus and hypophysis in relation to reproduction. Excerpta Medica Int. Congr. Ser. No. 219, 764–773.

Krulich, L., Dhariwal, A. P. S. & McCann, S. M. (1968). Stimulatory and inhibitory effects of purified hypothalamic extracts on growth hormone release from rat pituitary *in vitro*. *Endocrinology* **83**, 783–790.

Kwa, H. G., Feltkamp, C.A., van der Gugten, A. A. & Verhofstad, F. (1970). The rate of elimination of prolactin as a determinant factor for plasma levels assayed in rats. *J. Endocr.* **48**, 299–300.

Lee, K. L., Bowers, C. Y. & Miller, O N. (1968). Some studies of the effect of thyroid hormone on the metabolism of the anterior pituitary gland of rats. *Endocrinology* **83**, 754–762.

Lewis, U. J., Cheever, E. V. & VanderLaan, W. P. (1965). Alterations of the proteins of the pituitary gland of the rat by estradiol and cortisol. *Endocrinology* **76**, 362–368.

MacLeod, R. M. & Abad, A. (1968). On the control of prolaction and growth hormone synthesis in rat pituitary glands. *Endocrinology* **83**, 799–806.

MacLeod, R. M., Abad, A. & Eidson, L. L. (1969). *In vivo* effects of sex hormones on the *in vitro* synthesis of prolactin and growth hormone in normal and pituitary tumor-bearing rats. *Endocrinology* **84**, 1475–1483.

Matsuzaki, S. (1970). Effects of graded doses of L-thyroxine on ^{14}C-leucine incorporation into anterior pituitary proteins of thyroidectomized rats. *Endocr. jap.* **17**, 379–385.

Meites, J. & Fiel, N. J. (1965). Effect of starvation on hypothalamic content of " somatotropin releasing factor " and pituitary growth hormone content. *Endocrinology* **77**, 455–460.

Mishkinsky, J., Khazen, K. & Sulman F. G. (1968). Prolactin-releasing activity of the hypothalamus in post-partum rats. *Endocrinology* **82**, 611–613.

Neill, J. D. (1970). Effect of " stress " on serum prolactin and luteinizing hormone levels during the estrous cycle of the rat. *Endocrinology* **87**, 1192–1197.

Suzuki, M. & Shibasaki, K. (1970). Early effect of thyroxine on synthesis of growth hormone and prolactin in the adenohypophysis of thyroidectomized rat. *Endocrinol. Exp.* **4**, 187–199.

Tashijian, A. H., Jr., Barowsky, N. J. & Jensen, D. K. (1971). Thyrotropin releasing hormone : Direct evidence for stimulation of prolactin production by pituitary cells in culture. *Biochem. biophys. Res. Commun.* **43**, 516–523.

Tonoue, T. & Yamamoto, K. (1967*a*). Amino acid uptake and incorporation into rat pituitary as affected by thyroidectomy and thyroxine administration. *Biochem. biophys. Res. Commun.* **26**, 315–319.

Tonoue, T. & Yamamoto, K. (1967*b*). Effect of thyroidectomy and a thyroxine supplement on amino acid incorporation into proteins of rat anterior pituitary. *Endocrinology* **81**, 1029–1032.

Tonoue, T. & Yamamoto, K. (1968*a*). Inhibition by thyroxine of protein synthesis in the adenohypophysis of the rat. *Gunma Symp. Endocrinol.* **5**, 185–199.

Tonoue, T. & Yamamoto, K. (1968*b*). Effect of thyroxine on L-alanine incorporation into proteins of the secretory granules of rat anterior pituitary. *Jap. J. Physiol.* **18**, 481–488.

Winkler, B., Rathgeb, I., Steele, R. & Altszuler, N. (1971). Effect of ovine prolactin administration on free fatty acid metabolism in the normal dog. *Endocrinology* **88**, 1349–1352.

Yamamoto, K. (1971). Eighteen years of our thyroid research (in Japanese). *J. Physiol. Soc. Japan* **33**, 153–168.

Yamamoto, K. Thyroid hormones in regulation of protein biosynthesis. In *Methodicum Chimicum*, vol. 11, ed. Goto, M. Stuttgart: Georg Thieme (in press).

Yamamoto, K. & Ieiri, T. (1972). Effects of cold exposure on the synthesis and release of growth hormone and prolactin. *Jap. J. Physiol.* **22**, 491–503.

Yamamoto, K., Taylor, L. M. & Cole, F. E. (1970). Synthesis and release of GH and prolactin from the rat anterior pituitary *in vitro* as functions of age and sex. *Endocrinology* **87**, 21–26.

NEUROENDOCRINE CONTROL SYSTEMS AND SYSTEMS ANALYSES OF THE BRAIN-HYPOPHYSEAL-GONADAL MECHANISM

Shojiro Inoué and Takashi Sekiguchi

Characteristics of the nervous and endocrine systems are described with reference to their evolutionary aspects and complementary correlations. On the basis of patterns in communication among organ components, the kinds of organization of neuroendocrine and endocrine control systems in vertebrates are classified into seven types, and some examples are given. The significance of the systems approach to complex neuroendocrine regulatory mechanisms is discussed and attempts at mathematical modeling and computer simulation are briefly reviewed.

Our systems approach to the sex cycle control mechanism is dealt with. Each component in the system—the brain, which is composed of the neural, neurosecretory, and store subsystems; the hypophyseal portal system; the hypophysis; the general circulation system; the gonad; and the reproductive organs—is mathematically formulated into first-order differential equations. Time-course changes in information quantity are theoretically analyzed for the case of the 4-day estrous cycle in female rats. Two sets of digital computer simulations are shown for hormone secretions and ovarian structural changes.

Experimental approaches to controlling blood levels of hormones by means of continuous infusion yielded oscillatory responses of uterine weights to stepwise estradiol inputs in ovariectomized rats. Hence the uterus is assumed to be a fifth-order oscillatory system, and four different types of mathematical models are formulated. Analog computer simulation exhibits a good correspondence to the obtained results.

SYSTEMS APPROACHES TO NEUROENDOCRINE SYSTEMS

The nervous and endocrine systems and their correlation

Two different organ systems exist in metazoans for the integration of higher bodily functions, the nervous system and the endocrine system. These two control systems have evolved in the course of development of the organization, exhibiting some characteristics complementary to each other. Here the outline of their characteristics and correlation will be briefly dealt with.

The nervous system, observable in almost all multicellular members of the animal kingdom, even in such lower organisms as coelenterates, plays an essential role in the control of living processes to adapt an individual to changes in both internal and external environments. In accordance with phylogenetic development, the nervous system becomes more and more complicated and centralized. The central nervous system, composed of the brain and the spinal cord, is well established in vertebrates. It is specialized cells called neurons that make up the nervous system. They are connected with other neurons and cells of receptors or effectors by elongated nerve fibers. Roughly speaking, this may be called a " series connection." Accordingly, the nervous system is suited to transmitting information to and from specially restricted regions of the body directly through the fiber lines.

Originally neurons seem to have both excitatory and secretory natures. The majority of neurons have developed the former nature. They can generate electric spikes as a result of their excitation and very rapidly transmit a signal through their nerve fibers to other cells. Since a neuron generates one and the same kind of spike, neural information is encoded in time-series combinations of spikes and sequentially transmitted to effectors connected in series. The secretory nature of these neurons is not remarkable, although recognizable by the fact that neurotransmitters are liberated from their fiber endings.

Other neurons have become secretory and they are called neurosecretory cells and distinguished from the above excitatory nerve cells. They produce various kinds of specific information-carriers called neurohormones. Such chemical messengers are liberated from fiber endings into the body fluid and exert specific effects on other cells. The excitatory nature in neurosecretory cells still exists but is secondary. Since neurohormones are distributed humorally to effectors, considerably wide regions of effector cells can receive information equally and simultaneously, although it takes slightly longer. Therefore, this type of connection exhibits parallel in addition to series characteristics. Thus, the communication system of neurosecretory cells may be regarded as a " series-parallel connection." Sometimes these secretory neurons are distinguished from the nervous system, which is then solely composed of excitatory neurons, by the designation of the neurosecretory system (Kobayashi, 1968). The neurosecretory system shows intermediary characteristics between the nervous system in the narrow sense and the endocrine system, discussed below.

In invertebrates, which have relatively simple organizations and small body sizes, these two kinds of neurons mainly carry out the regulatory function. However, when the organization becomes much more complex and the body size much larger, expanded organ systems of a higher level are required to maintain and develop physiological functions. The nervous system itself develops further, for instance acquiring the ability of saltatory conduction, but it is now insufficient to maintain control of all the subtle regulatory processes. Furthermore, if a communication system were constructed only from " wire-relaying " nerve fibers, a vast quantity of materials might be necessary.

Thus a second control system appears, the endocrine system. Endocrine organs, which exclusively produce and liberate a number of specific chemical information-carriers, namely hormones, into the body fluid, are very well developed in vertebrates, in contrast to their poor state in invertebrates. It may be safely said that the participation of the endocrine control system in addition to the nervous system has allowed higher animals their higher state.

The endocrine system has developed in concert with the development of the circulatory system, distributed in a network throughout the body. Thus endocrine communication systems are combined in " parallel connection " with whole tissues by means of the body fluid, being suited to controlling all regions of the body homogenously as well as simultaneously. Since hormones are effective in extremely slight quantity (usually $10^{-6}-10^{-10}$ g/ml), many different kinds of information can be transmitted at one time through the same circulatory route, without the need of a separate transmission line. Each hormone possesses a specific chemical composition, which eliminates the contamination of information. Moreover, owing to differences in the sensitivity of tissues, restricted regions of the body may be controlled selectively. Although hormonal information is accompanied by a considerable time delay in comparison with rapid neural information, it is characterized by continuing effectiveness.

The nervous and endocrine systems are not always independent of each other. Hormonal information can be transformed into neural information, and neural spikes into chemical signals. Furthermore, hormones may modify the excitability of neurons, and in turn the secretion of hormones may sometimes be controlled by neural inputs to endocrine organs. A locus of neuroendocrine transducers is known to exist in the brain, important parts being assigned to hypothalamic neurosecretory cells. The pineal gland and the adrenal medulla are also similar neuroendocrine transducers. Thus in higher organisms, close complementary relations between the nervous and the endocrine systems form the neuroendocrine control system, enabling the establishment of higher regulatory performance.

Patterns in neuroendocrine control systems

Three factors may modify the activity of the nervous and endocrine systems. The first factor is endogenous information delivered by genes. This is especially important in ontogenetic development but is not dealt with here. The second factor is physico-chemical states of internal and external environments, such as changes in blood compositions of nutrients or metabolites or osmotic pressure and temperature inside or outside the body. The third factor is encoded information coming from either the nervous or the endocrine system, namely neural spikes, neurohormones, and hormones. In vertebrates, seven types of organizations can be distinguished in the neuroendocrine and endocrine control systems as schematically shown in Fig. 1.

Type I is the simplest system, in which an endocrine organ reacts to changes in blood composition and controls by itself the secretion of hormones. Examples are the control systems of parathormone (PTH) and thyrocalcitonin. If blood concentrations of phosphorus and calcium ions become lower, the parathyroid gland detects the state and releases PTH into the general circulation system, which stimulates the resorption and liberation of these ions from the intestine and bone, respectively, to recover a normal blood level. The recovery stops the secretion of PTH. Meanwhile, if concentrations of these ions are raised by PTH in the blood, the thyroid gland detects the state and releases thyrocalcitonin into the blood, thus exerting an antagonistic action on the effectors. Since no other control component participates in these actions, type I can be called an autoregulation system.

Type II is represented by the placenta and the ovary at the time of pregnancy. Chorionic gonadotropin secreted by the placenta reaches the ovary via the circulatory route to cause the secretion of

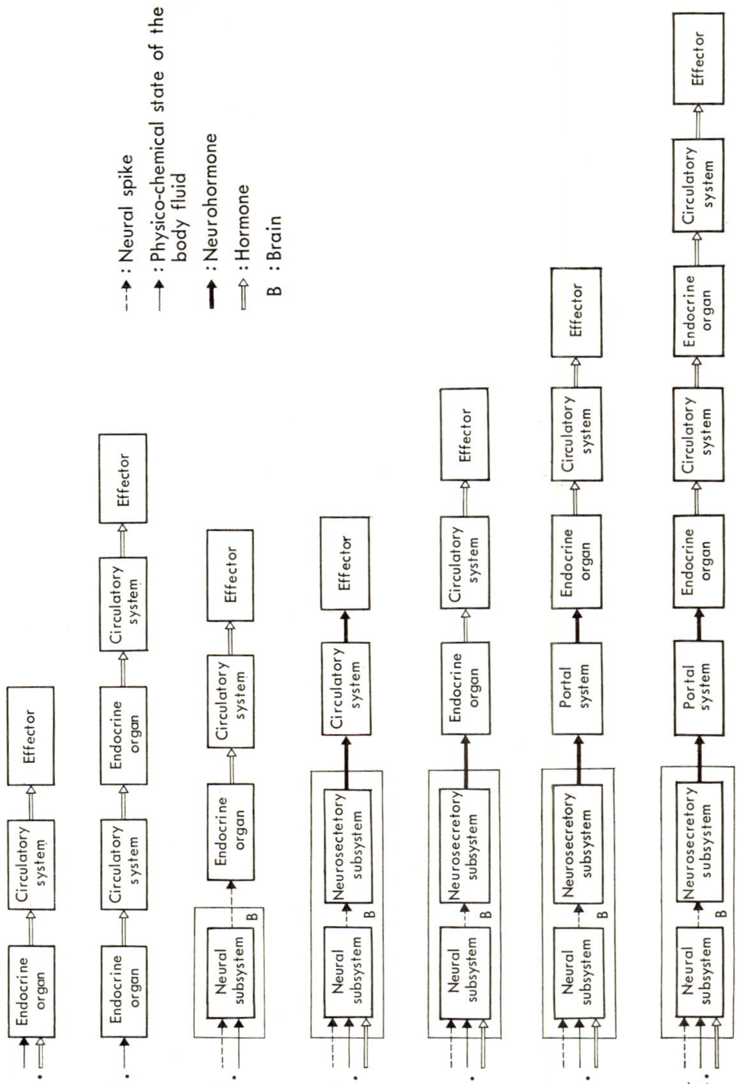

Fig. 1. Fundamental organizations in the neuroendocrine control system (from Inoué & Sekiguchi, 1971).

female sex hormones into the blood. Sex hormones then stimulate the reproductive organs and the mammary gland. Similarly, the ovarian-hypophyseal system may correspond to this type. Estrogen from the ovary influences the hypophysis directly to provoke the secretion of prolactin, which then stimulates the mammary gland.

In type III, neuroendocrine correlations are established. Control systems of the adrenal medulla and the pineal gland correspond to this type. For example, information about changes in internal and/or external environments are processed in the brain first and then transmitted through neural routes to the adrenal medulla to induce the release of adrenalin. In such a case, adrenalin acts not as a neurotransmitter nor as a tissue hormone, influencing a number of effectors in various regions of the body such as the skin, blood vessels, pupils, and so on.

As far as vertebrates are concerned, type IV is applicable to very restricted phenomena like the following. In a resting state, the osmotic pressure of the blood is detected by receptors in the brain and the periphery, which send neural spikes to hypothalamic neurosecretory cells. The neurosecretory cells steadily liberate antidiuretic hormone (ADH) into the cerebrospinal fluid; ADH reaches the kidney via the circulatory system to facilitate the resorption of water. There is no participation of an endocrine organ in this system. In contrast to vertebrates, this system is the most common in invertebrates.

However, if a great deal of ADH is required, as in dehydration for instance, this substance becomes released largely from the posterior hypophysis into the general circulation system. Since the secretory pattern in the posterior hypophysis shows characteristics of an endocrine gland, we may consider this organ as an endocrine component and ADH not as a neurohormone, although it is produced in the hypothalamus, but as a hormone. Consequently, such a control system corresponds to type V. Another hormone of the posterior hypophysis, oxytocin, is also controlled by a similar pattern. During the lactation period, sucking stimuli transmitted from the mammilla to the brain via neural pathways provoke the release of oxytocin, which is produced in the hypothalamic neurosecretory cells and stored in the posterior hypophysis, into the circulatory system to bring about milk secretion from the mammary gland.

Type VI is a neuroendocrine system representative in vertebrates. Neurohormones, which control the secretory activity of an effector endocrine organ (exclusively the hypophysis), are transported by the hypophyseal portal system. This circulatory system closely connects the hypothalamus with the hypophysis and transmits chemical messengers in concentrated conditions, ensuring high efficiency. Thus a close correlation between the two components is established, known as the brain-hypophyseal system. To this type belong the control systems related to the secretion of somatotropin (STH), melanophore-stimulating hormone, and luteotropin. For example, if blood glucose levels decrease, intracerebral neurons detect the change and send stimulatory signals to hypothalamic cells that produce somatotropin-releasing factor (SRF). Then SRF is liberated into the blood of the hypophyseal portal system and reaches the anterior hypophysis, provoking the secretion of STH. This hormone is distributed to all tissues of the body via the blood of the general circulation system, bringing about various changes in metabolic performance.

Type VII is a more complex neuroendocrine system, related to the control of the thyroid, the adrenal cortex, and the gonad by the brain-hypophyseal system. As for the control of the gonadal function, the hypothalamus and the hypophysis produce three kinds of neuro-hormones and hormones, respectively, and the gonad as well as the reproductive organs differ in structure, secretion, and function between the two sexes (*vide infra*). Hence the brain-hypophyseal-gonadal-reproductive organ system is the most complex neuro-endocrine control system.

In these seven types of control systems, shown in Fig. 1, no feedback information routes are indicated. Actually, however, there are some pathways of positive and/or negative feedback signals, forming complicated regulatory circuits. Furthermore, information of a control system is sometimes transmitted to another system, thus creating a more complex network. Consequently, for the purpose of understanding a neuroendocrine control mechanism, a systems approach seems to be of great significance.

Systems approaches

Some recent important attempts to study neuroendocrine control mechanisms embody systems approaches by means of mathematical modeling and computer simulation. In neuroendocrine systems, composed of multiorgan components, as shown in Fig. 1, many kinds of information coming from the internal and external environments dynamically integrates the system into a certain pattern of physiological state. Precise analyses of unit processes, for instance measurements of the blood level of a certain hormone at a certain moment, are not sufficient to understand such complex systems as a whole. Consequently, a synthetic method to gather all analytically obtained facts and put them into an integrative form, with some assumptions or under certain hypotheses, is required to recognize the total figure of the system.

This is a new branch of theoretical biology. Traditionally, such an aspect has remained at the stage of schematization, there being no further useful techniques. This is chiefly due to the lack of biomathematical treatments. However, significant progress is promised because extensive applications of systems engineering techniques to biological problems are now available. These techniques make it possible to deal with all components and information concerned with control performances at one time, no matter how complex the system may be. Of course, at present, the rather poor state

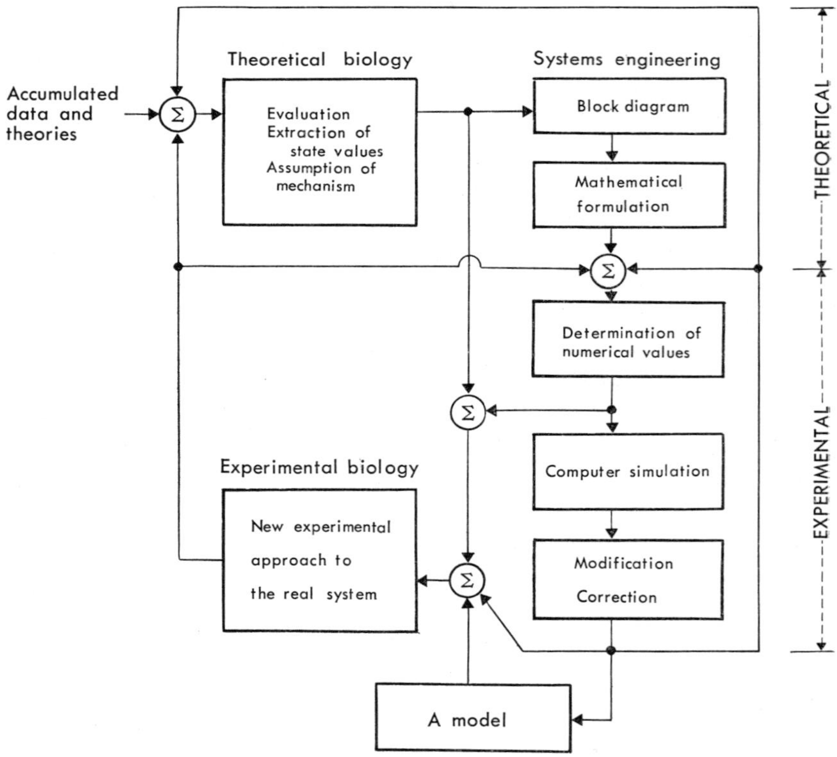

Fig. 2. Steps in systems approaches to neuroendocrine control mechanisms (from Inoué, Nakamura & Sekiguchi, 1970).

of knowledge may cause some comical misunderstanding, for sufficient materials to construct a good model are not available despite the vast quantity of accumulated data. The steps for modeling are schematically illustrated in Fig. 2.

Such attempts may bring about no direct answer as to what occurs in the real system. However, the modeling process can reveal what is lacking in our understanding of the real system and what should be done experimentally. Furthermore, suggestions as to how we should deduce the mechanism inside a logically safe range can be anticipated, and all possibilities as to what may occur in the real system may be predicted before experimental investigation (Nagumo & Inoué, 1971; Inoué, 1972). Complementary relations between such attempts and experimental research may become more and more important in understanding integrative control mechanisms in neuroendocrine phenomena.

As to modeling, especially in mathematical forms, extensive work has already been published since the classic paper of Lamport (1940). (For reviews, see Inoué & Sekiguchi, 1971; Inoué, 1972.) The thyroid hormone control system has been mathematically analyzed by Danziger and Elmergreen (1954, 1957), Roston (1959), and Norwich & Reiter (1965). Systems approaches have been done by DiStefano & Stear (1968, 1969); DiStefano (1969*a*, *b*); and DeGroot, Decostre & Phair (1971). The adrenal cortex hormone control system has been studied by Yates, Brennan, Urquhart, Dallman, Li & Halpern (1968); Gann, Ostrander & Schoeffler (1968); Yates & Brennan (1969); Li & Urquhart (1969); Urquhart (1970); Urquhart, Krall & Li (1970); and Dolkas & Leon (1970). The sex hormone control system, especially concerning female sex cycles in rats and women, has been studied by Thompson (1966); Thompson, Horgan & Delfs (1969); Schwartz & Hoffmann (1967); Schwartz (1969*a*, *b*, 1970); Inoué, Nakamura & Sekiguchi (1969, 1970); Inoué (1970*a*, 1973); Nakamura, Sekiguchi & Inoué (1969); Sekiguchi, Nakamura & Inoué (1969, 1970); Vande Wiele, Bogumil, Dyrenfurth, Ferin, Jewelewcz, Warren, Rizkallah & Mikhail (1970); and Shack, Tam & Lardner (1971). The insulin control system has been studied by Cerasi (1967); Charette, Kadish & Sridhar (1969), Gatewood, Ackerman, Rosevear & Molnar (1970); and Chorbajian, Coull & Coull (1971). The ADH control system has been studied by Roston (1959) and DeHaven & Shapiro (1970).

CONTROL MECHANISM OF THE BRAIN-HYPOPHYSEAL-GONADAL SYSTEM

The system and its components

The system concerned with the control of sex hormone secretion belongs to type VII of Fig. 1, the most complex assembly of organs. This brain-hypophyseal-gonadal system is functionally composed of three kinds of components: (1) information-processing components, such as the brain, the hypophysis, the gonad, and the reproductive organs; (2) information-transmitting components, such as the hypophyseal portal system, the general circulation system, and some sensory pathways; and (3) information-eliminating components, such as the submandibular gland, the liver, the kidney, and other tissues. Besides these, there are several associated components, such as the thyroid, the adrenal, and the neurohypophysis, which may exert minor effects on the regulatory function. The placenta is an important information-processing components at the time of pregnancy but is not dealt with here.

The functional combinations of the above components are schematically shown in Fig. 3. If external environmental conditions are stable and do not cause disturbances to the system, only photoperiodicity and copulatory stimuli may modify the state of activity. The former and the latter trigger the so-called biological clock and the state of pregnancy, respectively.

Characteristics and mathematical formulation of components

Brain

The brain can be divided into three different subcomponents—the neural, neurosecretory, and store subsystems (Fig. 3)—merely according to physiological functions, since intracerebral anatomical connections related with the hypophyseal-gonadal control are still not fully understood. The functional connections of the information flow in the brain are schematically illustrated in Fig. 4.

The neural subsystem contains detectors for all hormones and for sensory inputs coming from peripheries. Light-dark rhythms in the external environment adjust the intracerebral biological clock. Copulation interrupts the estrous cycle in most species, usually bringing about pregnancy. We propose that there are both stimulatory and inhibitory neurons (designated as $\mathcal{N}^s$ and $\mathcal{N}^i$, respectively, in Fig. 4), which control the production and release of neuro-

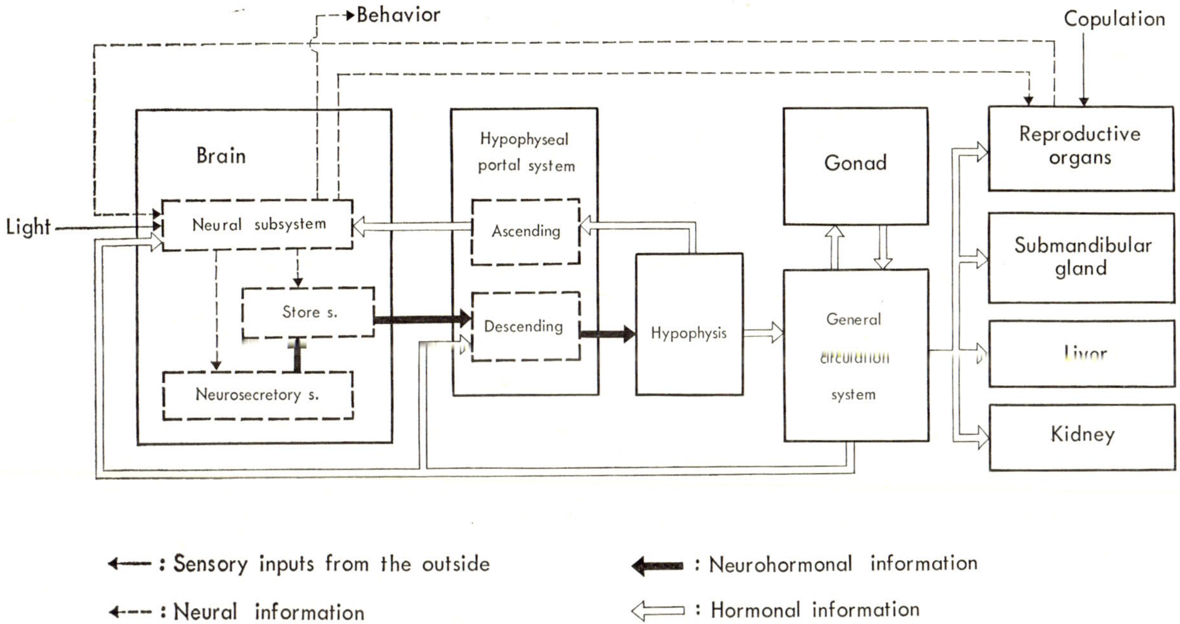

Fig. 3. Functional connections of components concerned with the control of reproductive functions. Arrows indicate the direction of information flows (modified from Inoué *et al.* 1970).

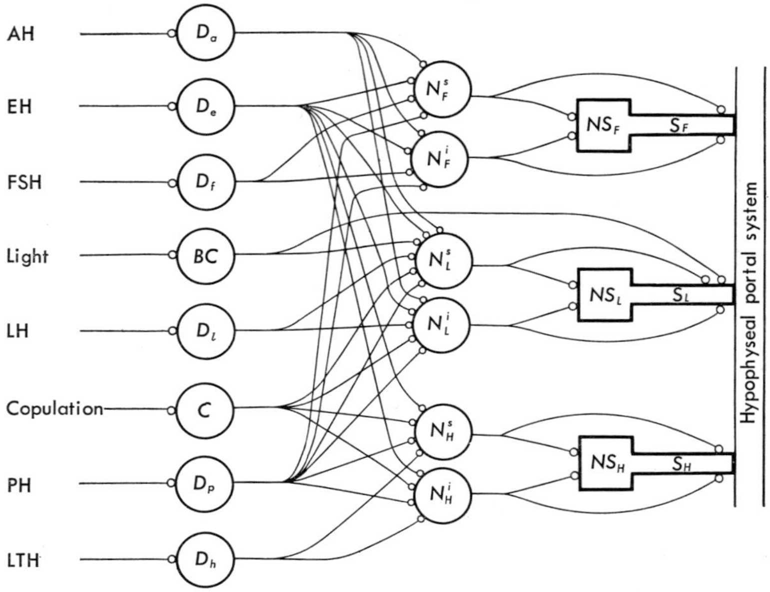

Fig. 4. Functional connections in the brain. Information is transmitted from the left to the right. White and black spots indicate excitatory and inhibitory "synaptic" junctions, respectively. *BC*, Intracerebral biological clock. *C*, Detector for copulatory stimulation. *D* with suffix, Detector for each hormone written at the left. *N* with suffixes, Neurons regulating the production and release of neurohormone (*s* and *i* indicate stimulatory and inhibitory, respectively, and *F*, *L*, and *H* relations to FSH–RF, LH–RF, and PIF, respectively). *NS* with suffix, Neurosecretory subsystem (each suffix indicates the same relation as *N*). *S* with suffix, Store subsystem (each suffix indicates the same relation as *N*). For other abbreviations, see text (from Inoué *et al.* 1970).

hormones. Another assumption is that the sexual difference in the regulatory function of the brain is not to be derived from structural differences, the same neural structures existing in both sexes. We assume that the sensitivity, working capacity, and/or time constant of the neural subsystem may develop differently in the male and female, and also in the spontaneous and reflex ovulators.

We formulate the neural subsystem into the following equation, considering that for example N_F (Table 1), both stimulatory and inhibitory inputs to NS_F, is modified by blood levels of androgen (AH), estrogen (EH), progestin (PH), and follicle-stimulating hormone (FSH) (Fig. 4), and that N_F is affected by the summation of the effects of each hormone.

TABLE 1. The symbols and definitions of state values extracted from the brain-hypophyseal-gonadal system*

Symbol	Definition (unit)
F, L, H, A, E, P, F_N, L_N, and H_N	Amount of secreted FSH, LH, LTH, AH, EH, PH, FSH-RF, LH-RF, and PIF, respectively, in the production site (μg/unit time)
$\mathcal{N}_F$, $\mathcal{N}_L$, and $\mathcal{N}_H$	Intensity of neural stimuli from the neural subsystem in the brain to cause the synthesis of FSH-RF, LH-RF, and PIF, respectively, in the neurosecretory subsystem (1/unit time)
$\mathcal{N}_f$, $\mathcal{N}_l$, and $\mathcal{N}_h$	Intensity of neural stimuli from the neural subsystem in the brain to cause the release of FSH-RF, LH-RF, and PIF, respectively, from the store subsystem (1/unit time)
$\mathcal{N}_b$	Intensity of neural stimuli from the intracerebral biological clock to cause the transient release of LH-RF from the store subsystem (1/unit time)
f_M, l_M, and h_M	Concentration of FSH-RF, LH-RF, and PIF, respectively, in the store subsystem (μg/ml plasma)
G_F, G_L, and G_H	Conductance of the boundary between the store subsystem and the descending hypophyseal portal system related to the release of FSH-RF, LH-RF, and PIF, respectively (1/unit time)
f_R, l_R, and h_R	Concentration of FSH-RF, LH-RF, and PIF, respectively, in the descending hypophyseal portal system (μg/ml plasma)
f_P, l_P, and h_P	Concentration of FSH, LH, and LTH, respectively, in the ascending hypophyseal portal system (μg/ml plasma)
α	Ratio in which hypophyseal outputs enter into the general circulation system
V_{HD}, V_{HA}, V_G, and V_M	Volume of the descending and ascending hypophyseal portal system, the general circulation system, and the store subsystem, respectively (ml or cm^3)
f_r, l_r, h_r, f, l, h, a, e, and p	Concentration of FSH-RF, LH-RF, PIF, FSH, LH, LTH, AH, EH, and PH, respectively, in the general circulation system (μg/ml plasma)
U, V, and W	Grade exhibiting the state of follicular growth, ovulation, and luteinization, respectively
Y and Z	Epithelial thickness (μ) and smear pattern, respectively, in the vagina

* Other symbols used in Eqs. (1)–(14) in the text stand for constants of nonnegative values. τ with suffixes are always time constants.

$$
\begin{aligned}
\mathcal{N}_F = {} & K_{Fa}U_{-1}(T_{Fa}-a)\cdot(\mathcal{N}_{Fa}-a)+K_{Fe}U_{-1}(T_{Fe}-e)\cdot(\mathcal{N}_{Fe}-e) \\
& +K_{Fp}U_{-1}(T_{Fp}-p)\cdot(\mathcal{N}_{Fp}-p)+K_{Ff}U_{-1}(T_{Ff}-f)\cdot(\mathcal{N}_{Ff}-f) \\
& +K_{Ffp}U_{-1}(T_{Ffp}-f_P)\cdot(\mathcal{N}_{Ffp}-f_P)
\end{aligned} \tag{1}
$$

where U_{-1} means unit step function, T and $\mathcal{N}$ with suffixes in parentheses are the threshold level to each hormone (for example, T_{Fa} is the threshold level to AH) and spontaneous activity of neurons (for example, $\mathcal{N}_{Fa}$ is the spontaneous activity of androgen-sensitive neurons), respectively. K with suffixes stands for a constant. The other symbols are defined in Table 1. All values are nonnegative. We assume that T, N, and K with suffixes may be functions of sexuality, physiological state, age, and species.

On the other hand, there are the possibilities that N_F is affected by the product of all the terms in Eq. (1) or the summation of products of some terms (for further details, see Sekiguchi *et al.* 1969). The latter possibility is most likely as judged by our computer analysis (Nakamura *et al.* 1969; Sekiguchi *et al.* 1970).

N_f, inputs to the store subsystem S_F (Fig. 4), is formulated in the same way. N_L and N_l, to which photic and copulatory stimuli are added, as well as N_H and N_h can be treated similarly.

The neurosecretory subsystem, which is known to be located in the so-called hypophysiotropic area of the hypothalamus, produces three kinds of neurohormones: follicle-stimulating hormone-releasing factor (FSH-RF), luteinizing hormone-releasing factor (LH-RF), and prolactin-inhibiting factor (PIF). They are transported through the axons of neurosecretory cells to the store subsystem, which is located in the median eminence of the basal hypothalamus and liberated into the blood of the hypophyseal portal system to cause the production and release of hypophyseal hormones.

The neurosecretory and store subsystems are formulated into Eqs. (2) and (3)–(4), respectively.

$$\frac{d}{dt}\begin{bmatrix} F_N \\ L_N \\ H_N \end{bmatrix} = -\begin{bmatrix} \tau_{FN} & 0 & 0 \\ 0 & \tau_{LN} & 0 \\ 0 & 0 & \tau_{HN} \end{bmatrix}\begin{bmatrix} F_N \\ L_N \\ H_N \end{bmatrix} + \begin{bmatrix} K_{FN} & 0 & 0 \\ 0 & K_{LN} & 0 \\ 0 & 0 & K_{HN} \end{bmatrix}\begin{bmatrix} N_F \\ N_L \\ N_H \end{bmatrix}, \tag{2}$$

$$V_M\frac{d}{dt}\begin{bmatrix} f_M \\ l_M \\ h_M \end{bmatrix} = -\begin{bmatrix} G_F & 0 & 0 \\ 0 & G_L & 0 \\ 0 & 0 & G_H \end{bmatrix}\begin{bmatrix} f_M \\ l_M \\ h_M \end{bmatrix} + \begin{bmatrix} F_N \\ L_N \\ H_N \end{bmatrix}, \tag{3}$$

$$\frac{d}{dt}\begin{bmatrix} G_F \\ G_L \\ G_H \end{bmatrix} = -\begin{bmatrix} \tau_{GF} & 0 & 0 \\ 0 & \tau_{GL} & 0 \\ 0 & 0 & \tau_{GH} \end{bmatrix}\begin{bmatrix} G_F \\ G_L \\ G_H \end{bmatrix} + \begin{bmatrix} K_{GF} & 0 & 0 & 0 \\ 0 & K_{GL} & 0 & G_{Gb} \\ 0 & 0 & K_{GH} & 0 \end{bmatrix}\begin{bmatrix} N_f \\ N_l \\ N_h \\ N_b \end{bmatrix}. \tag{4}$$

Hypophyseal portal system

This localized vascular system combines the brain and the hypophysis, transmitting humoral information from one to the other. The descending blood flow carries neurohormones from the hypothalamus to the anterior hypophysis, and the ascending flow conveys some of the hypophyseal hormones which exert short feedback effects on the brain. The physiological significance of this system is to send information in concentrated quantity directly as well as rapidly to the neighboring target organ.

Concentrations of neurohormones in the descending flow are expressed below.

$$V_{HD}\frac{d}{dt}\begin{bmatrix} f_R \\ l_R \\ h_R \end{bmatrix} = -\begin{bmatrix} \tau_{fR} & 0 & 0 \\ 0 & \tau_{lR} & 0 \\ 0 & 0 & \tau_{hR} \end{bmatrix}\begin{bmatrix} f_R \\ l_R \\ h_R \end{bmatrix} + \begin{bmatrix} G_F & 0 & 0 \\ 0 & G_L & 0 \\ 0 & 0 & G_H \end{bmatrix}\begin{bmatrix} f_M \\ l_M \\ h_M \end{bmatrix}. \tag{5}$$

Concentrations of hypophyseal hormones in the ascending flow are:

$$V_{HA}\frac{d}{dt}\begin{bmatrix} f_P \\ l_P \\ h_P \end{bmatrix} = -\begin{bmatrix} \tau_{fP} & 0 & 0 \\ 0 & \tau_{lP} & 0 \\ 0 & 0 & \tau_{hP} \end{bmatrix}\begin{bmatrix} f_P \\ l_P \\ h_P \end{bmatrix} + (1-\alpha)\begin{bmatrix} F \\ L \\ H \end{bmatrix}. \tag{6}$$

Hypophysis

This endocrine organ is regulated by hypothalamic neurohormones as well as gonadal hormones and in turn regulates the cerebral as well as gonadal activities by its own secretions. The anterior hypophysis secretes follicle-stimulating hormone (FSH), luteinizing hormone (LH), and prolactin (LTH). FSH and LH are synthesized and released by the actions of FSH-RF and LH-RF, respectively, while the production and release of LTH is inhibited by PIF. Estrogen is known to exert dual effects on gonadotropin secretion, either stimulatory or inhibitory. However, as far as the direct effect is concerned, we hypothesize that it acts only as an inhibitor in normal physiological conditions. Hypophyseal hormones are transported by the general circulation system and the ascending hypophyseal portal system.

Secretory activities of the hypophysis are approximated considering stimulatory and/or inhibitory effects of hypothalamic neurohormones and inhibitory effects of ovarian EH.

$$\frac{d}{dt}\begin{bmatrix} F \\ L \\ H \end{bmatrix} = -\begin{bmatrix} \tau_F & 0 & 0 \\ 0 & \tau_L & 0 \\ 0 & 0 & \tau_H \end{bmatrix}\begin{bmatrix} F \\ L \\ H \end{bmatrix} + \begin{bmatrix} K_{Ff} & 0 & 0 & -K_{Fe} \\ 0 & K_{Ll} & 0 & -K_{Le} \\ 0 & 0 & -K_{Hh} & -K_{He} \end{bmatrix}\begin{bmatrix} f_R \\ l_R \\ h_R \\ e \end{bmatrix}. \tag{7}$$

General circulation system

This vascular system is distributed broadly throughout the body and carries hormones from the site of production to the target organ.

The following equations are formulated for the circulating level of each hormone and neurohormone.

$$V_G \frac{d}{dt}\begin{bmatrix} a \\ e \\ p \end{bmatrix} = -\begin{bmatrix} \tau_a & 0 & 0 \\ 0 & \tau_e & 0 \\ 0 & 0 & \tau_p \end{bmatrix}\begin{bmatrix} a \\ e \\ p \end{bmatrix} + \begin{bmatrix} A \\ E \\ P \end{bmatrix}, \tag{8}$$

$$V_G \frac{d}{dt}\begin{bmatrix} f \\ l \\ h \end{bmatrix} = -\begin{bmatrix} \tau_f & 0 & 0 \\ 0 & \tau_l & 0 \\ 0 & 0 & \tau_h \end{bmatrix}\begin{bmatrix} f \\ l \\ h \end{bmatrix} + \alpha\begin{bmatrix} F \\ L \\ H \end{bmatrix} + \begin{bmatrix} \tau_{fP} & 0 & 0 \\ 0 & \tau_{lP} & 0 \\ 0 & 0 & \tau_{hP} \end{bmatrix}\begin{bmatrix} f_P \\ l_P \\ h_P \end{bmatrix}, \tag{9}$$

$$V_G \frac{d}{dt}\begin{bmatrix} f_r \\ l_r \\ h_r \end{bmatrix} = -\begin{bmatrix} \tau_{fr} & 0 & 0 \\ 0 & \tau_{lr} & 0 \\ 0 & 0 & \tau_{hr} \end{bmatrix}\begin{bmatrix} f_r \\ l_r \\ h_r \end{bmatrix} + \begin{bmatrix} \tau_{fR} & 0 & 0 \\ 0 & \tau_{lR} & 0 \\ 0 & 0 & \tau_{hR} \end{bmatrix}\begin{bmatrix} f_R \\ l_R \\ h_R \end{bmatrix}. \tag{10}$$

Gonad

The gonad shows sexual differences in structure and function. The testis, the male gonad, secretes mainly androgen (AH), and the ovary, the female gonad, secretes estrogen (EH) and progestin (PH).

Patterns of the production and release of ovarian hormones depend largely on the stage of the life cycle and on the estrous cycle. Close correlations exist between the morphological structure and the secretory activity in the ovary. We suppose that during the course of follicular growth the ovary secretes mainly EH under the stimulatory influence mainly of FSH, that at the time of follicular maturation both FSH and LH provoke ovulation, and that LH not only induces the successive luteinization but also alters the secretory state of the ovary from the EH phase to the PH phase (Fig. 5).

Choosing the rat as a model of the mammalian species, relatively small amounts of LTH are also involved in the luteinization process. If copulatory stimuli reach the brain, the production and release of PIF becomes markedly suppressed, and accordingly large amounts of LTH are secreted to activate corpora lutea for a certain duration. Activated corpora lutea secrete PH in larger amounts.

Since ovarian secretory activities are closely related with structural changes, we formulate the secretory activities (Eq. 11) as a function of states of the ovarian structure, such as U, V, and W, and in turn the structural states (Eq. 12) as a function of the secretory activities.

$$\frac{d}{dt}\begin{bmatrix} E \\ P \end{bmatrix} = -\begin{bmatrix} \tau_E & 0 \\ 0 & \tau_P \end{bmatrix}\begin{bmatrix} E \\ P \end{bmatrix} + \begin{bmatrix} 0 & K_{E'l} & 0 \\ 0 & K_{El} & 0 \end{bmatrix}\begin{bmatrix} f \\ l \\ h \end{bmatrix}$$
$$+\begin{bmatrix} K_{Ef}U & K_{El}U - K'_{El}e & 0 \\ 0 & K_{Pl}W & K_{Ph}W \end{bmatrix}\begin{bmatrix} f \\ l \\ h \end{bmatrix}, \tag{11}$$

$$\frac{d}{dt}\begin{bmatrix} U \\ V \\ W \end{bmatrix} = -\begin{bmatrix} \tau_U & 0 & 0 \\ 0 & \tau_V & 0 \\ 0 & 0 & \tau_W \end{bmatrix}\begin{bmatrix} U \\ V \\ W \end{bmatrix} + \begin{bmatrix} K_{Uf} & 0 & 0 \\ 0 & 0 & 0 \\ 0 & 0 & K_{Wh} \end{bmatrix}\begin{bmatrix} f \\ l \\ h \end{bmatrix}$$
$$+\begin{bmatrix} 0 & -K_{Ul}V & -K_{Uh}W \\ 0 & K_{Vl}U & -K_{Vh}W \\ -K_{Wf}U & -K_{Wl}V & 0 \end{bmatrix}\begin{bmatrix} f \\ l \\ h \end{bmatrix}, \tag{12}$$

where $0 \leq U \leq 1$, $0 \leq V \leq 1$, $0 \leq W \leq 1$, $0 \leq U+V+W \leq$ Constant.

The testis secretes AH under the stimulatory influences of FSH and

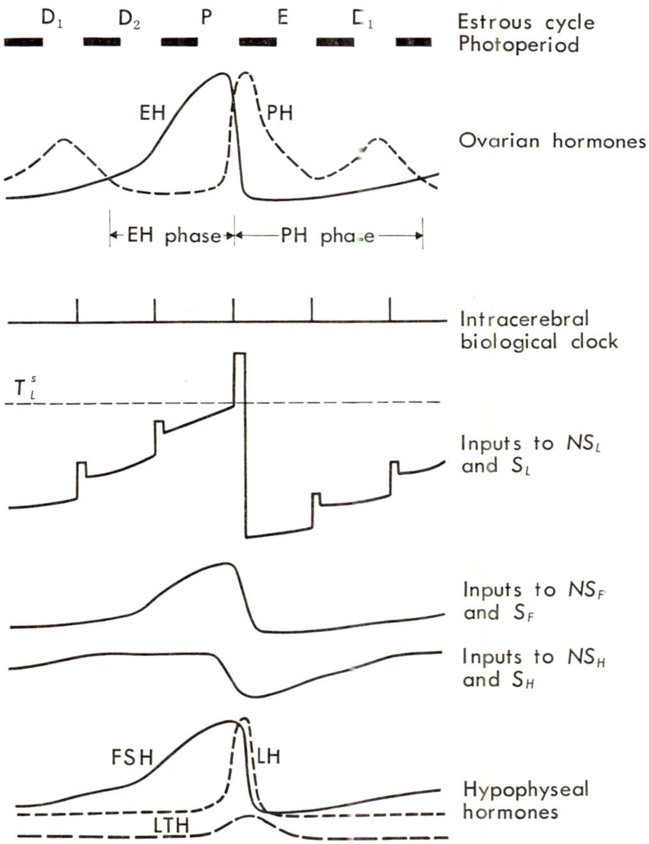

Fig. 5. Theoretically constructed time-course changes in information quantity during the 4-day estrous cycle of the rat. D_1 and D_2, The first and the second day of diestrus, respectively. E, The day of estrus. P, The day of proestrus. $T_L{}^s$, Threshold level to cause the LH surge. The diurnal photoperiodicity is 12-hr illumination alternated with 12-hr darkness (shown by black bar) (modified from Inoué *et al.* 1970).

LH. Differential effects of these hormones are not as clear as in case of the ovary, but LH is believed to stimulate the interstitial cells of the testis. The role of LTH in testicular regulation is uncertain. Structural changes in the testis are not so dynamic and drastic as in the ovary. Hence, we formulate the androgen secretion simply as follows.

$$\frac{d}{dt}A = -\tau_A A + K_{Af}f + K_{Al}l\,. \qquad (13)$$

Reproductive organs

Important reproductive organs are the uterus and the vagina in the female and the prostates and the seminal vesicles in the male. They react sensitively to sex hormones, changing their structures and functions. Most of the reproductive organs are merely targets of gonadal hormones and hence indicators of a state of the brain-hypophyseal-gonadal system. Here the vagina is mathematically formulated in the following form.

$$\frac{d}{dt}\begin{bmatrix} Y \\ Z \end{bmatrix} = -\begin{bmatrix} \tau_Y & \tau_{YZ} \\ \tau_{ZY} & \tau_Z \end{bmatrix}\begin{bmatrix} Y \\ Z \end{bmatrix} + \begin{bmatrix} K_{Ye} & K_{Yp} \\ K_{Ze} & K_{Zp} \end{bmatrix}\begin{bmatrix} e \\ p \end{bmatrix}. \qquad (14)$$

The utero-vaginal junction (vaginal cervix) is important as a sensory receptor for copulatory stimuli, which are transmitted to the brain via nervous pathways. In the state of pregnancy, the placenta develops in the uterus. This acts as an endocrine organ, playing a leading role during gestation.

Information-eliminating components

Hormones are more or less inactivated by various kinds of bodily tissues. Sex steroids are well known to be inactivated in the liver and excreted through the bile. Hypophyseal hormones are known to undergo inactivation by the submandibular gland. All neurohormones and hormones are excreted into the urine through the kidney. We suppose that the elimination of hormonal and neurohormonal information is proportional to their blood level.

Equations (1)–(14), except for (13), are summarized in Fig. 6 in the form of a block diagram. Here the neural subsystem is shown as a black box. Vectors and matrices are expressed as vector values which are represented by a symbol of the first element, for example:

$$\boldsymbol{f} = \begin{bmatrix} f \\ l \\ h \end{bmatrix}, \qquad \boldsymbol{\tau}_e = \begin{bmatrix} \tau_e & 0 \\ 0 & \tau_p \end{bmatrix}.$$

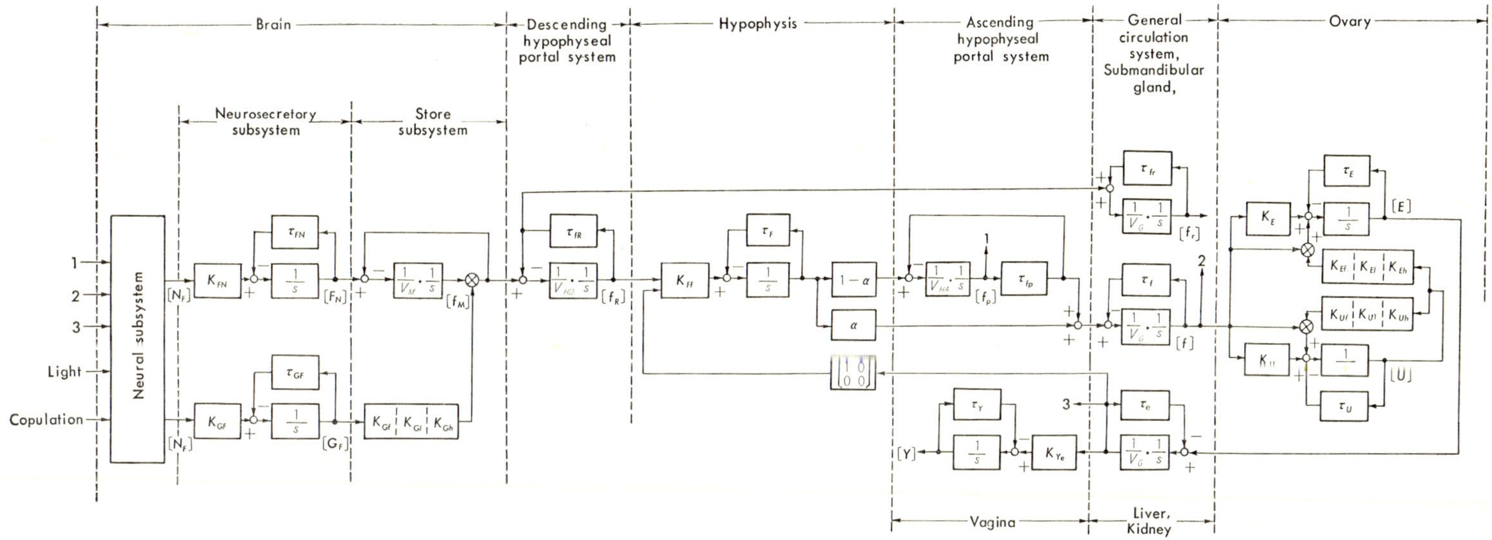

Fig. 6. Block diagram of the model of the brain-hypophyseal-ovarian system. For definition of symbols, see Table 1 and text. 's' stands for the Laplace operator. Vectors or matrices of state values or constants in equations are shown as vector values represented by the first element (modified from Inoué *et al.* 1970).

Control mechanism of the rat estrous cycle

EH phase

The control mechanism of the normal female rat with 4-day estrous cycles under constant environmental conditions is adopted here as an example. The diurnal photoperiodicity is an alternation of 12-hr illumination with 12-hr darkness. Figures 3–5 may help illustrate the following description.

The neural subsystem activating LH-RF production (N_L^s) is closely related to the intracerebral biological clock (BC). This clock sends daily a pulse-like signal to N_L^s regularly 9 hr after the initiation of illumination. This pulse becomes an input indirectly to the neurosecretory subsystem concerned with LH-RF (NS_L) via N_L^s and directly to the LH-RF store subsystem (S_L). Hence the production and release of LH-RF increases in proportion to the input.

On the other hand, EH secreted by the ovary reaches the detector in the neural subsystem (D_e) and modulates activities of both N_L^s and N_L^i (the neural subsystem inhibiting LH-RF production). If we define the thresholds for EH of N_L^s and N_L^i as T_{Le}^s and T_{Le}^i and their time constants as τ_{Le}^s and τ_{Le}^i, respectively, the following relations may be written.

$$T_{Le}^s < T_{Le}^i , \qquad \tau_{Le}^s < \tau_{Le}^i . \tag{15}$$

Consequently, only a positive feedback effect of EH appears at first, causing more elevated production and release of LH-RF. As a result, released LH-RF is transported to the hypophysis through the descending hypophyseal portal system, and induces LH secretion. LH can also exert a stimulatory effect at first on the neural subsystem and then an inhibitory effect later, known as the short feedback action. Hence the following relations may be formulated similarly.

$$T_{Ll}^s < T_{Ll}^i , \qquad \tau_{Ll}^s < \tau_{Lt}^i . \tag{16}$$

The combined effects of BC, EH, and LH on the neural subsystem increase inputs to NS_L and S_L gradually up to the onset of the PH phase, as illustrated in Fig. 5. When BC is triggered in the afternoon of the day of proestrus (P), which corresponds to the so-called critical period, the excitability of N_L^s is sharply raised to the maximum, exceeding the threshold level to cause the LH surge from the hypophysis (T_L^s). Then a large amount of LH-RF is transferred to the hypophysis.

Meanwhile, a rise in EH level during the EH phase similarly provokes a positive feedback effect on the brain, activating the $D_e \to N_F{}^s \to NS_F\text{-}S_F$ route at first. Consequently, FSH-RF and then FSH are released more and more, which in turn induces more EH secretion. FSH is also assumed to activate FSH-RF production and release through the short feedback pathway prior to its negative influence. As a result, when increased EH exceeds a certain level, the $D_e \to N_F{}^i \to NS_F\text{-}S_F$ route is activated, bringing about a decrease in FSH-RF and then FSH as well as EH (negative feedback effects). Hence relations of the threshold and the time constant between the stimulatory and inhibitory routes may be expressed as follows.

$$T_{Fe}{}^s < T_{Fe}{}^i, \qquad \tau_{Fe}{}^s < \tau_{Fe}{}^i. \tag{17}$$

The same relations may be applicable to each effect of FSH and PH, although the latter is almost negligible at the EH phase. The combined effects change chronologically as shown in Fig. 5.

As for the neural subsystem concerned with PIF, activities of the $D_e \to N_H{}^s$ or $N_H{}^i$ route may be very low, manifesting the change in levels as illustrated in Fig. 5. Here the same relations, as in the case of Relations (15)–(17), may exist:

$$T_{He}{}^s < T_{He}{}^i, \qquad \tau_{Fe}{}^s < \tau_{He}{}^i. \tag{18}$$

The effect of LTH and PH at the EH phase seems negligible. We hypothesize that the effects of EH on the neural subsystem as shown in Relations (15), (17), and (18) may be mutually related in the following manner.

$$\left.\begin{array}{ll} T_{Fe}{}^s < T_{Le}{}^s < T_{He}{}^s; & T_{Fe}{}^i < T_{He}{}^i \leq T_{Le}{}^i \\ \tau_{Fe}{}^s < \tau_{He}{}^s \leq \tau_{Le}{}^s; & \tau_{Fe}{}^{\cdot} < \tau_{He}{}^i \leq \tau_{Le}{}^i \end{array}\right\}. \tag{19}$$

The relation of $T_{Fe}{}^i < T_{Le}{}^i$ is shown clearly by experiments with chronic EH administration (Inoué, 1965).

Thus, gonadotropin secretion, under the above-described regulatory influences of the brain and the ovary (the direct effect of EH on the hypophysis), may proceed as shown in Fig. 5.

PH phase

A steep rise of LH level at P, induced by the sudden increase in LH-RF as described above, not only brings about PH secretion from the ovary but also suppresses EH secretion. The EH phase is then replaced by the PH phase. PH together with previously secreted EH induces estrous behavior. Matured follicles shed ova and change into corpora lutea. PH in turn acts on the brain, being detected by D_p, and activates at first $N_F{}^s$, $N_L{}^s$, and $N_H{}^s$ (positive feedback effects) and then $N_F{}^i$, $N_L{}^i$, and $N_H{}^i$ (negative feedback effects). Hence, the following relations may be written.

$$\left.\begin{array}{ll} T_{Fp}{}^s < T_{Fp}{}^i; & \tau_{Fp}{}^s < \tau_{Fp}{}^i \\ T_{Lp}{}^s < T_{Lp}{}^i; & \tau_{Lp}{}^s \ll \tau_{Lp}{}^i \\ T_{Hp}{}^s < T_{Hp}{}^i; & \tau_{Hp}{}^s \ll \tau_{Hp}{}^i \end{array}\right\}. \tag{20}$$

D_p of the female rat is so sensitive that $N_L{}^s$ and $N_L{}^i$ are greatly modified by PH levels. The results are then relayed to the hypophysis by means of neurohormones, to cause changes in secretory patterns of hypophyseal hormones (Fig. 5). In these processes too, the short feedback actions of each hypophyseal hormone may participate in the regulation of the neural activities, as described in Relation (16).

PH secretion continues even after the cessation of LH release, exhibiting the second peak on the first day of diestrus (D_1). This

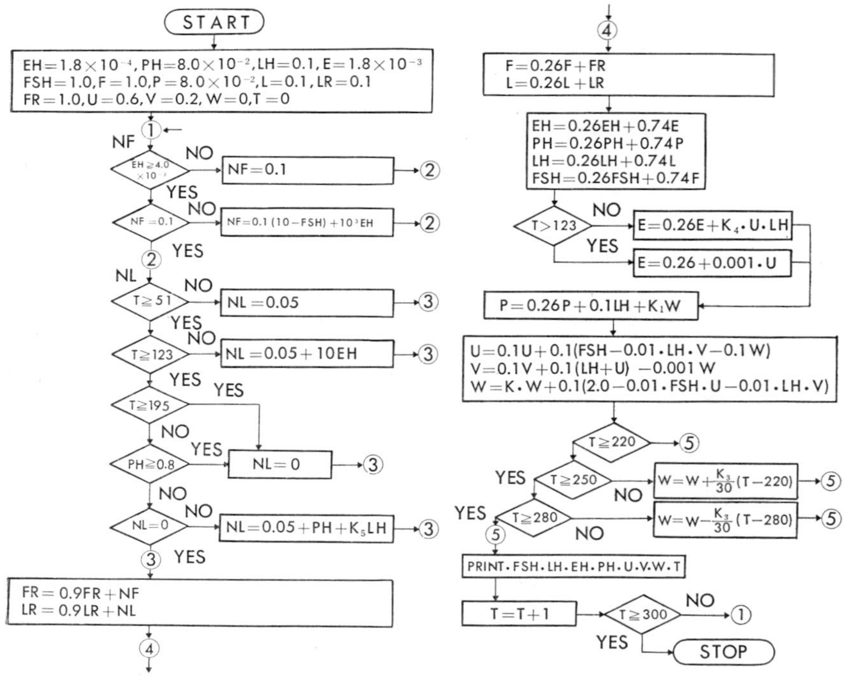

Fig. 7. Flow chart with some numerical values used for computer programming (from Nakamura *et al.* 1969).

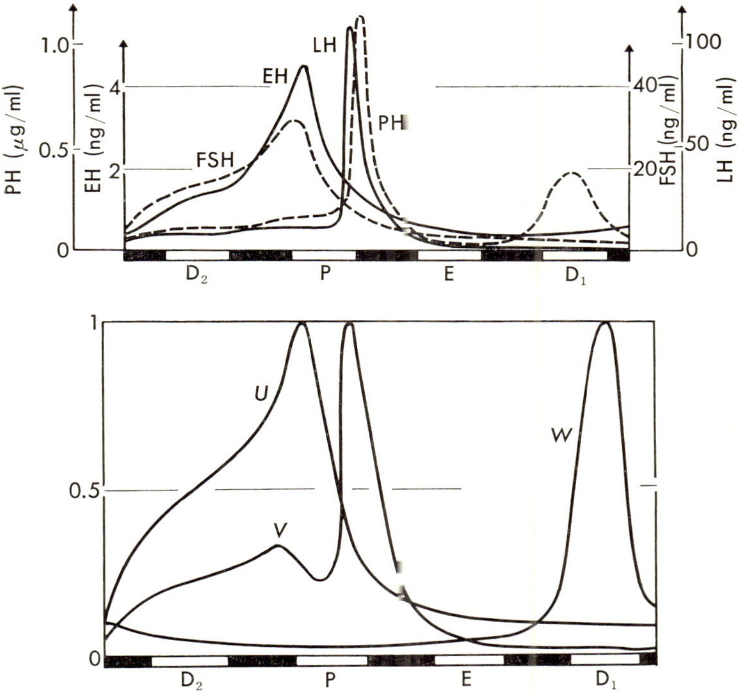

Fig. 8. Computer simulation of time-course changes in plasma concentration of hormones (above) and the ovarian structure (below) during the estrous cycle of the rat (modified from Inoué *et al.* 1970).

stage accords with the period in which the volume of fresh corpora lutea grows to the maximum. If copulatory stimuli reach the brain during the period from the onset of the PH phase to D_1, the neural subsystem concerned with LTH (the routes connected with C in Fig. 4) is triggered to suppress $\mathcal{N}_F{}^s$, $\mathcal{N}_L{}^s$, and $\mathcal{N}_H{}^s$ and activates $\mathcal{N}_F{}^i$, $\mathcal{N}_L{}^i$ and $\mathcal{N}_H{}^i$. Consequently, the production and release of neurohormones is inhibited, and only LTH becomes freely and abundantly liberated to cause successive secretion of PH in elevated levels. This state characterizes pregnancy or pseudopregnancy.

The suppressive effect of PH on the brain is especially important to the chronological sequences of performance at the PH phase. As expressed in Relation (20), after a short period of positive feedback action, the high level of PH exerts a strong negative effect on the neural subsystem. The second peak is assumed to enhance the effect further and elongate the duration. Hence, the neural activities concerned with the production and release of each neurohormone rise very slowly (Fig. 5). If these negative effects of PH are entirely removed, the diestrous stage might disappear. Contrarily, if such effects continue longer, the duration of diestrus might be longer.

Computer simulations of the rat 4-day estrous cycle

We have tried to simulate the dynamic performance of the brain-hypophyseal-gonadal system in the female rat with 4-day estrous cycles. The above-formulated differential equations are partially simplified and transformed into difference equations. An estrous cycle is divided into 288 sampling intervals. Since almost all numerical values for variables and constants are unknown, appropriate values are given in the trial-and-error manner. Initial values of EH, PH, and LH are adopted from Miyake (1968). Calculations start at D_1. The flow chart of decision functions with some numerical values used in a calculation is shown in Fig. 7.

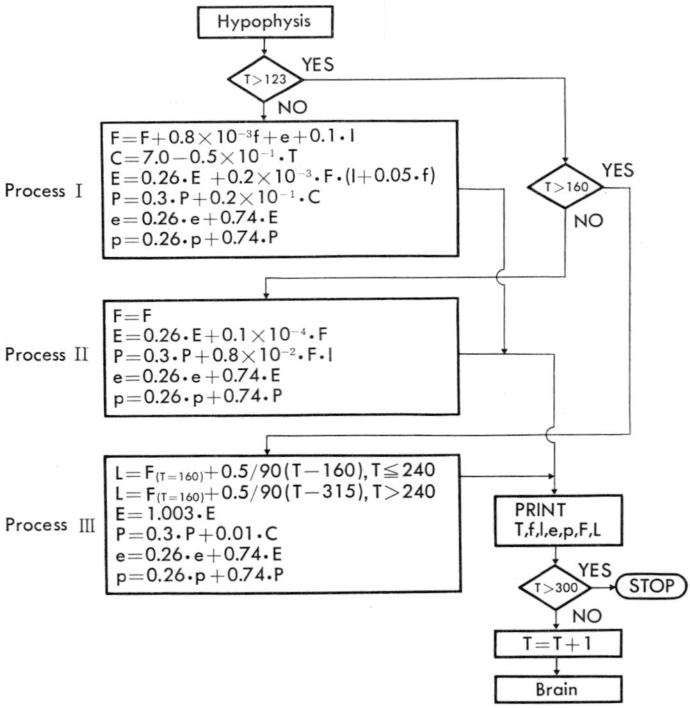

Fig. 9. Flow chart with some numerical values used for computation of ovarian structure and secretion in a revised model (from Sekiguchi *et al.* 1970).

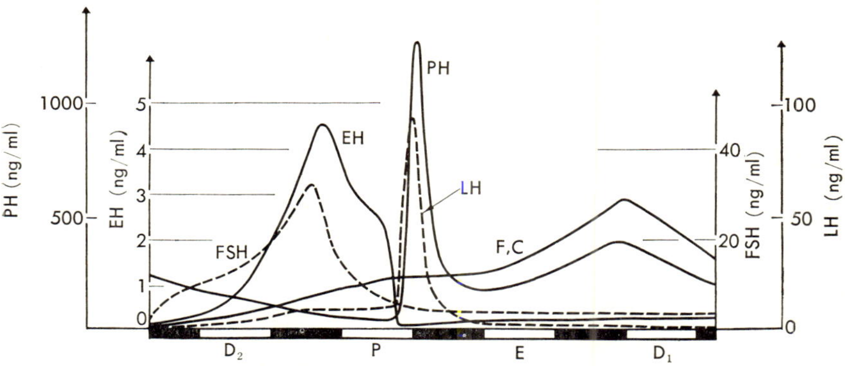

Fig. 10. Computer simulation of time-course changes in plasma concentration of hormones and the relative ovarian weight (F, C) during the estrous cycle of the rat (modified from Sekiguchi *et al.* 1970).

Figure 8 exhibits a set of calculated results. Structural changes in the ovary are considerably well simulated, although follicular maturation ($U=1$) and ovulation ($V=1$) take place a little earlier in this calculation than those in the real system. In addition, FSH and EH are secreted in accordance with the follicular growth, and ovulation occurs at the time of maximal LH secretion. Thus numerical connections between U and FSH or EH levels, and between V and LH levels prove to be too strong. Full development of corpora lutea ($W=1$) synchronizing with the second peak of the PH level fits the performance of the real system. It is clearly demonstrated that corpora lutea are not maintained further.

In our revised model, U, V, and W are replaced by processes I, II, and O, respectively, and the structural states are replaced by changes in ovarian weight using two state variables—F (follicular weight) and C (luteal weight). A set of numerical values and calculated results is shown in Figs. 9 and 10.

ANALYSIS OF DYNAMIC ESTROGEN-UTERINE WEIGHT INTERRELATIONSHIPS

Control of blood estrogen levels and its effect on the uterine weight in ovariectomized rats

We have developed a new chronic infusion technique for unrestrained animals, in which small amounts of hormone solutions are steadily sent into the external jugular vein through a thin tubing

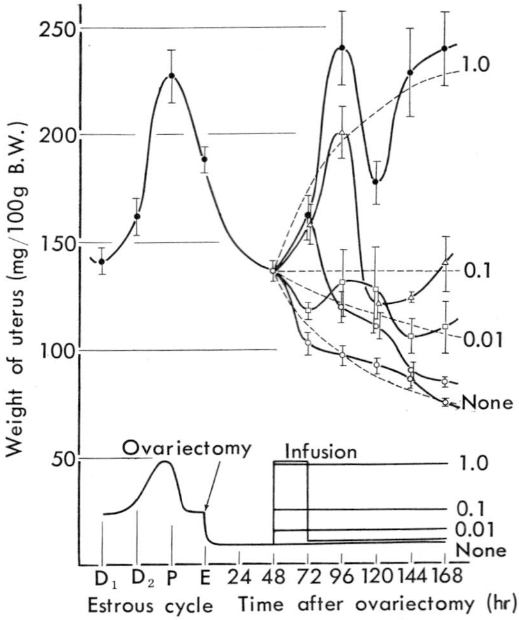

Fig. 11. Time-course changes in mean uterine weight during a normal estrous cycle and the following period of estradiol infusion, with reference to assumed blood levels of estrogen (bottom). Vertical lines indicate standard deviations. Daily doses of estradiol in micrograms are shown at the right. Broken lines are the asymptotic curves (from Inoué & Sekiguchi, 1970).

connected with an ultraslow rotary micropump (Inoué, 1970*b*). Using this technique, responses of the female reproductive organs to EH (estradiol only) were studied (Inoué & Sekiguchi, 1970; Inoué, 1973). Since the uterus reacts most sensitively to EH, analyses here are focused on uterine weight changes.

Young female rats of the Sprague-Dawley strain, 60–70 days of age and weighing 200–250 g, were ovariectomized under ether anesthesia. Ovariectomy was done in the afternoon of estrus (E) after observation of at least 2 normal estrous cycles. Hence, the uterus experienced the maximal estrogenic stimulation just before the operation. Forty-eight hours later, by which time the circulating EH level of ovarian origin might be nullified, EH infusion was initiated for 24–120 hr. Daily doses were 1.0, 0.1, or 0.01 μg, *i.e.*, 6.9×10^{-10} to 10^{-12} g/min. At exactly 24-hr intervals, 8 rats in each dosage group were killed, and weights of the uterus not including the luminal fluid were recorded. Normal rats with 4-day estrous cycles were killed at 13:00–

15: 00 (10 animals at each stage). These rats and ovariectomized but not EH-infused rats served as controls.

Figure 11 shows that continuous supply of EH at 1.0 μg/day caused a sharp rise in uterine weights in the first 48 hr, a sudden fall in the following 24 hr, and then a gradual rise again. The difference between the maximum and the minimum was highly significant. If EH infusion was stopped after the initial 24-hr supply, the uterus lost weight rapidly. The declining curve is largely parallel to that of ovariectomized not EH-infused rats, if it is shifted to the left by 3 gradations (72 hr).

EH infusion of 0.1 μg/day also brought about oscillatory changes in uterine weights. Although the height of the peak was lower than that of the preceding group, amplitude difference between the maximum and the minimum was much larger.

Contrarily, uterine weights of 0.01 μg/day-infused rats could not exceed or even maintain the value at the beginning of infusion. However, such a small dose was still effective to stimulate the uterus, inhibiting the weight loss occurring in noninfused ovariectomized rats. The difference between these two groups was highly significant. Slight oscillatory changes were also observable during the infusion period of 120 hr, although the amplitude differences were not significant.

Because of rapid hepatic inactivation and of urinary or biliary excretion, the blood level of EH in infused rats might be kept constant throughout the infusion period. Hence EH inputs to the uterus can be regarded as step functions (Fig. 11, bottom). However, transient responses of the uterus to physiological doses of EH were not stepwise but oscillatory. We can draw an asymptotic curve over each of the weight curves (Fig. 11, broken). The latter may converge with the former if the infusion period is prolonged further. The height of the asymptotic curve of 1.0 and 0.1 μg/day-infused groups may be nearly equal to that of the uterine weight at P and D_1, respectively, of normal rats. Consequently, we assume that EH levels of normal rats may be about 1.0 μg/day or 700 pg/min at the pre-estrous stage and 0.1 μg/day or 70 pg/min at the postestrous stage. We also assume that the minimal EH level in the blood of normal rats may be higher than 0.01 μg/day or 7 pg/min, since this dose of EH never brought about uterine weights in the normal range.

Mathematical models of the uterus

On the basis of the above-described oscillatory responses of the uterus to stepwise EH inputs, we formulated mathematical models of the uterus. On the analogy of oscillatory transient reactions of physical systems, the uterus may belong fundamentally to a nonlinear first-order system combined either in series or in parallel with a second-order oscillatory system. The latter may be linked with a second-order parameter variation of either frequency modulation (FM) or order-decrease type.

Thus the uterus can be mathematically described by five state variables, x_1–x_5, although their biological or physical entity is not exactly known. At least, weight and diameter of the whole uterus, proliferation rate and thickness of endometrial cells, and electrolyte concentrations in the luminal fluid may be involved. However, no further attempt is made to analyze them except for correspondences between u and EH input and between y and uterine weight.

There are the following four possibilities in combinations of the above systems.

(1) Series combination of the nonlinear first-order system and the second-order oscillatory system which is linked with the second-order parameter variation of the FM type.

$$\begin{cases} \dot{x}_1 = x_2 \\ \dot{x}_2 = -cx_1 - bx_2 + (x_3 - x_1)(x_4 + x_5 + A - f(u)) \\ \dot{x}_3 = -ax_3^2 + f(u) \\ \dot{x}_4 = -kx_4 + gf(u) \\ \dot{x}_5 = -lx_5 - hf(u) \\ y = x_1 \end{cases} \tag{21}$$

(2) Series combination of the nonlinear first-order system and the second-order oscillatory system which is linked with the second-order parameter variation of the order-decrease type.

$$\begin{cases} \dot{x}_1 = (x_4 + x_5 + A - f(u))x_2 \\ \dot{x}_2 = x_3 - dx_2 - [x_1 + \{B - (x_4 + x_5 + A - f(u))\}x_2] \\ \dot{x}_3 = -ax_3^2 + f(u) \\ \dot{x}_4 = -kx_4 + gf(u) \\ \dot{x}_5 = -lx_5 - hf(u) \end{cases} \tag{22}$$
$$y = x_1 + \{B - (x_4 + x_5 + A - f(u))\}x_2$$

(3) Parallel combination of the nonlinear first-order system and the second-order oscillatory system which is linked with the second-order parameter variation of the FM type.

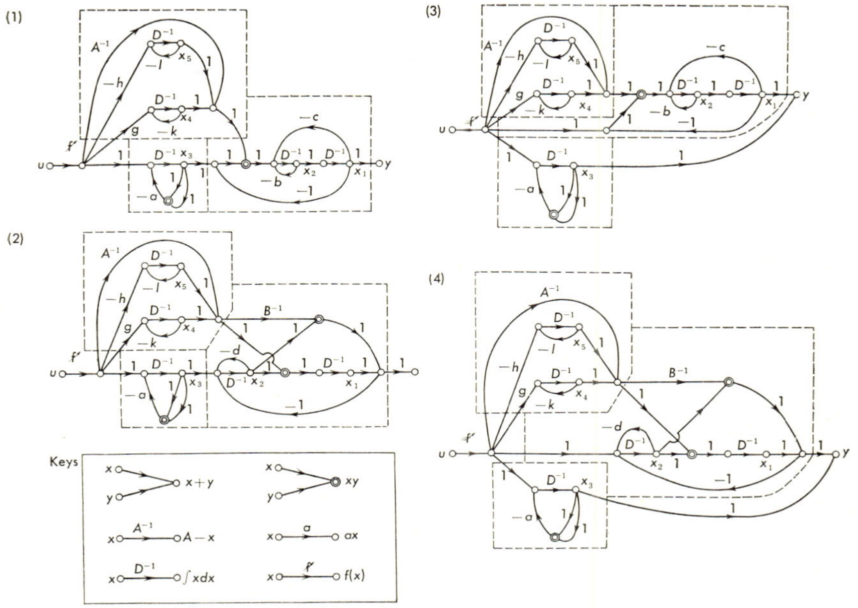

Fig. 12. State diagrams of four possible types of mathematical models for the rat uterus.

$$\left\{\begin{aligned} \dot{x}_1 &= x_2 \\ \dot{x}_2 &= -cx_1 - bx_2 + (f(u) - x_1)(x_4 + x_5 + A - f(u)) \\ \dot{x}_3 &= -ax_3^2 + f(u) \\ \dot{x}_4 &= -kx_4 + gf(u) \\ \dot{x}_5 &= -lx_5 - hf(u) \\ y &= x_1 + x_3 \end{aligned}\right. \tag{23}$$

(4) Parallel combination of the nonlinear first-order system and the second-order oscillatory system which is linked with the second-order parameter variation of the order-decrease type.

$$\left\{\begin{aligned} \dot{x}_1 &= (x_4 + x_5 + A - f(u))x_2 \\ \dot{x}_2 &= f(u) - dx_2 - [x_1 + \{B - (x_4 + x_5 + A - f(u))\}x_2] \\ \dot{x}_3 &= -ax_3^2 + f(u) \\ \dot{x}_4 &= -kx_4 + gf(u) \\ \dot{x}_5 &= -lx_5 - hf(u) \end{aligned}\right. \tag{24}$$

$$y = x_1 + \{B - (x_4 + x_5 + A - f(u))\}x_2 + x_3$$

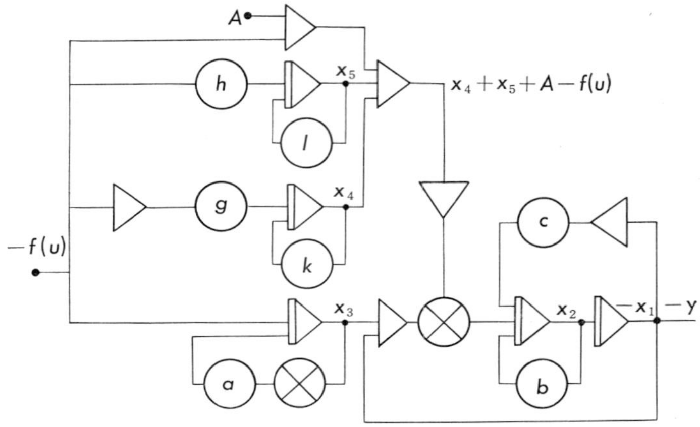

Fig. 13. Analog computer circuit for solution of Eq. (21) (from Inoué, 1973).

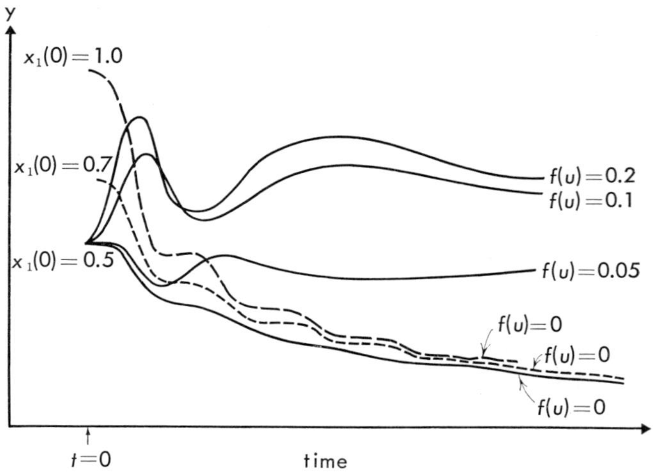

Fig. 14. Computer simulation of the responses of rat uterine weights to various levels of stepwise estrogen inputs. This is calculated from Eq. (21). Numerical values used here are: $a=0.5$, $b=0.1$, $c=0$, $g=0.5$, $h=0.1$, $k=1$, $l=0.2$, and $A=2.16$ (from Inoué, 1973).

where A, B, a, b, c, d, g, h, k, and l are nonnegative constants, and $f(u)$ is a zero-memory nonlinear element. State diagrams of the above models are shown in Fig. 12.

Computer simulation of estrogen-uterine weight dynamics

Among the four possibilities, the first case (1) is calculated by means of an analog computer. Figure 13 shows the analog computer circuit for solution of Eq. (21). Numerical values are chosen through the trial-and-error manner. Computations yield very satisfactory results. For example, curves shown in Fig. 14 largely correspond to those of the experimentally obtained results shown in Fig. 11: $x_1=0.5$ corresponds to the uterine weight of 48 hr ovariectomized rats; $f(u)=0.2$, 0.1, 0.05, and 0 to EH=1.0, 0.1, 0.01, and 0 μg/day, respectively. Moreover, it is theoretically suggested how a uterus, which has undergone estrogenic stimulation, weighing much heavier ($x_1=0.7$ and 1.0, Fig. 11), may lose its weight after the cessation of EH supply. The weight curve of 24-hr infused rats may correspond to the case of $x_1=0.7$.

REFERENCES

Cerasi, E. (1967). An analogue computer model for the insulin response to glucose infusion. *Acta Endocrinol.* **55**, 163–183.

Charette, W. P., Kadish, A. H. & Sridhar, R. (1969). Modeling and control aspects of glucose homeostasis. *Math. Biosci.* Suppl. 1, 115–149.

Chorbajian, T., Coull, B. & Coull, J (1971). An *in numero* study of glucose-insulin interaction. *Bull. math. Biophys.* **33**, 451–462.

Danziger, L. & Elmergreen, G. L. (1954). Mathematical theory of periodic relapsing catatonia. *Bull. math. Biophys.* **16**, 15–21.

Danziger, L. & Elmergreen, G. L. (1957). Mathematical models of endocrine systems. *Bull. math. Biophys.* **19**, 9–18.

DeGroot, L. J., Decostre, P. & Phair, R. (1971). A mathematical model of human iodine metabolism. *J. clin. Endocr. Metab.* **32**, 757–765.

DeHaven, J. C. & Shapiro, N. Z. (1970). Simulation of the renal effects of antidiuretic hormone (ADH) in man. *J. theor. Biol.* **28**, 261–286.

DiStefano, J. J., III. (1969*a*). Hypothalamic and rate feedback in the thyroid hormone regulation system: A hypothesis. *Bull. math. Biophys.* **31**, 233–246.

DiStefano, J. J., III. (1969*b*). A model of the normal thyroid hormone glandular secretion mechanism. *J. theor. Biol.* **22**, 412–417.

DiStefano, J. J., III. & Stear, E. B. (1968). Neuroendocrine control of thyroid secretion in living systems A feedback control system model. *Bull. math. Biophys.* **30**, 3–26.

DiStefano, J. J., III & Stear, E. B (1969). Modeling and control aspects of thyroid function. *Math. Biosci.* Suppl. 1, 161–192.

Dolkas, C. B. & Leon, H. A. (1970). A nonlinear time-varying mathematical analog of the gluco-corticoid control. *IEEE Trans. Bio-Med. Eng.* BME- **17**, 1–10.

Gann, D. S., Ostrander, L. E. & Schoeffler, J. D. (1968). A finite state model for the control of adrenal cortical steroid secretion. In *Systems Theory and Biology*, ed. Mesarović, M. D., pp. 185–200. Berlin, Heidelberg, New York: Springer-Verlag.

Gatewood, L. C., Ackerman, E., Rosevear, J. W. & Molnar, G. D. (1970). Modeling blood glucose dynamics. *Behav. Sci.* **15**, 72–87.

Inoué, S. (1965). Studies on the hypophyseal-gonadal mechanism in parabiotic rats; II. Transfer of injected estrogen from one animal to the other and its differential effects on the gonadotrophin secretion. *Endocr. jap.* **12**, 29–35.

Inoué, S. (1970*a*). Systems theory of the regulation of sex hormone secretion (in Japanese). In *Seitai no Seigyokiko*, ed. Igaku no Ayumi, pp. 206–213. Tokyo: Ishiyaku Shuppan.

Inoué, S. (1970*b*). A technique for continuous intravenous infusion of small amounts of hormone solutions in unrestrained rats. *Reports Inst. Med. Dent. Eng.* **4**, 140–142.

Inoué, S. (1971). Regulatory mechanisms in endocrine systems. *Jap. J. Med. Electronics Biol. Eng.* **9**, 225–230 (in Japanese).

Inoué, S. (1972). Modeling approaches to endocrine phenomena. *Jap. J. Clin. Med.* **30**, 503–504 (in Japanese).

Inoué, S. (1973). Regulatory mechanism in the sex cycle. *Adv. Biophys.* vol. 5, ed. Kotani, M. Tokyo: Univ. Tokyo Press, in press.

Inoué, S., Nakamura, T. & Sekiguchi, T. (1969). Control mechanism involved in the cerebro-hypophysio-gonadal system. *Reports Inst. Med. Dent. Eng.* **3**, 88–96.

Inoué, S., Nakamura, T. & Sekiguchi, T. (1970). A theoretical approach to the regulatory mechanism in the brain-hypophyseal-gonadal system with reference to its mathematical modeling and computer simulation. *Endocr. jap.* **17**, 567–583.

Inoué, S. & Sekiguchi, T. (1970). Oscillatory responses of uterine weights to constantly infused estrogen in ovariectomized rats. *Proc. Japan Acad.* **46**, 1022–1025.

Inoué, S. & Sekiguchi, T. (1971). Endocrine control systems (in Japanese). In *Seitai Shisutemu*, ed. Nagumo, J., pp. 101–149. Tokyo: Nikkan Kogyo Shimbunsha.

Kobayashi, H. (1968). Neurosecretory phenomena. *Kagaku* **38**, 608–614 (in Japanese).

Lamport, H. (1940). Periodic changes in blood estrogen. *Endocrinology* **27**, 673–680.

Li, C. C. & Urquhart, J. (1969). Modeling of adrenocortical secretory dynamics. In *Concepts and Models of Biomathematics. Simulation Techniques and Methods*, ed. Heinmets, F., pp. 185–210. New York: Marcel Dekker.

Miyake, T. (1968). Interrelationship between the release of pituitary luteinizing hormone and the secretions of ovarian estrogen and progestin during estrus cycle of the rat. *Hokkaido Univ. Med. Library.* Series 1, 139–149.

Nagumo, J. & Inoué, S. (1971). Biological systems theory (in Japanese). In *Seitai Shisutemu*, ed. Nagumo, J., pp. 1–12. Tokyo: Nikkan Kogyo Shimbunsha.

Nakamura, T., Sekiguchi, T. & Inoué, S. (1969). Computer simulations of the hypothalamo-hypophyseal-ovarian system. *Reports Inst. Med. Dent. Eng.* **3**, 107–115.

Norwich, K. H. & Reiter, R. (1965). Homeostatic control of thyroxin concentration expressed by a set of linear differential equations. *Bull. math. Biophys.* **27**, 133–144.

Roston, S. (1959). Mathematical representation of some endocrinological systems. *Bull. math. Biophys.* **21**, 271–282.

Schwartz, N. B. (1969*a*). A model for the regulation of ovulation. *Recent Prog. Horm. Res.* **25**, 1–55.

Schwartz, N. B. (1969*b*). Modeling and control in gonadal function. *Math. Biosci.*, Suppl. 1, 229–255.

Schwartz, N. B. (1970). Cybernetics of mammalian reproduction. In *Mammalian Reproduction*, ed. Gibian, H. & Plotz, E. J., pp. 97–111. Berlin, Heidelberg, New York: Springer-Verlag.

Schwartz, N. B. & Hoffmann, J. C. (1967). A model for the control of the mammalian reproductive cycle. Excerpta Med. Int. Congress Series 132, 997–1003.

Sekiguchi, T., Nakamura, T. & Inoué, S. (1969). Modeling of the hypothalamo-hypophyseal-ovarian system. *Reports Inst. Med. Dent. Eng.* **3**, 97–106.

Sekiguchi, T., Nakamura, T. & Inoué, S. (1970). A model for the control of ovarian structure and secretion (in Japanese). In *Seitaikogaku to Seigyokogaku Shimpojumu Koenronbunshu*, ed. Nihon Jidoseigyo Kyokai, pp. 21–26. Kyoto: Nihon Jidoseigyo Kyokai.

Shack, W. J., Tam, P. Y. & Lardner, T. J. (1971). A mathematical model of the human menstrual cycle. *Biophys. J.* **11**, 835–848.

Thompson, H. E. (1966). Hormonal control of ovulation. Simulations of the anterior pituitary-ovarian endocrine control system. M. Sc. Thesis, Marquette Univ., Milwaukee.

Thompson, H. E., Horgan, J. D. & Delfs, E. (1969). A simplified mathematical model and simulations of the hypophysis-ovarian endocrine control system. *Biophys. J.* **9**, 278–291.

Urquhart, J. (1970). Endocrinology and the systems paradigm. *Behav. Sci.* **15**, 57–71.

Urquhart, J., Krall, R. L. & Li, C. C. (1970). Adrenocortical secretory funcion—communications and control aspects. *Automatica* **6**, 193–205.

Vande Wiele, R. L., Bogumil, J., Dyrenfurth, I., Ferin, M., Jewelewcz, R., Warren, M., Rizkallah, T. & Mikhail, G. (1970). Mechanisms regulating the menstrual cycle in women. *Recent Prog. Horm. Res.* **26**, 63–103.

Yates, F. E. & Brennan, R. D. (1969). Study of the mammalian adrenal glucocorticoid system by computer simulation. *Math. Biosci.* Suppl. 1, 20–87.

Yates, F. E., Brennan, R. D., Urquhart, J., Dallman, M. F., Li, C. C. & Halpern, W. (1968). A continuous system model of adrenocortical function. In *Systems Theory and Biology*, ed. Mesarović, M. D., pp. 141–184. Berlin, Heidelberg, New York: Springer-Verlag.